Developmental
Neurobiology of Vision

NATO ADVANCED STUDY INSTITUTES SERIES

A series of edited volumes comprising multifaceted studies of contemporary scientific issues by some of the best scientific minds in the world, assembled in cooperation with NATO Scientific Affairs Division.

Series A: Life Sciences

Recent Volumes in this Series

The series is published by an international board of publishers in conjunction with NATO Scientific Affairs Division

A	Life Sciences	Plenum Publishing Corporation
B	Physics	New York and London
C	Mathematical and Physical Sciences	D. Reidel Publishing Company Dordrecht and Boston
D	Behavioral and Social Sciences	Sijthoff International Publishing Company Leiden
E	Applied Sciences	Noordhoff International Publishing Leiden

Developmental Neurobiology of Vision

Edited by
Ralph D. Freeman
University of California
Berkeley, California

PLENUM PRESS • **NEW YORK AND LONDON**
Published in cooperation with NATO Scientific Affairs Division

Library of Congress Cataloging in Publication Data

Nato Advanced Study Institute on Developmental Neurobiology of Vision, Réthim-
non, Greece, 1978
Developmental neurobiology of vision.

(NATO advanced study institutes series: Series A, Life sciences; v. 27)
"Lectures presented at the NATO Advanced Study Institute on Developmental
Neurobiology of Vision, held in Rethymnon, Crete, Greece, September 18–27,
1978."
Includes index.
1. Vision—Congresses. 2. Developmental neurology—Congresses. I. Freeman, Ralph
D. II. Title. III. Series
QP475.N33 1978   599'.03'3   79-19389
ISBN 0-306-40306-4

Lectures presented at the NATO Advanced Study Institute on Developmental
Neurobiology of Vision, held in Rethymnon, Crete, Greece, September 18–27, 1978.

© 1979 Plenum Press, New York
A Division of Plenum Publishing Corporation
227 West 17th Street, New York, N.Y. 10011

Preface

This volume contains summaries of most of the invited presentations given by lecturers and participants at the NATO Advanced Study Institute, "Developmental Neurobiology of Vision," held in Rethymnon, Crete, Greece 18-27 September 1978. The purpose of this meeting was to enable a relatively small international group of scientists and students to hold discussions and to present their views on current problems in the field. It was intended that the results of the exchanges would be conveyed to others in the native countries of the participants. An attempt was made to obtain broad representation of disciplines within the area of the Institute and this is reflected in the diversity of the chapters in this book.

Considerable interest has been generated in recent years concerning the development and plasticity of vision. Perhaps not unexpectedly, along with the high level of activity in this field, there have been some controversial findings. In certain respects, the Institute was an ideal forum for the consideration of these different areas. The level of participant interest throughout the meeting was very high and discussions were invariably ended only because of time constraints. This is not to suggest that there was universal agreement about problematic issues or that there was widespread feeling that most major questions had been answered. On the contrary, the links are still tenuous at best between molecular events, neurochemical mechanisms, neuronal activity, and perceptual behavior. Hopefully, much of the enthusiasm displayed at the Institute in attempting to bridge these areas will be evident in this book.

The co-director of the Institute was Dr. Wolf Singer. His participation at every stage, including decisions about topics, lecturers, and participants, was crucial to the success of the meeting. Only those who have organized similar meetings can fathom the variety of details that require attention and many of these were handled by Ikuko Nakao. She is also largely responsible for the production of this book into which she has poured enormous effort. Ruth Suzuki helped with typing and Sarah Miyazaki proofread the chapters.

Ralph D. Freeman
University of California
Berkeley, California USA
1979

Lecturers and Participants

ATKINSON, JANETTE Psychological Laboratory, University of Cambridge, Cambridge CB2 3EG, England.

BAGDONAS, EILEEN 216-76 Caltech, Pasadena, California 91125 USA.

BAGNOLI, PAOLA Istituto di Fisiologia, Umana Via S. Zeno 31, 56100 Pisa, Italy.

BARLOW, H. B. Physiological Laboratory, University of Cambridge, Cambridge CB2 3EG, England.

BAUGHMAN, ROBERT W. Department of Neurobiology, Harvard Medical School, 25 Shattuck Street, Boston, Massachusetts 02115 USA.

BERARDI, NICOLETTA Laboratorio di Neurofisiologia del CNR, Via S. Zeno 56, 56100 Pisa, Italy.

BERKLEY, MARK Department of Psychology, Florida State University, Tallahassee, Florida 32306 USA.

BISTI, SILVIA Laboratorio di Neurofisiologia del CNR, Via S. Zeno 51, 56100 Pisa, Italy.

BLAKEMORE, COLIN Physiological Laboratory, University of Cambridge, Cambridge CB2 3EG, England.

BONDS, A. B. School of Optometry, University of California, Berkeley, California 94720 USA.

BRADDICK, OLIVER Psychological Laboratory, University of Cambridge, Cambridge CB2 3EG, England.

BONHOEFFER, FRIEDRICH Laboratorio di Neurofisiologia del CNR, Spemannstr. 35/I, Tübingen, Germany.

CATTANEO, ANTONINO Laboratorio di Neurofisiologia del CNR, Via S. Zeno 51, 56100 Pisa, Italy.

CYNADER, M. Department of Psychology, Dalhousie University, Halifax, Nova Scotia B3H 4J1, Canada.

DI STEFANO, MARIA Istituto Fisiologia, Umana Via S. Zeno 29-31, 56100 Pisa, Italy.

DISTEL, HANSJURGEN Institut für Medizinische Psychologie der Universität München, Pettenkoferstr. 12, München 2, Germany.

DURSTELER, MAX R. Neurologische Universitätsklinik, Ramistr. 100, CH-8091 Zürich, Switzerland.

EYSEL, ULF Th. Institut für Physiologie, Universitätsklinikum Essen, Hüfelandstr. 55, D-4300 Essen 1, Germany.

FINLAY, BARBARA L. Department of Psychology, Cornell University, Ithaca, New York 14850 USA.

FLANDRIN, JEAN-MARC Laboratoire de Neuropsychologie, INSERM U94, 16 ave. doyen Lepine, 69500 Bron, France.

FRASER, SCOTT E. Jenkins Department of Biophysics, Baltimore, Maryland 21218 USA.

FREGNAC, YVES Laboratoire de Neurophysiologie, Collège de France, 75231 Paris, Cedex 05, France.

FREEMAN, RALPH D. School of Optometry, University of California, Berkeley, California 94720 USA.

GARDNER, JILL G. Department of Psychology, Dalhousie University, Halifax, Nova Scotia B3H 4J1, Canada.

GAREY, L. J. Institut d'Anatomie, Université de Lausanne, Rue du Bugnon 9, CH-1011 Lausanne, Switzerland.

GAZZANIGA, MICHAEL S. Department of Neurology, Cornell Medical Center, 1300 York Avenue, New York, N.Y. 10021 USA.

GENIS-GALVEZ, JOSE M. Catedra de Anatomia, Facultad de Medicina, Sevilla, Spain.

GODEMENT, PIERRE Laboratoire de Neurophysiologie, Collège de France, 75231 Paris, Cedex 05, France.

GRUSSER, O.-J. Department of Physiology, Freie Universität Berlin, Arnimallee 22, 1 Berlin 33, Germany.

HAMMOND, P. Department of Communication and Neuroscience, University of Keele, Keele, Staffordshire ST5 5BG, England.

HANNY, PAUL-ERNST Neurologische Klinik, Kantonspital, CH-8091 Zürich, Switzerland.

HEITLANDER, HELGA D. Institut für Zoologie der Joh. Gutenberg-Universität, P.B. 3580, 65 Mainz, Germany.

HOCHSTEIN, SHAUL Neurobiology Unit, Institute of Life Sciences, Hebrew University of Jerusalem, Jerusalem, Israel.

HOFFMANN, K.-P. Abteilung für Biologie IV der Universität, Oberer Eselsberg, Postfach 4066, D-7900 Ulm, Germany.

HUBEL, DAVID H. Department of Neurobiology, Harvard Medical School, 25 Shattuck Street, Boston, Massachusetts 02115 USA.

IMBERT, MICHEL Laboratoire de Neurophysiologie, Collège de France, 75231 Paris, Cedex 05, France.

INNOCENTI, GIORGIO M. Institute of Anatomy, University of Lausanne, 1011 Lausanne-CHUV, Switzerland.

JOHNS, PAMELA RAYMOND Neurosciences Laboratory, University of Michigan, Ann Arbor, Michigan 48109 USA.

KILLACKEY, HERBERT P. Department of Psychobiology, University of California, Irvine, California 92717 USA.

KREUTZBERG, GEORG W. Max-Planck-Institut für Psychiatrie, Kraepelinstr. 2, 8000 München 40, Germany.

MAFFEI, L. Laboratoire di Neurofisiologia del CNR, Via S. Zeno 51, 56100 Pisa, Italy.

MARTIN, K. A. C. University Laboratory of Physiology, Parks Road, Oxford OX1 3PT, England.

MORRONE, MARIA C. Laboratorio di Neurofisiologia CNR, Via S. Zeno 51, 56100 Pisa, Italy.

MUSCHAWECK, LEONHARD G. Max-Planck-Institut für Biophyskalische Chemie, D-3400 Göttingen-Nikolausberg, Germany.

NEUMANN, GUNTHER Max-Planck-Institut für Psychiatrie, Kraepelinstr. 2, 8000 München 40, Germany.

PALM, GUNTHER Max-Planck-Institut für biol. kybernetik, Spemannstr. 38, 74 Tübingen, Germany.

RAKIC, PASKO School of Medicine, Yale University, New Haven, Connecticut 06510 USA.

RAMACHANDRAN, V. S. Trinity College, Cambridge CB2 3EG, England.

RAUSCHECKER, J. P. Max-Planck-Institut für Psychiatrie, Kraepelinstr. 2, 8000 München 40, Germany.

RUBINSON, KALMAN New York University School of Medicine, 550 First Avenue, New York, N.Y. 10016 USA.

RUSOFF, ANNE C. Division of Biological Sciences, University of Michigan, Ann Arbor, Michigan 48109 USA.

SCHMIELAU, FRITZ Agnos Bernauer Str. 99A, 8000 München 21, Germany.

SCHWARZ, ULI Max-Planck-Institut für Virusforschung, Spemannstr. 35/II, 74 Tübingen, Germany.

SHERMAN, S. MURRAY Department of Physiology, University of Virginia Medical School, Charlottesville, Virginia 22908 USA.

SINGER, W. Max-Planck-Institut für Psychiatrie, Kraepelinstr. 2, 8000 München 40, Germany.

SPINELLI, D. Laboratorio di Neurofisiologia, Via S. Zeno 51, 56100 Pisa, Italy.

SWINDALE, N. Physiological Laboratory, University of Cambridge, Cambridge CB2 3EG, England.

TIMNEY, BRIAN Department of Psychology, University of Western Ontario, London, Ontario N64 5C2, Canada.

VERNADAKIS, ANTONIA Departments of Psychiatry and Pharmacology, University of Colorado, School of Medicine, Denver, Colorado 80262 USA.

VITAL-DURAND, F. Laboratoire de Neuropsychologie, INSERM U94, 16 ave. doyen Lepine, 69500 Bron, France.

YINON, URI Electrophysiological Laboratory, Goldschleger Eye Institute, Tel-Aviv University Medical School, Tel-Hashomer, Israel.

Contents

Three Theories of Cortical Function

H. B. BARLOW
Physiological Laboratory
Cambridge, England

Ralph Freeman asked me, as an introduction to this meeting, to try to give you some kind of review of the present status of knowledge in the field of cortical development. The more I looked into it, the more confused I became. There has been a lot of papers published, and these, of course, all constitute local advances; but overall the situation has not, in my view, changed very significantly since I tried to review it three years ago (Barlow, 1975), or even before that date. It is as if we were winning all the battles, but losing the war, and when that sort of thing happens, it's a sure sign that you don't know what the war is all about. I began to wonder if the trouble with work on the development of the cortex is that we have not identified the problem or problems correctly.

If we were discussing the development of the heart, it would be quite easy to recognize certain landmarks, such as the onset of rhythmic contractions, the formation of separate chambers, the beginning of vigorous blood circulation, and the closing of the pulmonary circulation short circuits. That is because we know the function of the adult heart and the mechanisms by which it is achieved, and this knowledge gives very useful landmarks in describing development. In spite of the fact that it is almost 20 years since the first paper from Hubel and Wiesel (1959), I feel we are in a pre-Harvey-an state with regard to our knowledge of cortical function, and this is what is holding us up.

Imagine a NATO summer school taking place 400 years ago on the subject of cardiac development. One person might say that the heart is underdeveloped at birth but matures very soon after, because we know the function of the heart is to warm the blood, and only newborn infants chill easily. Another might say it only matures at adolescence, because it is the heart that falls in love. Of course, there may be sensible things to say about development without any knowledge of function, but they won't be as sensible, and not nearly as interesting, as the things you can say about development when you know how the system works.

It is interesting, by the way, to recall Harvey's comparative lack of success in describing the physiology of reproduction. He made many absolutely valid observations, but failed to interpret them correctly, probably because he did not realize that actions could be produced in distant organs by means of hormones (e.g. the ovary acting on the uterus). For this reason I don't think we should feel too ashamed of our ignorance of cortical function. Sound knowledge of elementary hydraulics was available in Harvey's day, and this is what was required to understand the circulation. But, although we have knowledge of communication engineering which is relevant to understanding how our brains are fed with sensory information, it isn't clear that we have the knowledge that will give us insight into the kinds of things our brains do with that information.

Following this line of thought, I decided to use a few minutes of your time to put forward three theories of cortical function. I shall call them the interpolation theory, the jigsaw puzzle theory, and the association theory. However, I don't want you to consider them as mutually exclusive rivals, for I think there is some truth in all of them. Also, recall that the cortex is a surprisingly uniform structure and must be the dominant one responsible for man's intellectual, artistic, social, and moral qualities; it would be silly to opt for a theory of visual cortex that allowed no possibility at all of explaining, by analogy or extension, these other aspects of cortical function. I am by no means confident that any or all of these theories do this, so you must leave room for other theories.

Interpolation: Reconstruction of the Visual Image in Space and Time

It is an old idea that the projection from LGN to area 17 recreates on the cortex a copy of the retinal image; let us take this old idea seriously.

The image arriving from the retina suffers two serious defects that can be identified, and could perhaps be ameliorated by neural mechanisms in the cortex or earlier. First, what was a continuous distribution of light intensity on the retina is represented by a set of impulse frequencies in a limited number of nerve fibers in the optic radiation. What can, and what cannot, be reconstructed from such sampled data is well understood, and this is the first point to discuss. The second defect, to be considered later, is that the eye is almost always moving relative to the scene it surveys, and, consequently, the image being transmitted to the cortex is also moving.

There is a well-known theory that says a continuous function can be completely represented by a finite set of values sampled at intervals $1/2F_{max}$, where F_{max} is the highest frequency contained in the continuous function (see, for instance, Bracewell, 1965). This can be applied to a two-dimensional function, such as the retinal image, although there are minor problems, which will not be considered here, arising from the necessity of arranging the sample points in a triangular, or some other regular, array. Because the image is not perfect there is an upper limit to the spatial frequencies it contains, and because of diffraction that pupil diameter enables one straight away to calculate the highest possible value of this frequency: expressed in radians, it is λ/D, where

λ is the wavelength of light and D is the pupil diameter measured in the same unit of length. If D is taken as 2 mm, the smallest value it normally reaches in humans, and λ as 560 nm, then λ/D is 2.8×10^{-4} radians or nearly 1 minute; this corresponds closely to the limiting resolution of humans, and it has long been recognized that the sampling interval of the most densely packed foveal cones is close to half this period. Of course, for larger pupil diameters the optimum performance is not approached so closely, and in the peripheral field of vision resolution and sampling interval, though they are probably concordant with each other, fall orders of magnitude below the upper limit of spatial frequencies contained in the image.

Nothing can be done in the cortex to replace the spatial frequencies lost by the pupillary cut-off, or by the limited band pass of peripheral receptive fields (Enroth-Cugell and Robson, 1966). This information has gone forever, and in this sense the point sampled representation of the optic radiation is as good as it can be, for it cannot be improved. But, for some purposes it may be an inconvenient representation. Many people must have encountered the problem of judging the position of the maximum in a curve when it lies between two sample points—for instance, specifying the λ_{max} of a spectral sensitivity curve which has only been measured at widely spaced intervals. Surprisingly, this can be done with complete accuracy by the process of interpolation, provided the intervals are not too coarse, and the process for doing this is shown in Fig. 1 (from Cherry, 1957). The bars in the lower figure show the values of the continuous waveform above, sampled at τ_1, τ_2, τ_3, etc., separated by intervals $1/2F_{max}$. For the sample at τ_5, the "interpolation function," $\sin(x)/x$, of appropriate peak amplitude, has been drawn. If similar functions of appropriate amplitude were placed at τ_1, τ_2, τ_3, etc., and all these curves added up, then the upper curve would be recreated *exactly* at every point, including those intermediate between τ_1, τ_2, etc. If, in the cortex, each incoming optic radiation fibre distributed its terminals with a spatial pattern reproducing the two-dimensional interpolation function, then the pattern of synaptic activity would also recreate the retinal image *at every point.*

Now the relevance of this for visual, and cortical, physiology is that there are some tasks the eye can perform which seem to exceed the 1 minute resolution limit. Westheimer and McKee (1977) call this phenomenon "hyperacuity" and the best known examples are vernier and stereo acuity. In both cases, psychophysical results show that the relative positions of objects can be judged with an accuracy of 6 sec of arc, or better, for stereoacuity. This is an order of magnitude less than 1 minute of arc or 60 cycles per degree, which is the pupillary cut-off frequency and the highest attainable grating resolution. Of course, it has been realized that the attainment of such positional accuracy does not break any physical laws, but it certainly raises the physiological question of how a position can be judged with greater accuracy than the distance separating two foveal cones. It is, furthermore, a serious problem if, as I do, you like to explain perceptual phenomena in terms of single neurons (Barlow, 1972), for how can you explain seeing something placed, say, one-fifth of the way between the positions of two neurons?

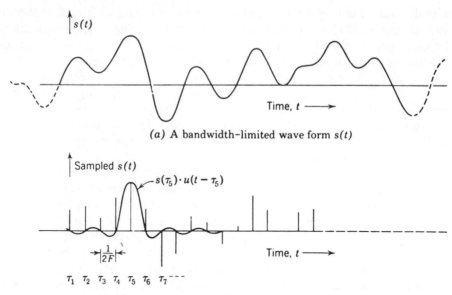

(a) A bandwidth–limited wave form $s(t)$

(b) The sampled $s(t)$ and the sin x/x interpolation function

FIGURE 1 A continuous band-limited signal can be fully reconstructed from a finite number of samples. The interpolation function $(sin(x)/x)$ of appropriate height is placed at each sample point, and the values simply added at all points. Similarly, a fine-grained version of the retinal image could be reconstructed in the geniculate and cortex from the limited number of samples provided by the retinal ganglion cells. (From Cherry, 1957).

Now, it so happens that the geniculo-cortical fibres terminate in a layer of granule cells which are extremely densely packed, so densely that there are 30 to 100 times as many granule cells per unit area as there are optic radiation terminals (L. Garey, personal communication; Rockel, Hiorns, and Powell, 1974). Along any straight line there are five to ten times as many granule cells as there are sample points from the retinal image, and a finer-grained version of the retinal image could be created by having connectivity from input fibres to granule cells vary as $sin(x)/x$, x being position in the granule cell layer. The position of the most active granule cell would then represent a peak in the image with five to ten times the accuracy possible by doing the same operation on the sampled data presented in the optic nerve. Thus, hyperacuity might be explained, and the single-unit doctrine preserved at the same time.

The second aspect of the cortical representation to be considered is the problem of the movement of the image resulting either from unintended eye movements, or failure of the eye to track a moving object accurately. Marshall and Talbot (1942) suggested many years ago that movements of the eye, in particular the rapid tremor known as micronystagmus, actually aided resolution, and this received some initial support from the observation that stabilizing the image causes rapid fading (Ditchburn and Ginsberg, 1952; Riggs, Ratliff, Cornsweet and Cornsweet, 1953). But the idea that rapid tremor helped resolution was always unattractive, because it neglected the fact that the eye is slow even in photopic conditions, and in a system that integrates temporally, movement will inevitably blur the image. Further evidence against Marshall and Talbot accumulated when it was shown that the micronystagmus was not constant (Barlow, 1952), that resolution was better when the eye was still than when it was moving (Ratliff, 1952), and that resolution was as good under stabilized image conditions as with a moving eye (Tulunay-Keesey, 1960). The question therefore changed. Westheimer and McKee (1975) have recently asked "How well does the eye resist the degradation of performance by image movement?", and their measurements show that resolution stays remarkably good provided the rate of movement is below 2 to 3 $deg \cdot sec^{-1}$. Considering the integration time must be of the order of 1/50 second, this means that the eye can move about 3 minutes within 1 integration time; that is 3 times the grating resolution, and at least 30 times the hyper-resolution value for vernier acuity!

Meanwhile, evidence of a quite different kind has been accumulating which suggests that the eye has a well-developed capacity for temporal interpolation: that is, the position of an image is represented as an almost continuous function of time, even when the image itself is presented at discrete positions at separate instants.

Burr (1975), Morgan (1975, 1976), and Ross and Hogben (1975) more or less simultaneously made a very interesting observation while working on the well-known stereoscopic phenomenon originally described by Pulfrich: a pendulum swinging in the frontal plane appears to move in a circular orbit when the image seen by one eye is dimmed by interposing a neutral filter. The new observation was that the effect works for an image in apparent motion produced by flashing the image at a set of discrete positions. Furthermore, the effect can be imitated, or counteracted (Burr and Ross, 1979), by introducing a delay in the view of the object through one eye. Delay by dimming is, of course, the explanation that has often been given of the original Pulfrich effect (see Morgan and Thompson, 1975, for an interesting account of the history), and this was confirmed, but the observation that it works for apparent, stroboscopic, movement implies that a delay is interpreted by the visual system as a shift of position when the object is moving, for the stations at which the object actually appears have zero disparity. Unfortunately, the possibility that the eye was tracking the moving object mars this interpretation, but Morgan and Turnbull (1978) had shown that movement does not account for it. Burr and Ross (1979) have shown that the stereoacuity for "virtual" disparity (i.e., the angle through which the object would have moved during the delay) is as good as true stereoacuity, and it is hard to believe that their results only occur with tracking. Burr (1979) has an elegant demonstration that the illusion of a moving vernier can result from displaying a line stroboscopically at discrete stations, but with a short delay between upper and lower halves, and he used Westheimer and McKee's trick of curtailing the duration of motion to 180 msec so that reflex following could not occur, and randomizing its direction to avoid any possibility of anticipatory movements. Finally, Cynader, Gardner, and Douglas (1979) have observations on the disparity-selective cells of cat cortex which suggest that delay of a moving object may cause a shift of its apparent position.

Phenomena like these, of course, form the basis of cinematography, and the explanation usually given is in terms of temporal integration and apparent movement (phi-phenomenon). But, temporal *interpolation* is a better term for it, because the essence of the process is to recreate a continuous function from sampled data. In the case of temporal interpolation the process is a linear one, like estimating the position of a car from the instants at which it passes two marks. But the most remarkable aspect of the process is not just that it occurs, but that it can apparently be done with nearly the full accuracy of the mechanism underlying hyperacuity. This is probably a conclusion implied by Westheimer and McKee's measurements on the resistance of vernier acuity to movement, and by the observations of Morgan and Thompson (1975) and Burr and Ross (1979) on the Pulfrich phenomenon, but Burr's experiment (1979) with moving verniers makes the conclusion quite explicit. He finds that resolution for the illusory offset caused by a temporal delay is very nearly as good as it is with a real offset. What this seems to mean neurophysiologically is that the peak of a wave of disturbance in the granule cell layer moves almost continuously, even when it is caused by stimuli applied at discrete points and times in the image.

Such a reconstitution of the image by spatial and temporal interpolation is remarkable. I wish I could present to you a neural model suggesting how it is done, but instead all I can do is to relay to you this psychophysical evidence on the accuracy of the position sense, which is maintained even when the image is moving, and to suggest that the very large numbers of granule cells in layer IV may underlie these capacities. Reconstitution of the visual image by interpolation in space and time is a neat trick, and one should also recall that the reconstituted image is not continuous, but sliced into thin strips alternating for the two eyes (Hubel and Wiesel, 1977).

Jigsaw Puzzles: Schemes for Disseminating the Information in an Image Fragment

One function that must be performed by each small part of area 17 can be deduced with reasonable confidence from current knowledge of anatomy and physiology: each part must disseminate to other parts of the brain information about the small fragment of the visual image that it has access to. A given point in 17 receives direct connections from only a small part of the image (about 1/1000 of it; Hubel and Wiesel, 1977), and it has limited connections with closely neighbouring regions of 17 itself (Fisken, Garey and Powell, 1973). The outgoing fibres go to distant regions, and fibres to different destinations come predominantly from cells in different layers; layer VI goes mainly back to the thalamus; layer V goes mainly to midbrain visual centres; and layers II, III, and IV go to other distant areas of the cerebral cortex, including area 17 on the other side.

Thus the primary visual cortex certainly acts as a redistribution centre, but what information does it redistribute? This appears to be different in different layers; layer V, for instance, contains a large proportion of "special complex" cells that are especially sensitive to small objects moving in a particular direction over quite a large receptive field (Gilbert, 1977). These project to the superior colliculus and are presumably mainly responsible for the directionally selective properties of units recorded there. Also, layer VI contains a special type of large, orientation-selective simple cell, and these presumably have something to do with the cortico-geniculate projection, but the functional role of this pathway is mysterious. Layers II, III, and IV, projecting to other areas of the cortex, contain the bulk of the simple, complex, and hypercomplex cells, and the type of analysis performed on the visual image by the visual system as a whole must be determined in part by the nature of the information that is disseminated elsewhere by these cells, in part by the exact pattern of dissemination to parastriate and other areas, and in part by what goes on in these areas. Perhaps some idea of the analysis that must be done can be given by describing the following scene for solving a giant jigsaw puzzle.

Imagine a room like an examination hall with a thousand or so clerks sitting at desks in it. Take an enormously enlarged picture of the retinal image and cut it up into a thousand pieces, small pieces in and around the foveal area, and progressively larger pieces towards the edges. Give one piece to the clerk at each desk, whose job it is to fill in a limited number of entries on cards which are then consigned elsewhere and used to facilitate the solution of the puzzle. How would you design these cards and their entries? Where would you send them? What would you do with them? I think this is the point where our situation is like that of Harvey dealing with the reproductive system while ignorant of hormones, and not like Harvey looking at the hydraulics of the circulation. But we do know a little about how to transform and code images, and we can make some guesses about what might go on.

The purpose of the analogy is to bring out the fact that the cells of the cortex, like the clerks, can only do local analysis and they can only pass on messages about these local properties. The analogy is a bad one in many ways; unlike the jigsaw, only the picture is important, for all the pieces have the same shape. Also, the analogy could probably be improved by making the fragments allotted to each desk overlap the neighbouring fragments given to other desks. But it is more important that the aim of the cortex, which is what we are now considering, is not to solve the puzzle in the sense of reconstructing the picture, though people sometimes talk as if that was the only task for the cortex. Probably what the cortex needs to do first is to distinguish foreground from background and to recognize which fragments belong to the same objects. It is difficult to say what is required for doing this, but as with the problem of resolution and interpolation there are some general principles that may act as a guide.

First, note that the cards could be planned so that no information was lost and the whole picture could, in principle, be reconstructed from the cards. For instance, the clerk might simply be instructed to measure the luminance at each resolvable point and punch the value, suitably coded, at the appropriate point on the card. That would not seem to help the process of analysis at all, and a much more interesting possibility would be to instruct him to do a Fourier Analysis of his fragment and enter the amplitudes of, say, the first half dozen sine and cosine terms. The reason this is more interesting is that a "language" of properties that might extend across fragments is thereby introduced. For instance one can imagine one clerk saying "I've got a lot of components of various frequencies, all oriented at 45°", other clerks would say "So have I", and thus the fragments containing the outline of an object (assumed to be continuous and locally straight) could be identified. In contrast, there is no reason why high values at the same X, Y coordinates in different fragments should have any relation to each other. So the universal language introduced by Fourier analysis gives a potential advantage, but it is not necessarily the best form of universal language for analysing picture fragments. What are the other possibilities?

Elsewhere, Sakitt and I (in preparation) present a theoretical scheme for decomposing a picture fragment into orientation and spatial frequency selective components, and we think that it has various merits. It is economical and efficient, it bears a good resemblance to the operations of simple cortical neurons, it can probably account for the psychophysical detectability of sinusoids (Wilson, 1979; Robson and Graham, in preparation, 1979), and perhaps for the effects of adaptation to gratings (Gilinsky, 1967; Pantle and Sekuler, 1968; Blakemore and Campbell, 1969). There is not room for more detail here, but one additional point needs to be made.

It would be a very desirable feature of a coding scheme if the cards submitted by the clerks very often required only a single entry or none at all. For example, on the very first scheme, in which the picture was coded point by point, if the pictures to be analysed all consisted simply of photographs of the sky at night, the cards for the many fragments containing no stars would be blank, there would be a good number with a single entry, and no doubt also several with multiple entries; but the total number of entries would be quite small compared with the number of resolvable image points. On the other hand, the same pictures presented for Fourier analysis would yield a highly complex set of cards. Now consider photographs of scenes with strongly periodic elements such as railings, wire meshes, and repetitive decorative components. The cards resulting from local Fourier analysis would now be relatively simple, whereas those resulting from point-by-point representation would be complex. So, on those grounds alone, one would design the cards differently if one knew what kinds of pictures were to be presented. Most visual scenes would not be simple for either of the codes above, and anyway we do not have very much control over what is presented to our eyes, but the fact that different coding schemes suit different populations of targets suggests there is something to be said for designing the cards so that features that are important for further analysis yield simple and prominent card entries. Hence, possibly, the use of edge and line detectors in the cortex.

Those who know Marr's (1976) work will recognize that he has been dealing with a related problem in considering what the first step in the computer analysis of a picture should be, once it is available as a list of digitized luminance values. His "primal sketch" is a list of quantities calculated locally from the digitized image, but more useful for subsequent analysis than the digitized quantities themselves. This list includes quantities rather like what, one supposes, the outputs of cortical neurones represent. It is probably from sources like this that we can hope to pick up interpretive keys like hydraulics and hormones. We may then understand the requirements for designing good forms for the clerks solving the jigsaw puzzle, and perhaps thereby see what primary visual cortex does.

Testing the Occurrence of Paired Associations

This is the theory for which there is least evidence and which is hardest to explain, but it is also the one that is mostly likely to have implications outside vision and in that sense it is the most interesting. Let me give you the evidence first, because it is on the one hand so scanty that it will not take long to present it, while on the other hand it will suggest to you why I have been led to consider the cortex as the organ for detecting paired associations.

Figure 2 shows three patterns produced by pairing two dots in a regular way (Glass, 1969; Glass and Perez, 1973), the first dot of each point being positioned entirely at random. For these randomly placed dots (top left) there is, of course, no pattern, except that implied by the circular outline. For the top right, each of the randomly placed dots is repeated in a position mirror symmetric about the vertical midline. For bottom left, the pair is placed a short distance up and to the right of its mate, and for bottom right, it is displaced outwards from the centre and rotated clockwise round the centre.

The perceptual prominence of these Glass figures makes one suspect at once that the visual system is specially adapted to detecting pairs placed according to a regular strategy, but it could, of course, be an accidental by-product of some other process such as line-detecting. Reeves and I have, however, made measurements of how well these patterns are detected (Barlow and Reeves, 1979). The answer appears to be that when the characteristics of the regular feature are well adapted to those of the symmetry detecting mechanism, a high proportion of the available information about the pattern is made use of; this conclusion follows from the small number of errors made in distinguishing samples of these patterns from samples lacking the regular feature—the number of errors is not much greater than the minimum possible. The flavour of these experiments is given by Fig. 3.

In this experiment a subject had to classify unknown patterns as symmetric or not, and the errors he made were used to calculate d' values in the usual way.

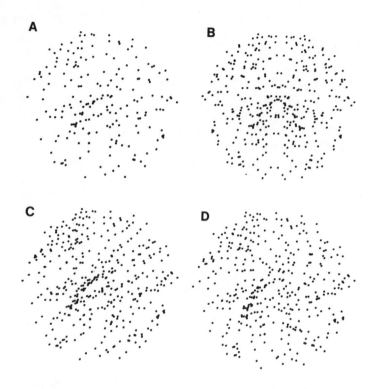

FIGURE 2 Examples of structure produced by pairing dots. In (B) every dot has a pair at the position mirror-symmetric about the vertical mid-line. In (C) the pair is placed at a constant distance up and to the right. In (D) it is displaced a constant distance along a spiral path. In all these patterns it is the regular feature produced by pairing that is perceived readily. The irregular arrangement shown in (A) was used as the basis for all the other patterns, but it requires careful inspection to detect or confirm this.

These values are plotted as ordinates. Half of the patterns were entirely random, and the other half had half of their dots placed at random, each of these having a mate placed near the mirror symmetric position. The accuracy of this placing was varied, and this is what is plotted on the abscissa. Obviously, as the accuracy gets worse, the d' scores diminish, but it is interesting to note that the task can be well performed even in the presence of considerable inaccuracy.

In order to account for this kind of performance, we were led to consider the kind of model shown in Fig. 4. Here the information on which symmetry is

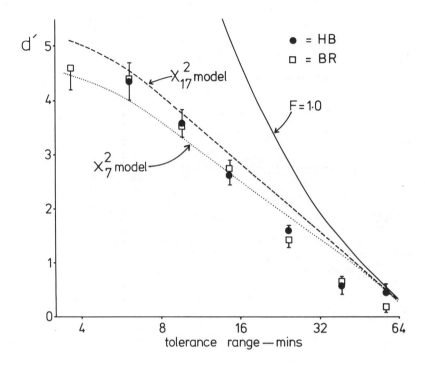

FIGURE 3 If mirror-symmetric pairs are placed inaccurately the detectability is impaired, but surprisingly large inaccuracy (± 8 minutes) can be tolerated. The d' values are about half those obtainable ideally (curve marked $F = 1.0$), which implies utilization of 25% of the statistical information available for the task. The dotted line shows the predicted behaviour of the model illustrated in Fig. 4, and the dashed line the behaviour for a similar model with 36 smaller squares.

thought to be judged consists of the numbers of dots contained in each of the 16 squares. Because the system tolerates inaccurate placing of the mirror pairs, it is unnecessary to suppose that the positions of individual dots are important, and the regions within which dots are counted can be large. The next step consists of the pairwise comparison of squares which would contain equal numbers of dots if the pattern was symmetric. This could be done a number of ways, but we chose a χ^2 test, as indicated. The model classifies a given test pattern as symmetric or not according to the value of χ^2 obtained.

Going back to Fig. 3, the lines here show the performance of this model, and also another one in which the squares were smaller. Clearly, the models describe the performance quite well. I do not know of any method of detecting symmetry that would not require pairwise comparisons of regions that would be

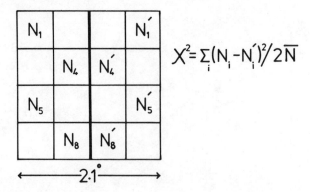

FIGURE 4 Detection of symmetry might be based on the number of dots in fixed regions rather than the exact positions of individual dots. In the model shown diagrammatically here, the pattern is subdivided into 16 squares, and the difference between numbers in symmetric squares are compared; low values of χ^2 are expected if the arrangement is symmetric, not random. Predictions of this model are shown in Fig. 3 (dotted line). A similar model uses 36 squares instead of 16 (dashed line).

alike if symmetry was present. The fact is, our visual system does it well, and this is the evidence (admittedly not overwhelming), that the cortex can test for and detect paired associations.

Testing for paired associations is not a trivial task. In the first place, to do it well requires implementation of a substantial part of the contents of a statistical textbook, but that is the easy part of the problem. The textbooks assume that the associations to be tested for are self-evident and there is no difficulty in deciding upon them, but that is not the case at all. The skill of science lies in suspecting that two things *might be* associated, and this is where the real difficulty lies: some 3 million sensory fibres relay their messages to the sensory cortex, so between them there are $n(n-1)/2$ possible paired associations, about 5×10^{12} of them. If we possess a cortex for the purpose of testing the occurrence of pairs it is no surprise that a large neural organ is required for the task!

There is another point to be made here. One often thinks of the sensory cortex as a place where sensory information is simply represented. The idea is elaborated a little by saying "features" are represented, or hierarchies of features, and I have indulged in this myself (e.g. Barlow, Narasimham, and Rosenfeld, 1972). But there is a great deal of difference between representing facts, or features, and acquiring knowledge, and what our senses surely do for us is to acquire new knowledge of the world about us, not simply represent it. Now, for the acquisition of knowledge you need statistical tests (Fisher, 1935; Barlow, 1974), and especially, tests of association, so again it seems that testing for paired associations is a job that has got to be done, and it would be a good idea to explore the possibility that cortex does it.

Of course, there must be restrictions on the pairs of sensory messages whose association could be detected, and it is not clear that it would be advantageous to test them all in pairs, even if it could be done. So, what might these restrictions be? As with computers, wiring is likely to be the problem, so a first guess about a restriction might be that the pair of sensory messages must be represented in the cortex within a short distance of each other, say 1 mm, before their association can be detected. Thus, in primary visual cortex only the association of local events close to each other in the visual field could be found. Could one perhaps regard the interleaving of right and left eye representations as a device enabling associations between them to be found? Two points define a line, so to detect orientation, do you not require evidence of excitation at *two*? Could orientation columns represent the systematic exploration of the occurrence of such pairs?

Let us briefly look at the requirements for symmetry detection. By analogy with Hubel and Wiesel's micro-columnar structure within a topographical map, one might suppose the first step would be to reproject to one small area all the pairs of regions lying about a particular position of the axis of symmetry in the visual field, and then, as with orientation columns, test for the occurrence of pairs: if enough pairs were found, that would be evidence for symmetry about the axis.

I do not know if these will prove to be fertile notions to pursue further, but ideas about the sensory cortex are certainly needed, and the proposal that it converts a representation of the environment to *knowledge* about it by detecting the occurrence of paired associations seems a worthy task for it to perform.

Summary and Conclusions

There is a need to consider what the visual cortex *does;* otherwise, we are likely to lose our bearings in describing its development. I am suggesting three ideas about this.

The first is that the cortex is able to interpolate between the samples of the retinal image provided by the ganglion cells and fibres of the optic radiation, thus providing a representation of finer grain size on which judgements of high positional accuracy can be performed (though high spatial frequencies are not, of course, restored). When the scene is also represented by samples in time, as in a movie, the system can interpolate in time, interpreting a delay as a change in apparent position, and it does this with very high accuracy. It is thought this fine-grained representation must occur in the granule cells of layer IV, which are numerous enough to account for vernier and stereoacuity.

The second idea concerns the function of area 17 as a distributor of information about fragments of the visual image to other parts of the brain. It is suggested that the messages relayed contain rather complete information about the image fragment, coded in a form intermediate between a point-by-point representation and the coefficients of a local fourier analysis. The scheme will be described fully elsewhere (Barlow and Sakitt, 1979).

The third idea arose from considering how symmetry could be detected. The suggestion is that sensory cortex converts a mere representation of sensory messages into new knowledge of the sensory environment by testing for the associated occurrence of pairs of events. The detection of the paired regions that give evidence of symmetry is likely to occur in parastriate cortex, but the idea may also apply to local associations detected in area 17, and it is, of course, a general task that could be the essential one performed in other cortical regions.

Acknowledgements A review like this owes much to individuals who have discussed their unpublished work with the reviewer, and I am particularly indebted to D. Burr, L. Garey, S. McKee, M. Morgan, and G. Westheimer for explaining their published work and for giving me access to more awaiting publication (though I fear they may not all agree with my interpretations). The theoretical work with Dr. Sakitt that is briefly described was started under her grant from NIH, EY01336.

References

Barlow, H. B. (1952). Eye movements during fixation. J. Physiol. (Lond.) 116:269-306.

Barlow, H. B. (1972). Single units and sensation: A neuron doctrine for perceptual psychology? Perception 1:371-394.

Barlow, H. B. (1974). Inductive influence, coding, perception and language. Perception 3:123-134.

Barlow, H. B. (1975). Visual experience and cortical development. Nature 258:199-204.

Barlow, H. B., R. Narsimhan, and A. Rosenfeld (1972). Visual pattern analysis of machines and animals. Science 177:567-575.

Barlow, H. B., and B. C. Reeves (1979). The versatility and absolute efficiency of detecting mirror symmetry in random dot displays. Vision Res. (in press).

Blakemore, C., and F. W. Campbell (1969). On the existence of neurones in the human visual system selectively sensitive to the orientation and size of retinal images. J. Physiol. (Lond.) 203:237-260.

Bracewell, R. (1965). The Fourier Transform and Its Applications. McGraw Hill, New York.

Burr, D. C. (1975). A second binocular depth perception system. B.Sc. thesis, University of Western Australia.

Burr, D. C. (1979). Acuity for apparent vernier offset. Vision Res. (in press).

Burr, D. C., and J. Ross (1979). How binocular delay gives information about depth. Vision Res. (in press, August 1978).

Cherry, C. (1957). On Human Communication. Science Editions, Inc., New York; M.I.T. Press and John Wiley & Sons, Ltd.

Cynader, M., J. Gardner, and R. Douglas (1979). Neural mechanisms underlying stereoscopic depth perception in cat visual cortex. In: Frontiers of Vision Research, S. J. Cool and E. L. Smith, III (eds.), Springer, New York.

Ditchburn, R. W., and B. L. Ginsberg (1952). Vision with a stabilized retinal image. Nature 170:36-37.

Enroth-Cugell, C., and J. G. Robson (1966). The contrast sensitivity of retinal ganglion cells of the cat. J. Physiol. (Lond.) 187:517-552.

Fisher, R. A. (1935). The Design of Experiments. Oliver and Boyd, Edinburgh.

Fisher, R. S., L. J. Garey, and T. P. S. Powell (1973). Patterns of degeneration after intrinsic lesions of the visual cortex (area 17) of the monkey. Brain Res. 53:208-213.

Gilbert, C. D. (1977). Laminar differences in receptive field properties of cells in cat primary visual cortex. J. Physiol. (Lond.) 268:391-421.

Gilinsky, A. S. (1967). Masking of contour-detectors in the human visual system. Psychon. Sci. 8:395-396.

Glass, L. (1969). Moiré effect from random dots. Nature 223:578-580.

Glass, L., and R. Perez (1973). Perception of random dot interference patterns. Nature 246:360-362.

Hubel, D. H., and T. N. Wiesel (1959). Receptive fields of single neurones in the cat's striate cortex. J. Physiol. (Lond.) 148:574-591.

Hubel, D. H., and T. N. Wiesel (1977). Ferrier Lecture: Functional architecture of macaque monkey visual cortex. Proc. Roy. Soc. B 198:1-59.

Marshall, W. H., and S. A. Talbot (1942). Recent evidence for neural mechanisms in vision leading to a general theory of sensory acuity. Biological Symposia (J. Cattell, ed.) 7:117-164.

Marr, D. (1976). Early processing of visual information. Phil. Trans. Roy. Soc. B 275:483-519.

Morgan, M. (1975). Stereoillusion based on visual persistence. Nature 256:639-640.

Morgan, M. J. (1976). Pulfrich effect and the filling in of apparent motion. Perception 5:187-195.

Morgan, M. J., and P. Thompson (1975). Apparent motion and the Pulfrich effect. Perception 4:3-18.

Morgan, M. J., and D. F. Turnbull (1978). Smooth eye tracking and the perception of motion in the absence of real movement. Vision Res. 18:1053-1054.

Pantle, A., and R. Sekuler (1968). Size-detecting mechanisms in human vision. Science 162:1146-1148.

Ratliff, F. (1952). The role of physiological nystagmus in monocular acuity. J. exp. Psychol. 43:163-172.

Riggs, L. A., F. Ratliff, J. C. Cornsweet, and T. N. Cornsweet (1953). The disappearance of steadily fixated visual test objects. J. Opt. Soc. Am. 43:495-501.

Rockel, A. J., R. W. Hiorns, and T. P. S. Powell (1974). Numbers of neurons through full depth of neocortex. Proc. Anat. Soc. Gt. Brit. Ire. 118:371.

Ross, J., and J. H. Hogben (1975). The Pulfrich effect and short-term memory in stereopsis. Vision Res. 15:1289-1290.

Sakitt, B., and H. B. Barlow (1979). An economical model for the cortical encoding of the visual image. In preparation.

Tulunay-Keesey, U. (1960). Effects of involuntary eye movements on visual acuity. J. Opt. Soc. Am. 50:769-774.

Westheimer, G., and S. P. McKee (1975). Visual acuity in the presence of retinal image motion. J. Opt. Soc. Am. 65:847-850.

Westheimer, G., and S. P. McKee (1977). Integration regions for visual hyperacuity. Vision Res. 17:89-93.

Wilson, H. R. (1979). Quantitative characterization of two types of line spread function near the fovea. Vision Res. (in press).

STUDIES OF THE KITTEN'S VISUAL SYSTEM

Development of Ganglion Cells in the Retina of the Cat

ANNE C. RUSOFF

Division of Biological Sciences
University of Michigan
Ann Arbor, Michigan USA

Abstract The eye of a cat grows significantly between birth and adulthood. Part of this growth occurs after the kitten has begun to exhibit visually guided behavior. Both the neural retina and the optical components of the eye participate in the growth. Unless these two components of the eye grow at the same rate, individual retinal neurons will receive spatial information from different amounts of the visual world at different times. Measurements of dendritic fields of retinal ganglion cells, specifically beta cells, show that many beta cells have reached their adult size at three weeks after birth, many weeks before the optical components of the eye are mature. Thus, the amount of visual world from which a ganglion cell receives spatial information must gradually decrease as the optical components of the eye grow. Measurements of the receptive-field center size of ganglion cells from kittens of various ages provide additional support for this idea.

The components of a cat's eye grow greatly between birth and adulthood. The surface area of the retina of a newborn kitten is only about 40% as large as it will be in the adult cat (M. A. Ransford, M. W. Dubin, and A. C. Rusoff, unpublished observations). Some of the growth of the retina is accomplished by adding new cells—the inner and outer nuclear layers add new neurons up to three weeks after a kitten is born. However, all the ganglion cells are present in the retina within 24 hours after birth (Johns, Rusoff, and Dubin, in preparation). Essentially all the growth of the ganglion cell layer after birth must then be accomplished by redistribution of existing cells. The retina grows rapidly during the first few weeks of life, but at three weeks after birth, its surface area is still only about 65% of the area of the adult retina (M. A. Ransford, M. W. Dubin, and A. C. Rusoff, unpublished observations). Since neurogenesis ceases at about this time, all the later growth must come from movement of existing cells (Johns, Rusoff, and Dubin, in preparation).

The growth of the retina occurs in conjunction with growth and maturation of the optical components of the eye. At birth the kitten's eye is not just a miniature adult eye. The anterior chamber occupies a much smaller fraction of the eye than it will in the adult so that most of the kitten's eye is posterior chamber. The different optical components of the eye then grow at different rates; the eye changes shape as well as size with age (Thorn, Gollender, and Erickson, 1976). One result of this growth is that the posterior nodal distance increases with age. Three weeks after birth the posterior nodal distance of the kitten's eye is approximately 8 mm (F. Thorn and M. Gollender, personal communication); that of the adult cat is approximately 12 mm (F. Thorn and M. Gollender, personal communication; Vakkur, Bishop, and Kozak, 1963). The posterior nodal distance determines the size of the retinal image of an object in the visual world. Using these values of posterior nodal distance, one can calculate that an object which subtends 1° of visual angle will have a retinal image 140 μm long in the 3-week-old kitten and 210 μm long in the adult cat. (These values are approximate and apply only to the central part of the retina. Hughes (1976) found a large variation in magnification factor between central and peripheral retina in the adult cat.) As the kitten's eye matures, a larger area on the retina receives input from a piece of the visual world subtended by 1° of visual angle. Since kittens begin making placing responses to purely visual stimuli by 3 weeks of age (Norton, 1974), the phase of retinal and optical maturation occurring after 3 weeks of age happens while the kitten is using its eyes.

Change in the retina during this period is not limited to stretching of the retina to increase its surface area. Neuronal cell bodies increase in size (A. C. Rusoff, unpublished observations; Donovan, 1966; Tucker, 1978; Vogel, 1978) and the plexiform layers increase in thickness (Tucker, 1978; Vogel, 1978), indicating that the dendrites and axons of cells are growing. New synapses are also being added (Cragg, 1975; Morrison, 1977). The pattern of this neuronal growth and formation of interconnections is of interest because it affects the kitten's ability to perceive the visual world. Individual neuronal elements in the retina may grow and make connections with other neurons such that their spatial information from the external world does not change during the period of growth. This result could be accomplished if the retina grows like a balloon being blown up, with the cell bodies and their processes simply expanding as the optical components of the eye inflate. Each neuron would then continuously receive information from both the same number of degrees of visual space and from the same part of visual space. Another possibility is that individual neurons achieve their adult size early in life and then the optical components of the eye grow independently. Increases in retinal surface area must then be accomplished by moving adult-sized cells apart. These two possibilities predict different changes in receptive-field measurements and measurements of the anatomical extent of retinal cells. The former possibility predicts that (1) *receptive-field centers* and *surrounds* of retinal cells should subtend the *same* number of degrees of visual angle in kittens and in adult cats, and (2) *anatomical correlates* of receptive-field center and surround should *increase* in size with age. The second possibility predicts that (1) *receptive-field centers* and *surrounds* of retinal cells should *decrease* in degrees of visual angle with age and (2) their anatomical correlates should be the *same* size in kittens and in adult cats. One might measure these parameters on any of the classes of cells in the retina. Since all visual information is funneled through the retinal ganglion cells, their pattern of development and of connections with the rest of the retina determines the visual information available to the rest of the visual system. Therefore, I have measured both the size of receptive-field centers of retinal ganglion cells and the extent of their reputed anatomical correlate, the dendritic fields of ganglion cells (Brown and Major, 1966), in an attempt to determine the pattern of retinal growth. (These measurements have been discussed previously in a different context (Rusoff and Dubin, 1977; Rusoff and Dubin, 1978).)

Anatomical Measurements

Brown and Major (1966) suggested that the dendritic-field size of a ganglion cell was related to the size of its receptive-field center. Although the exact equation does not appear valid (Nelson, Famiglietti, and Kolb, 1978), a strong correlation between dendritic-field size and the size of the receptive-field center does appear to exist (Boycott and Wässle, 1974; Levick, 1975). Therefore, the extent of the dendrites of ganglion cells from retinas of 3-week-old kittens and of adult cats was measured. Retinas were stained by a Golgi technique, and cells which appeared fully stained, that is, their dendrites either tapered down to a fine point or ended in a swelling, were drawn with a Leitz drawing tube. The mean of the longest diameter of each drawing and of the diameter orthogonal to it was used to calculate a diameter for each cell in microns on the retina.

All the ganglion cell types described by Boycott and Wässle (1974) in the adult cat retina were also found in the kitten retina. Figure 1 shows examples of alpha and gamma ganglion cells from a retina of an adult cat and from that of a 3-week-old kitten. A beta cell from a kitten retina is shown next to the alpha cell from a kitten to demonstrate the large differences in size and branching patterns of the dendrites between these two classes of cells. All cells are shown at the same magnification. Often the dendrites of the large alpha and gamma cells appeared chopped off as if they had failed to fill with the stain; it was difficult to determine if the entire dendritic field was visible.

Figure 2 shows examples of delta and beta ganglion cells from a retina of an adult cat and from that of a 3-week-old kitten. Note that Fig. 2 is at a higher magnification than Fig. 1. Only a few delta ganglion cells were found in the retinas of either adult cats or kittens. However, many beta ganglion cells were found with dendrites which appeared fully stained. Therefore, a comparison of the diameters of the dendritic fields of beta cells from the retinas of 3-week-old kittens and of adult cats was made; the results are shown in Fig. 3.

The diameters of the dendritic fields of the beta cells from the retinas of adult cats (diamonds) agree fairly well with the diameters of beta cells measured by Boycott and Wässle (1974) on retinas stained by a similar Golgi technique. The solid lines on Fig. 3 show the range of their measurements. It is

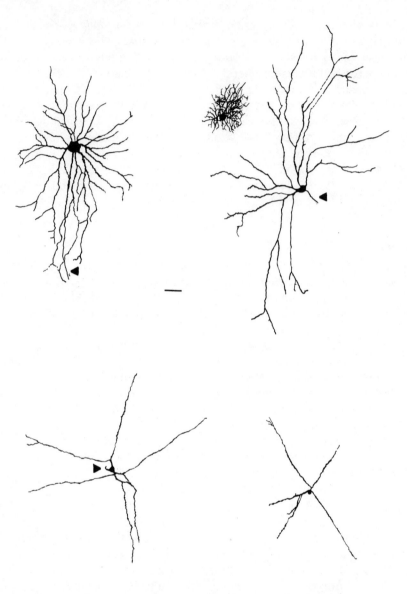

FIGURE 1 Drawing of ganglion cells from adult cat (left) and 3-week-old kitten (right) retinas stained by a Golgi technique. Top: Alpha cells with a kitten beta cell for comparison. The adult alpha cell is 7.7 mm from the area centralis and has a dendritic field diameter of 515 μm; the kitten alpha cell is 7.9 mm from the area centralis and has a dendritic-field diameter of 682 μm; the kitten beta cell (shown at higher magnification in Fig. 2) is 5.1 mm from the area centralis and has a dendritic-field diameter of 147 μm. Bottom: Gamma cells. The adult gamma cell is 10.8 mm from the area centralis and has a dendritic-field diameter of 463 μm; the kitten gamma cell is 6.8 mm from the area centralis and has a dendritic-field diameter of 380 μm. Scale: 50 μm. Arrowheads indicate the axon of each cell.

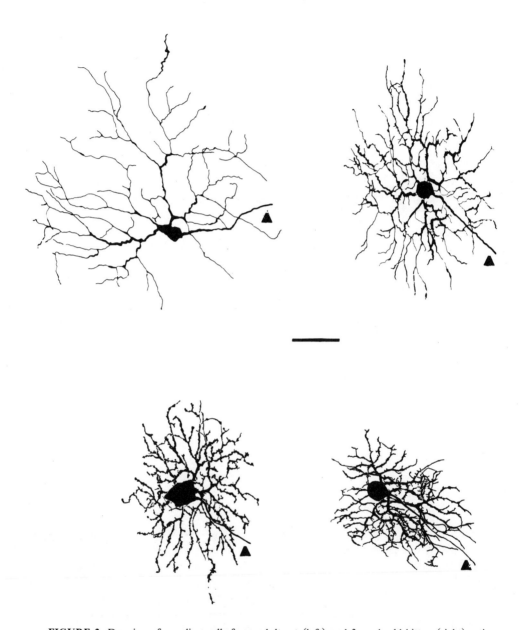

FIGURE 2 Drawing of ganglion cells from adult cat (left) and 3-week-old kitten (right) retinas stained by a Golgi technique. Top: Delta cells. The adult delta cell is 13.3 mm from the
area centralis and has a dendritic-field diameter of 269 μm; the kitten delta cell is 9.2 mm
from the area centralis and has a dendritic-field diameter of 213 μm. Bottom: Beta cells
(reprinted with permission from Invest. Ophthalmol. Visual Sci. 17:819-821 (1978)). The
adult beta cell is 4.9 mm from the area centralis and has a dendritic-field diameter of 177 μm;
the kitten beta cell is 5.1 mm from the area centralis and has a dendritic-field diameter of 147
μm. Scale: 50μm. Arrowheads indicate the axon of each cell.

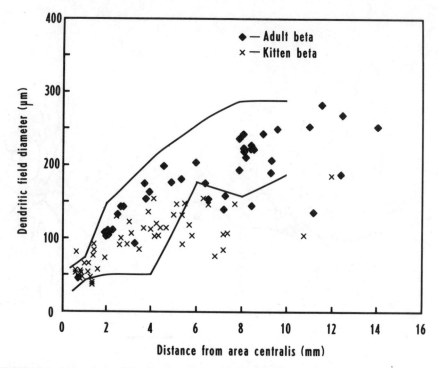

FIGURE 3 Comparison of the dendritic-field diameters of beta ganglion cells from the retinas of adult cats and of 3- week-old kittens. Dendritic-field diameter is defined here as the arithmetic mean of the longest diameter of the dendritic field and of the diameter orthogonal to it. (Reprinted with permission from Invest. Ophthalmol. Visual Sci. 17:819-821 (1978)). Solid lines encompass the range of dendritic-field diameters of beta ganglion cells from the retinas of adult cats measured by Boycott and Wässle. (Redrawn from Fig. 7, J. Physiol. 240:397-419 (1974).)

greater than the range of my measurements at most distances from the area centralis, probably because their sample of cells was larger than mine. Most of the dendritic-field diameters of adult cells in my sample are within the range of their measurements. Near the area centralis the dendritic fields are quite small, below 50 μm in diameter; with increasing eccentricity, the dendritic-field diameters increase, as noted by Boycott and Wässle (1974), approaching, but not exceeding, 300 μm in diameter at the far periphery.

Near the area centralis the beta cells from kittens (crosses) and from adult cats appear to have dendritic fields of the same size, although my beta cell sample from adult retinas is too small to be certain. However, the dendritic-field sizes measured by Boycott and Wässle form an envelope which encompasses the sizes of both my adult and kitten cells up to 5 mm from the area centralis. At eccentricities greater than 5 mm, the beta cells from adult retinas, both in my sample and in that of Boycott and Wässle, are larger than the beta cells from kittens. At some eccentricities the diameters of the adult cells are twice as large as those of the cells from 3-week-old kittens. Thus, there is a gradient of development of the dendritic fields across the retina; beta cells in central retina have achieved their adult size three weeks after birth, but those in peripheral retina are still growing at that time. Further studies are necessary to determine when the beta cells of peripheral retina reach their adult size and whether the other classes of ganglion cells follow this same pattern of development.

Since the ganglion cell layer of the retina of the kitten must still increase in area without adding any new neurons, at least some of the kitten cells will be farther from the area centralis in the adult retina than they are at 3 weeks of age. If the kitten cells could be shown in Fig. 3 in their final adult position with respect to the area centralis, many cells with diameters within the adult range in their present position would probably be moved to a part of the graph where all the adult cells are larger than the kitten cells. However, the extent of this effect cannot be determined until the pattern of movement of ganglion cells as the retina grows to its adult size is understood. At present this is not known.

There are at least two possible ways for the ganglion cell layer to achieve its final state. The retina may stretch symmetrically across its entire extent with each ganglion cell moving away from its neighbors by a constant amount. Or only the peripheral retina, where dendrites are still growing, may stretch. The mature-sized cells in the central retina could then maintain their interactions with neighbors without having their neighbors change. Measurements of the increase in distance between the optic disc and the area centralis with age provide some evidence that growth is symmetrical. Tucker (1978) found an increase of 17% in this length between three weeks and adulthood in the cat. If the total length of the retina, measured from one retinal margin to the other through the retina (Johns and Easter, 1977), is 23 mm in the 3-week-old kitten (M. A. Ransford, M. W. Dubin, and A. C. Rusoff, unpublished observations) and 28 mm in the adult cat (Hughes, 1976), the increase in total length is about 23%; the two percentages are essentially equivalent, suggesting that each part of the retina stretches about 20% in the linear dimension. However, Tucker did not measure the total length of the retina in the eyes she studied. Since different histological procedures produce different retinal lengths in eyes from animals of the same age (Johns, Rusoff, and Dubin, in preparation), the pattern of retinal growth cannot be determined by comparing the measurements made by Tucker with retinal lengths measured by others. Comparisons of the positions of landmarks within the retina at different ages will only be useful in determining the pattern of stretching when both the total retinal length and the distances between the landmarks are measured on the same retinas.

To summarize the anatomical evidence, the beta ganglion cells in the central retina mature first and have reached their adult size at three weeks after birth, that is, at about the time the kitten begins making visually guided placing responses. Thus, the kitten has mature-sized ganglion cells in its area of greatest visual acuity when it is learning to view the world. However, the kitten's eye is still only about two-thirds of its mature size and, therefore, has a different magnification factor from the adult eye. If these ganglion cells are using the full extent of their dendrites to form adult-like connections, the amount of the world from which each of these cells receives visual information must be greater than it will be in the adult. The receptive fields of these ganglion cells should receive information from one and one-half times as many degrees of visual angle as they will when the optical components of the eye grow to their adult size. The next section discusses measurements of these receptive fields.

Electrophysiological Measurements

The size of the receptive-field center was measured while recording extracellularly from individual ganglion cells from the retinas of kittens 3-7 weeks old and of adult cats. The technique has been described previously (Cleland, Levick, and Sanderson, 1973; Rusoff and Dubin, 1977). All ganglion cells studied were at least 7° from the center of the area centralis but within 30° of it. In anatomical terms these positions correspond approximately to the region between 1 and 4.5 mm from the center of the area centralis, that is, in the part of the retina where the dendritic fields of the ganglion cells appear to have reached their adult size by three weeks after birth (Fig. 3). The responses of each cell to a battery of tests were used to assign the cell to one of the classes of ganglion cells—brisk X, brisk Y, or sluggish; the cells which did not fit these categories were classified as unknown. Then the size of the receptive-field center of each cell was measured using an area-threshold technique (Cleland, Levick, and Sanderson, 1973; Rusoff and Dubin, 1977). The size of the receptive-field center of each ganglion cell is shown in Fig. 4, plotted as center diameter in degrees of visual angle versus the age of the kitten; cell type is indicated by the different symbols, as noted on the figure.

As predicted, many of the ganglion cells measured in the retinas of kittens have receptive-field centers larger than those of any of the adult ganglion cells. Many cells do have centers within the adult range. However, there are clear differences in the distribution of center size within this range. No centers as small as the smallest adult centers are found until seven weeks after birth, and the median center diameter is larger in kittens than in adults. These generalizations apply to the measurements on cells in the different classes as well as to measurements on the population as a whole. Many of the kitten cells responded clearly to visual stimuli; these responses were used to divide the cells into the same classes as the adult cells. Within a ganglion cell class the average receptive-field center is larger in the kitten retinas than in the adult retinas. As an example, the average brisk X center is 1.3° in diameter in 4-week-old kittens, 1.0° in diameter in 7-week-old kittens, and 0.8° in diameter in adult cats. (The average adult value was calculated from individual center sizes, not from their range.) Brisk X cells are probably the electrophysiological

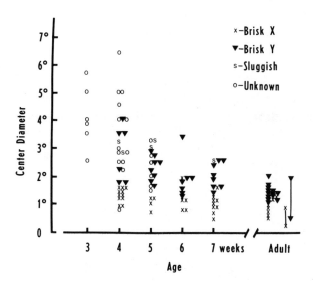

FIGURE 4 Change in receptive-field center diameter with age. Each point indicates the center diameter determined for a single cell using the area-threshold measurement technique. Cell type is indicated by the point markers illustrated on the figure. The adult center measurements came from two sources: points indicate individual center diameters measured on cells in three adult cats; bars indicate the range of center diameters of brisk X and brisk Y cells measured by Cleland, Levick, and Sanderson (1973). (Figure reprinted with permission from J. Neurophysiol. 40:1188-1198 (1977).)

equivalents of the anatomical class of beta cells (Boycott and Wässle, 1974; Levick, 1975). Thus, for at least this one class of cells in the area of the retina studied, both the electrophysological and anatomical evidence suggest that cells attain their mature anatomical size early and the optical components of the eye grow later, causing the part of the visual world viewed by each cell to shrink.

The gradual shrinkage of the size of the receptive-field centers is also apparent in Fig. 4. The median center diameter decreases between three and seven weeks after birth. During this four-week period the whole eye grows; but at 7 weeks of age it is still significantly smaller than the adult eye, having a posterior nodal distance of less than 9.5 mm compared to the adult value of 12 mm (F. Thorn and M. Gollender, personal communication). Thus, center diameters above the adult size are still expected and are found, both in the total range of measurements and within the measurements on each class of ganglion cells, but these centers are not so much larger than the adult centers as they were four weeks after birth.

One other set of cells must be discussed. These are the cells found in 3- and 4-week-old kittens, and classified as unknowns, whose receptive-field centers are even *larger* than predicted. The difference in posterior nodal distance alone between the eyes of kittens and of adult cats predicted receptive-field centers for the 3-week-old cells as much as one and one-half times the adult size, or up to 3.5° in diameter. However, some centers up to 6.5° in diameter were found. These very large centers were not found in kittens older than 4 weeks. These very large centers may be the result of real immaturities in retinal connections, causing them to be more extensive in kittens than in adults, but they may also be an artifact of the poor optical quality of the eye. There is a membrane, the *tunica vasculosa lentis,* over the lens in the eye of a 3-week-old kitten; this membrane probably scatters light extensively (Thorn, Gollender, and Erickson, 1976; Freeman and Lai, 1978). One must consider the possibility that this scattered light affected the measurement of center size, causing the centers to appear larger than they really were. By 4 weeks of age the *tunica vasculosa lentis* is largely gone (Thorn, Gollender, and Erickson, 1976; Freeman and Lai, 1978), although its remnants or other optical imperfections may have affected measurements in some eyes, causing the appearance of the few very large (greater than 3.5° diameter) centers measured in 4-week-old kittens. Linespread functions suggest that the optical ability of the kitten approaches that of the adult between four and five weeks after birth (Bonds and Freeman, 1978). Therefore, the measurements of center diameter made after 4 weeks of age, and most of those made in the 4-week-old kittens, were probably not severely affected by scattered light; the center sizes measured at these ages then are probably a true reflection of the compromise between dendrites which are adult in extent and optical components of the eye which are smaller than they will be in the adult. Additional support for this idea comes from the clear responses of many of these cells to visual stimuli, including gratings. The ability of these cells to detect many of these stimuli suggests that they were viewing the stimuli through relatively clear optics.

In summary, some of the ganglion cells mature early in the kitten's life—those within the central part of the retina achieve dendritic fields of adult size by three weeks after birth. As the optical components of the eye grow, these cells view a gradually decreasing portion of the visual world. The rest of the kitten's visual system must then cope with this slow change in the meaning of incoming information. Whether the ganglion cells also change their relationships with their neighbors is dependent on the manner in which the retina attains its final size—either pulling these cells away from their neighbors as it expands symmetrically, or leaving the central cells in place and moving only the more peripheral cells. Since there is a gradient of maturation across the retina, the peripheral cells are still growing at this time. What these cells "see" must go through an extensive period of change as their dendrites and the optical components of the eye grow, and the retina stretches.

Acknowledgements The work summarized in this paper was done in the laboratory of Dr. Mark Wm. Dubin and was supported by National Science Foundation Grants BNS76-00506 to M. W. Dubin. I also thank Dr. Pamela Johns and Dr. S. S. Easter, Jr. for criticism of the manuscript and Ms. DeAnn Madrid for typing the manuscript. I was supported by National Eye Institute Research Fellowship Award 1F32 EY05294-01 during the preparation of the manuscript.

References

Bonds, A. B., and R. D. Freeman (1978). Development of optical quality in the kitten eye. Vision Res. 18:391-398.

Boycott, B. B., and H. Wässle (1974). The morphological types of ganglion cells of the domestic cat's retina. J. Physiol. 240:397-419.

Brown, J. E., and D. Major (1966). Cat retinal ganglion cell dendritic fields. Exp. Neurol. 15:70-78.

Cleland, B. G., W. R. Levick, and K. J. Sanderson (1973). Properties of sustained and transient ganglion cells in the cat retina. J. Physiol. 228:649-680.

Cragg, B. G. (1975). The development of synapses in the visual system of the cat. J. Comp. Neur. 160:147-166.

Donovan, A. (1966). The postnatal development of the cat retina. Expt'l. Eye Res. 5:249-254.

Freeman, R. D., and C. E. Lai (1978). Development of the optical surfaces of the kitten eye. Vision Res. 18:399-407.

Hughes, A. (1976). A supplement to the cat schematic eye. Vision Res. 16:149-154.

Johns, P. R., and S. S. Easter (1977). Growth of the adult goldfish eye. II. Increase in retinal cell number. J. Comp. Neur. 176:331-342.

Levick, W. R. (1975). Form and function of cat retinal ganglion cells. Nature 254:659-662.

Morrison, J. D. (1977). Electron microscopic studies of developing kitten retina. J. Physiol. 273:91-92P.

Nelson, R., E. V. Famiglietti, and H. Kolb (1978). Intracellular staining reveals different levels of stratification for on- and off-center ganglion cells in cat retina. J. Neurophysiol. 41:472-483.

Norton, T. T. (1974). Receptive-field properties of superior colliculus cells and development of visual behavior in kittens. J. Neurophysiol. 37:674-690.

Rusoff, A. C., and M. W. Dubin (1977). Development of receptive-field properties of retinal ganglion cells in kittens. J. Neurophysiol. 40:1188-1198.

Rusoff, A. C., and M. W. Dubin (1978). Kitten ganglion cells: dendritic field size at 3 weeks of age and correlation with receptive field size. Invest. Ophthalmol. Visual Sci. 17:819-821.

Thorn, F., M. Gollender, and P. Erickson (1976). The development of the kitten's visual optics. Vision Res. 16:1145-1150.

Tucker, G. S. (1978). Light microscopic analysis of the kitten retina: postnatal development in the area centralis. J. Comp. Neur. 180:489-500.

Vakkur, G. J., P. O. Bishop, and W. Kozak (1963). Visual optics in the cat, including posterior nodal distance and retinal landmarks. Vision Res. 3:289-314.

Vogel, M. (1978). Postnatal development of the cat's retina. Adv. Anat. Embryol. Cell Biol. 54:1-66.

Development of Orientation Tuning in the Visual Cortex of Kittens

A. B. BONDS

School of Optometry
University of California
Berkeley, California USA

Abstract Orientation tuning of single units in striate cortex of kittens aged 2-6 weeks was measured quantitatively. Results are based on 227 visually responsive units from 36 kittens of which 9 were reared in complete darkness. Cells were classed as non-oriented (N.O.), orientation-biased (O.B.), and orientation-selective (Or.) using objective criteria. Specificity of O.B. and O.S. cells was gauged using the half-width of the orientation tuning curve at half the maximum response amplitude.

In 2-week-old kittens, O.B. and O.S. cells comprised 41% and 38% of the total (responsive) sample ($N = 39$), respectively. By 5 to 6 weeks of age, O.B. represented 10% and O.S. 86% of the sample ($N = 61$), with intermediate ages showing a roughly linear progression between the two levels. In contrast, dark-reared animals of up to 6 weeks of age retained distributions nearly identical with that seen at 2 weeks. Orientation specificity of cells in normally reared kittens improved from a mean of 29° at 2 weeks to essentially adult performance (19°) at 5 and 6 weeks. For the dark-reared kittens this figure remained essentially static through 6 weeks of age.

Discounting such factors as optical blur (the justification of which will be discussed), it may be concluded that visual experience plays a major role in refining the neural mechanisms responsible for orientation selectivity. Moreover, total visual deprivation "freezes" both the number and specificity of oriented cells at the level found in very young kittens.

Introduction

Single units in striate visual cortex (area 17) generally respond well only to a narrow range of stimuli. Hubel and Wiesel (1962) first showed that in the normal adult cat these cells, unlike more distal elements of the visual pathway, require a bar or edge stimulus to be oriented at or near a specific azimuth to drive the cell effectively. This property is one of the major consequences of the

31

processing of visual information within striate cortex; the degree of orientation specificity of single units can therefore be useful as an indicator of the level of sophistication of cortical mechanisms during development.

Orientation specificity in visually naive (newborn or dark-reared) and growing kittens has been studied several times, but in the current literature results are inconsistent. Hubel and Wiesel (1963) found cells in the cortex of young kittens to be sluggish and to fatigue easily, but reported nonetheless almost adult performance with regard to orientation specificity. Since this property was present in animals who had no prior visual experience, they concluded that such specificity was genetically determined. Pettigrew (Barlow and Pettigrew, 1971; Pettigrew, 1974), on the other hand, found very few orientation-selective cells in young kittens and suggested rather that such a capability was acquired via visual experience. More recent work suggests a compromise, with some cells in young kittens having distinct preferences for orientation and others being totally non-specific (e.g., Blakemore and Van Sluyters, 1975; Buisseret and Imbert, 1976). There remains some disagreement over the proportions of these cells. The consequences of dark-rearing are also not consistently described. One study shows all cells in (6-week-old) dark-reared animals to be "non-specific" (Buisseret and Imbert, 1976), while another reports over 90% of such cells to have at least some sensitivity to stimulus orientation (Sherk and Stryker, 1976).

A probable cause for these differences is that, in most of these studies, orientation specificity was assessed by categorization into loosely specified classes based on subjective judgments of the experimenter. The variability and weakness of responses of cells in young or deprived kittens can be a confounding factor in such observations. The present study is an attempt to overcome these difficulties by describing the progressive maturation of the response characteristics of these cells in an objective manner, using explicit quantitative criteria. By comparing response properties of cells in very young and growing kittens, it can be shown by just how much and over what time period orientation selectivity improves. Adding to this sample a group of cells from kittens reared in total darkness enables the description of whether (and, if so, by how much) the improvements in orientation selectivity are dependent on visual experience.

The strength of a strictly quantitative approach as taken here lies not in the absolute accuracy with which specificity can be measured, but rather in the consistency and reliability of the answers. By counting nerve impulses under a given test condition, one arrives at a number. The significance of this number is up to the experimenter, but it can be used to calculate a performance index which is consistent between experiments (and experimenters). Also, single cells in striate cortex are notorious for variation of their excitability over time. By spreading a measurement over several minutes and interleaving test configurations (Henry, Bishop, Tupper, and Dreher, 1973) the data reflect more reliably the performance of a cell for all time as opposed to a single moment.

Methods

This study is based on quantitative recording from 233 cells in 36 kittens aged 12 days to 6 weeks. Nine of these kittens were reared in total darkness. The kittens were prepared for recording using standard methods (e.g., Blakemore and Van Sluyters, 1975). Brevital anesthesia was used for surgery, and nitrous oxide (75% N_2O/25% O_2) during recording. For the entire project, several different response properties were measured for each cell—response strength and variability, orientation, direction, and velocity selectivity, response planes (to determine cell type) and response habituation. The present discussion will be limited to results concerning direction and orientation selectivity.

As pointed out by Henry, Bishop, and Dreher (1974) and reiterated by Hammond earlier this week, selectivity of direction of motion and orientation of a stimulus may well involve two different mechanisms, and in the strict case these can be separated during testing. Directional selectivity is tested by observing changes in response for different directions of motion of a radially symmetric (spot) stimulus traversing a receptive field. Selectivity for orientation is resolved by flashing a stationary extended bar or edge at different azimuths. Unfortunately, neither of these configurations constitute an effective stimulus for many cells in very young kittens. The general test used here was simply to drift a bar in various directions (orthogonal to its orientation) across the receptive field; directionality was only considered along (to and fro) the favored axis of motion.

Testing was initiated by acquiring a single unit and manually assessing, with a bar stimulus, the location of its receptive field in visual space together with a best guess at its preferred orientation. Control was then passed to the computer, which drifted the bar across the receptive field at each of 12 orientations (30° intervals), one of them coincident with the best orientation found by plotting manually. Each stimulus was repeated at least ten times in quasi-random order (interleaving as per Henry et al., 1973), with intervals of up to 20 seconds between sweeps to compensate for response habituation, if present. When finished, a plot of response (spikes/sweep less the maintained discharge) vs. orientation was generated.

In many cases the orientation tuning of the cells was so broad that use of the usual coefficient for selectivity (see below) was not suitable. Instead, all results were first subject to the following categorization: the ratio of the largest response (at the preferred orientation) to the smallest response (usually orthogonal to the preferred orientation) was taken. If this ratio was 10:1 or better, the cell was classed as orientation-selective (Or.). If the ratio was less than 10:1 but greater than 2:1, the cell was judged to be orientation-biased (O.B.), and if the ratio was less than 2:1, the cell was considered non-oriented (N.O.). Similar criteria were used to class direction-selective (or unidirectional), direction-biased, and non-directional cells, with the ratio being taken between the response along the favored direction of motion and its exact opposite.

In instances when the cells proved orientation-selective, additional responses were measured around the favored orientation using smaller intervals (10° or less) between orientations. This resulted in a clearer picture of the tuning curve and allowed the derivation of a figure of merit for tuning, using half the width of the tuning curve at half the maximum response amplitude (Henry, Dreher, and Bishop, 1974; Rose and Blakemore, 1974).

Results

Orientation specificity Distributions of the three classes of orientation specificity (see Methods) in normally reared kittens aged 2 to 6 weeks are shown in Fig. 1(a). As the kittens mature, a clear progression is evident. The fraction of oriented cells, designated by the filled bar segments, increases from 38% of the sample (at 2 weeks) to 85% (at 5 and 6 weeks). Both N.O. and O.B. cells are

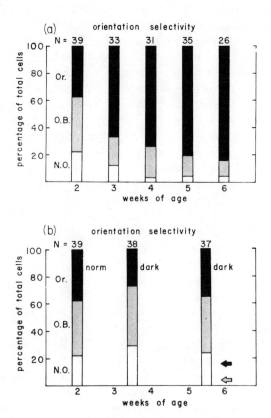

FIGURE 1 Proportions of classes of orientation selectivity within population of responsive cells as a function of age. Or.: orientation-selective; O.B.: orientation-biased; N.O.: non-oriented; definitions as per text. (a) Normally reared kittens, 2 to 6 weeks of age. (b) Normally reared kittens, 2 weeks of age, compared with dark-reared kittens (combined 3-4 week and 5-6 week).

decreasing in proportion, so it seems likely that with age and visual experience both of these classes are increasing their selectivity to orientation. These data of necessity represent proportions of visually responsive cells. This fraction (of all cells encountered) also increases with age and visual experience, from about 60% at 2 weeks to over 85% at 6 weeks.

Maturation of the kittens without visual experience produces distributions among the cell classes which are significantly different from normal (Fig. 1(b)). Data pooled from 3- and 4-week-old as well as 5- and 6-week-old dark-reared kittens show no large change in the proportion of orientation-selective units (about 35%); overall, the distributions resemble that of the (essentially visually naive) 2-week-old normal group. It thus appears that maturation without visual experience (at least to 6 weeks of age) "freezes" cortical orientation selectivity at the state found shortly after eye-opening. We also see from Fig. 1(a) that the benefits of allowing vision during rearing are apparent by 3 weeks of age, where there is a clear divergence from the two-week (or dark-reared) distribution.

For those cells judged to be orientation-selective, it was possible to measure their orientation tuning. This is plotted (half-width at half-height) as a function of age in Fig. 2. The normally reared kittens (filled circles) show a monotonic improvement (decrease) of the mean value for tuning from about 29° at 2 weeks to an asymptote of 20° at 5 and 6 weeks. This latter value is near the adult mean of 18°-19° reported by Rose and Blakemore (1974) and Henry, Dreher, and Bishop (1974), so the adult level of performance is very nearly

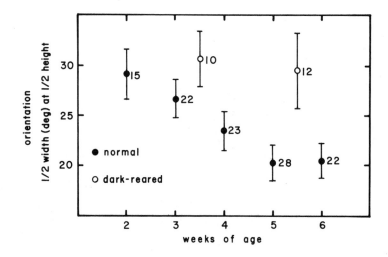

FIGURE 2 Orientation specificity (of orientation-selective cells) as a function of age. The ordinate represents the half-width of the tuning curve at half the maximum response (see text). Filled circles: normally reared kittens; open circles: dark-reared kittens. Bars represent ± one standard error of the mean.

achieved by 5 weeks of age. The points for the dark-reared kittens (open circles) do not fit in the normal progression but rather remain overlapping with the sample from the normally reared 2-week-old kittens, again emphasizing that dark-rearing just maintains the minimal performance seen in the youngest animals.

Direction selectivity The testing of directional selectivity (within the limits discussed earlier) yields substantially the same message. Figure 3(a) shows the improvement of directional selectivity with age in normal kittens. In this case, a stabilization appears by the fourth week. The apparent retrogression in the fifth week may be attributable to a large proportion of complex cells (which are generally less direction-selective than simple cells; Goodwin and Henry, 1975) in that sample. Dark-rearing again shows little change in the distribution of classes up to 6 weeks of age, as shown by the comparison of the youngest normal animals with the dark-reared animals (Fig. 3(b)).

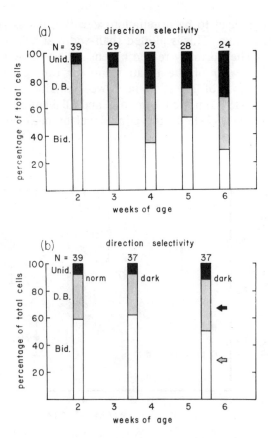

FIGURE 3 Proportions of classes of direction selectivity within population of responsive cells as a function of age. Unid.: unidirectional; D.B.: direction-biased; Bid.: bidirectional; definitions as per text. (a) and (b) represent the same groups as described in Fig. 1.

Possible complications The results above, taken at face value, show that while some cells are initially orientation-selective, maturation with visual experience is necessary for normal development of the cortical mechanisms subserving orientation and direction selectivity. However, at least with regard to stimulus orientation, improvement of selectivity may not stem from changes in the neural substrate specific for that property but from other causes. Two possibilities are considered below.

First, it might be argued that the weakness of responses from very young animals may obscure inherently fine orientation tuning. The improvement with age might then reflect simply an enhancement of responsiveness rather than of orientation selectivity *per se*; the failure to improve in the dark would then be interpreted as a degeneration. Figure 4 is a scatter plot describing cells in 2-week-old kittens relating the degree of orientation specificity with the largest response attainable with a bar stimulus. In contrast with the suggestion above, the cells with the best orientation selectivity show in fact rather weak responses relative to the total sample, and many vigorously responding units had poor specificity. A similar result was found for dark-reared animals.

A second problem arises from the fact that the optical quality of the eyes of young kittens is initially quite poor (Bonds and Freeman, 1978). Blurring the image of a test stimulus might exaggerate the coarseness of tuning, which

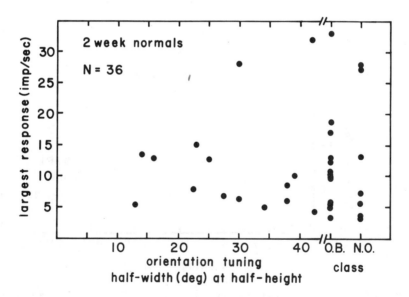

FIGURE 4 Scatter plot of the best response elicited from a cell compared with its orientation selectivity; 2-week-old kittens. This figure shows that fine selectivity is not simply a result of good responsiveness (better signal/noise ratio). Larger responses seem in fact to be associated with broader selectivity.

would improve with the clearing of the optical media. If this were true, the poor specificity found in older dark-reared kittens (where optics do not present a problem) would again, as suggested above, reflect a degeneration of performance. Figure 4 shows, however, that the effect of poor optics on orientation tuning cannot be great. Some of the units in 2-week-old kittens had tuning narrower than the adult mean (of about 19°), and these were found in the same eyes which had a great many poorly tuned units.

As a final check I simply tested the effect of optical blur on orientation tuning in a mature cat. A cell with reasonably tight tuning (17°-18°) was acquired, and tuning curves were measured (Fig. 5) with several optical perturbations: addition (to the best refraction for the viewing distance) of +4, +8, and +12 diopter lenses, a diffusing plate (sandblasted plexiglass) and the diffusing plate combined with a +8 diopter lens. After all of these tests, a measure under normal conditions was again made to assure stability of the tuning curve. Figure 5 shows that no large or systematic change in the tuning is caused by these kinds of optical degradation. As an indicator of the suitability of this test, a +4 diopter lens alone is sufficient to attenuate the optical transfer function of the adult cat eye to resemble that of a 2-week-old kitten (Bonds and Freeman, 1978). The additional perturbations serve to degrade the image still further, but do not seem to alter the orientation tuning.

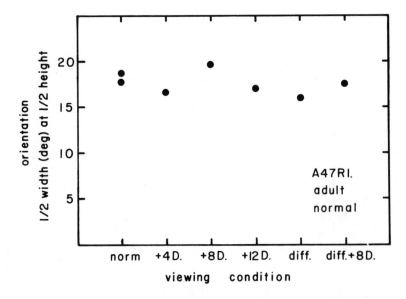

FIGURE 5 Change of orientation tuning with optical degradation. Tuning of a cell in an adult cat was measured in the usual way (norm) and with interposition of several optical perturbations: +4, +8 and +12 diopter auxiliary lenses, a diffusing plate, and a +8 diopter lens in combination with the diffusing plate. No large or systematic change is seen in the half-width of the tuning curve at half-height.

Discussion

These experiments demonstrate clearly that a substantial proportion (38%) of units in very young kittens are selective to the orientation of a bar stimulus. Allowing maturation with visual experience, this proportion approaches adult levels (85%) by the time the kitten reaches 5-6 weeks of age. With total light deprivation, the characteristics of cortical cells (at least with regard to orientation and direction selectivity) remain static.

Experiments with similar intent have been performed before many times (e.g., Hubel and Wiesel, 1963; Barlow and Pettigrew, 1971; Pettigrew, 1974; Sherk and Stryker, 1976; Blakemore and Van Sluyters, 1975; Buisseret and Imbert, 1976) but with rather disparate and often contradictory conclusions. One purpose of the present series of experiments is to serve as a guide for the rationalization of previous results. Use of objective methods for data collection and classification separates the complicating factors of sluggishness and response habituation (which may influence subjective judgments) from the measurement of orientation selectivity, and a consistent progression of development emerges.

As found by Hubel and Wiesel (1963), some cells in extremely young kittens possess "adult-like" properties, especially with regard to orientation selectivity. The primary difficulty with their report is the suggestion that since "preference in stimulus orientation was common to all of the units isolated" in large part all cortical connections were present at birth. Although their conclusion may have been troubled by the small sample of cells in their study, it should be remembered that what they found remarkable (and therefore emphasized) was the clear existence of such neural mechanisms at an early stage of development. The strongest disagreement with this concept, from Barlow and Pettigrew (1971) and Pettigrew (1974) is based primarily on studies of units selective for binocular disparity. The report that few, if any, cells in young kittens are selective for orientation (Pettigrew, 1974) makes use of a much more strict definition of orientation selectivity than is used normally, and does not invite a direct comparison. Subsequent studies (Blakemore and Van Sluyters, 1975; Buisseret and Imbert, 1976) estimate about 25% of the cells in young kittens to be orientation selective. While this is rather less than the 38% reported here, many of the finely tuned cells in this study had small responses (Fig. 4) and these might have been classed differently using less strict criteria.

Although judgments of the initial conditions vary, there is general agreement on the rate at which improvement of orientation selectivity proceeds under normal rearing conditions. Pettigrew (1974), Blakemore and Van Sluyters (1975) and Buisseret and Imbert (1976) all suggest adult-like specificity is achieved by 4 weeks of age. The present data (Figs. 1(a),(b)) show there is improvement between the fourth and fifth week, but this is primarily an enhancement of the specificity of individual cells rather than large changes in the proportions of selective cells, and might therefore have been overlooked in the other studies.

There remains some disagreement over the consequences of dark-rearing. Sherk and Stryker (1976), who tested 4-week-old dark-reared kittens, found remarkable (as did Hubel and Wiesel for the youngest kittens) the degree of cortical sophistication in these animals. They report 90% of the cells studied "responded selectively to the orientation of a moving bar stimulus." They did not, however, discriminate between orientation-based and orientation- selective cells and as their data are presented without reference to a maintained discharge this separation cannot be made *post hoc.* In the present study, combination of both the orientation-biased and orientation-selective cells of dark-reared kittens constitutes about 80% of the total sample, which is sufficiently close to the finding of Sherk and Stryker to remove any significant basis for disagreement. The results of Buisseret and Imbert (1976) are not so easy to reconcile. They found nothing but "non-specific" cells in the cortex of 6-week-old dark-reared kittens. How their interpretation of "non-specific" compares with other definitions is problematic. What is certain is that, in the present study, 59% (13/22) of the orientation-selective cells (or 17% of all responsive units) in dark-reared cats had tuning specificity within one standard deviation of the adult norm of Rose and Blakemore (1974), and it would be misleading to represent these cells as "immature" or "non-specific." Although Blakemore and Van Sluyters (1975) suggest there "may be progressive degradation following prolonged pattern deprivation," they also found 10-20% orientation-selective units in 3-to-6-week-old dark-reared kittens. On the basis of the presence of such cells, cortex of dark-reared kittens (at least to 6 weeks of age) must retain some degree of orientation specificity. As suggested by Michel Imbert earlier in this meeting, cells which are intrinsically specified may be more resistant to experimental modification, and it seems reasonable to extend this concept to include resistance to degeneration as a result of visual deprivation as well.

References

Barlow, H. B., and J. D. Pettigrew (1971). Lack of specificity in the visual cortex of young kittens. J. Physiol. (Lond.) 218:98-100P.

Blakemore, C. B., and R. C. Van Sluyters (1975). Innate and environmental factors in the development of the kitten's visual cortex. J. Physiol. (Lond.) 248:663-716.

Bonds, A. B., and R. D. Freeman (1978). Development of optical quality in the kitten eye. Vision Res. 18:391-398.

Buisseret, P., and M. Imbert (1976). Visual cortical cells: Their developmental properties in normal and dark-reared kittens. J. Physiol. (Lond.) 255:511-525.

Goodwin, A. W., and G. H. Henry (1975). Direction selectivity of complex cells in comparison with simple cells. J. Neurophysiol. 38:1524-1540.

Henry, G. H., P. O. Bishop, R. M. Tupper, and B. Dreher (1973). Orientation specificity and response variability of cells in the striate cortex. Vision Res. 13:1771-1779.

Henry, G. H., P. O. Bishop, and B. Dreher (1974). Orientation axis and direction as stimulus parameters for striate cells. Vision Res. 14:767-778.

Henry, G. H., B. Dreher, and P. O. Bishop (1974). Orientation specificity of cells in cat striate cortex. J. Neurophysiol. 37:1394-1409.

Hubel, D. H., and T. N. Wiesel (1962). Receptive fields, binocular interaction and functional architecture in the cat's visual cortex. J. Physiol. (Lond.) 160:106-154.

Hubel, D. H., and T. N. Wiesel (1963). Receptive fields of cells in striate cortex of very young, visually inexperienced kittens. J. Neurophysiol. 26:994-1002.

Pettigrew, J. D. (1974). The effect of visual experience on the development of stimulus specificity by kitten cortical neurones. J. Physiol. 237:49-74.

Rose, D., and C. B. Blakemore (1974). An analysis of orientation selectivity in the cat's visual cortex. Exp. Brain Res. 20:1-17.

Sherk, H., and M. P. Stryker (1976). Quantitative study of cortical orientation selectivity in visually inexperienced kittens. J. Neurophysiol. 39:63-70.

Maturation of Visual Cortex
with and without Visual Experience

MICHEL IMBERT

Laboratoire de Neurophysiologie
Collège de France
Paris, France

Abstract The evolution of the response properties of visual cortical cells was studied in two groups of kittens between 1 and 7 weeks of age—one group normally reared, the other reared in complete darkness. Four classes of striate neurones were defined: (a) non-activatable cells, (b) non-specific cells, (c) immature cells, (d) specific cells that are as selective for orientation as the simple or complex cells of the adult cat. The results confirm that as soon as neurones become visually activated, about 25% of the recorded visual units are definitely specific in terms of orientation selectivity. These neurones are present in earliest stages, even in the absence of any visual experience. However, active visual experience (visuomotor interaction) is necessary to maintain and develop these specific cells after the third week of postnatal life. Polar diagrams of the orientation encoded by specific cells show that for kittens under 3 weeks of age and whatever the rearing condition, there are more specific cells coding horizontal or vertical orientation than those coding oblique orientations. These horizontally and vertically oriented cells are preferentially driven by the contralateral eye. Thus, the ocular dominance distribution and the orientation selectivity appear as two linked parameters characterizing visual specificity in very young kittens independently of any visual experience. An hypothesis of "differential modifiability" is proposed: contralateral monocular "horizontal and vertical detectors" are supposed to be stable. They would remain so until they become binocular. Binocular cells, for which competition between two inputs occurs, are the labile units which can be despecified or be specified under the control of visual experience.

Since the discovery by Hubel and Wiesel (1962) of specific visual analysers at the cortical level in the adult cat, numerous workers have shown that restricted visual experience can lead to selective modification in the response properties of these visual cortical neurones. This restricted visual experience is effective only when it occurs during a limited critical period of postnatal development and when the kitten has had no previous visual experience. In order to understand the main effect of visual experience on the development of cortical specificity it is important to investigate the receptive field arrangement of neurones in very young visually inexperienced kittens. Recording a small sample of cells, Hubel and Wiesel (1963) described neurones with orientation selectivity and ocular dominance distribution of adult standard; these neurones are present in naive cortices by 8 days, the age of eyelid opening, and thus, before any visual experience. They concluded that the response properties of the neurones in the kitten's visual cortex are innately determined. At variance, according to Barlow and Pettigrew (1971) and Pettigrew (1974), "orientation selectivity" would appear to require an appropriate visual experience: before the fourth postnatal week, neurones in both normal and binocularly deprived kittens show directional properties but lack proper orientation selectivity and disparity selectivity. Blakemore and Van Sluyters (1975) agreed with some of Pettigrew's results, but noticed a considerable greater degree of inherent organization in deprived cortex. Sherk and Stryker (1976) have emphasized this property and Singer and Tretter (1976) have confirmed that after long-term binocular deprivation the main deficits were in the synaptic security of afferents and the strength of intracortical inhibition. The experiments I wish to describe were undertaken in order to compare the receptive fields of visual cortical cells in normally reared kittens and in dark-reared kittens at a whole series of different ages.

Most of these results have already been published (Imbert and Buisseret, 1975; Buisseret and Imbert, 1976; Buisseret et al., 1978; Fregnac and Imbert, 1978; Fregnac et al., 1978).

Methods

The activity of single cells were recorded extra-cellularly with metallic microelectrodes stereotaxically located in the medial edge of the post-lateral gyrus in kittens which were anesthetized with Nembutal or Penthotal and curarized. The characteristics of the receptive field were mapped by manually projecting small spots or slits of light upon a wide tangent screen diffusely illuminated at mesopic level. On completion of this preliminary examination, the neurones were studied using computer programmed visual stimuli. The activity of about 1200 single cells were recorded in 33 normally reared kittens (NR) and in 29 dark-reared kittens (DR) aged between 9 and 50 days (Fregnac and Imbert, 1978). For details of the recording procedure, see earlier reports (Imbert and Buisseret, 1975; Buisseret and Imbert, 1976).

Results

Functional classification of visual cortical neurons The cortical units were classified into four types of neurones: (a) non-activatable cells that cannot be excited by any visual stimulation, (b) non-specific cells that are visually activated by a circular stimulus moving in any direction across their receptive fields which are characteristically circular, (c) immature cells that are preferentially activated by a rectilinear stimulus but are broadly tuned around an optimal orientation, and (d) specific cells that appear to be as selective for orientation as the simple or complex cells in the adult cat. Orientation tuning curves of visually activated cells recorded in deprived kittens are illustrated in Fig. 1B (Fregnac, this volume).

The results confirm that cells with some of the highly specific response properties of the adult cortical visual neurones, especially in relation to orientation selectivity, are present in the earliest stages of postnatal development, independent of visual experience. However, visual experience is necessary to maintain and develop these specific cells after the third week of postnatal life (Buisseret and Imbert, 1976).

The distribution of the different cell types in three age groups (12-17, 18-28, and 29-50 days) is compared in NR and DR kittens. In the first age group (12-17 days), about 25% of the units were found clearly and positively activated by oriented stimuli, whatever the rearing conditions; the proportion of non-specific, immature, and specific cells for each rearing condition are no longer comparable between 18 and 28 days. After 28 days a clear-cut effect appears: in NR kittens the proportion of specific cells increases, while in DR kittens the proportion of non-specific cells becomes predominant.

TABLE 1 Properties of different types of visually responsive units at different ages.

AGE GROUP (days)	12-17		18-28		29-42	
REARING CONDITIONS	NR	DR	NR	DR	NR	DR
Nonspecific cells	22.5%	40%	18%	53%	9%	90%
Immature cells	39.0	37	46	28	22	6
Specific cells	38.5	23	36	19	69	4
Number of cells	n = 185	n = 94	n = 131	n = 246	n = 234	n = 160
Number of kittens	8	7	8	14	11	8

Properties of specific cells in kittens under 3 weeks of age Before 3 weeks of age, whatever the rearing condition, most of the specific cells are monocularly activated by the contralateral eye and have horizontal or vertical orientation preferences (Fregnac and Imbert, 1978) (Fig. 1).

Beyond 4 weeks of age, ocular dominance is independent of orientation preference, of the functional type of neurones considered, and of rearing conditions. It has been proposed by Fregnac and Imbert (1978) "that binocularly driven cells are the modifiable cells which may be specified or despecified in terms of orientation selectivity under the control of visual experience." The monocular "horizontal and vertical detectors" recorded in very young kittens

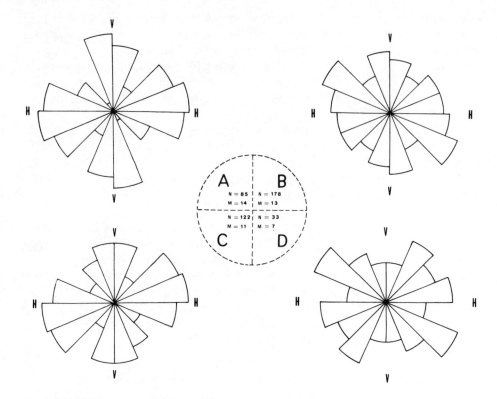

FIGURE 1 Polar diagrams of orientation preference. The number of specific cells recorded from comparable kittens is plotted versus the mean polar orientation class of 22.5° width. The length of each portion indicates the number of specific cells for one class of orientation. (A) NR and C: DR kittens under 3 weeks of age; (B) NR kittens over 4 weeks of age; (D) 6-week-old DR kittens after six hours of visual exposure.

More specific cells coding horizontal and vertical orientations than responding to oblique are found in very young kittens (A) and (C). There is no asymmetry in the distribution of orientation preference in NR kittens after four weeks (B) or in 6-week-old DR after recovery by six hours of visual experience (D).

form a special subpopulation...and are stable and resistant to the absence of visual input or to the selective exposure to an orientation to which they do not respond as long as they are influenced only by one eye." This hypothesis of "differential modifiability" has been tested (Fregnac et al., 1978) by recording cells in the primary visual cortex of 6-week-old dark-reared kittens unilaterally enucleated at birth. Such a surgical procedure, which disrupts binocular interaction, seems to stabilize that particular contingent of oriented cells, which are no longer recorded in the absence of visual experience in the intact DR kitten of the same age.

Restoration of specificity in 6-week-old DR kittens The repartition of the three types of units, non-specific, immature, and specific is given in Fig. 2 for normally reared (left column NR) and dark-reared (right column DR) kittens older

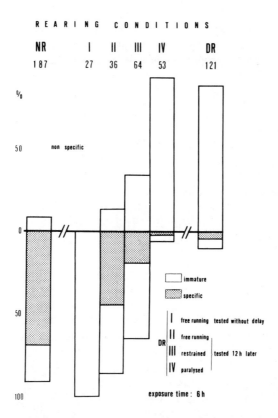

FIGURE 2 Distribution in percentage of the three types of visual cortical neurones (non-specific upper part, immature, and specific lower part) recorded after different visual exposures in 6-week-old DR kittens (see text for details).

than 4 weeks of age. In 6-week-old kittens, 6 hours of visual experience followed by 12 hours of rest in darkness are sufficient to increase rapidly the proportion of orientation specific cells (column II). A consolidation time of 12 hours between the end of the experience and the electrophysiological exploration does not seem to be necessary for the acquisition of orientation selectivity (column I). Moreover, when visuo-motor interaction is restricted during exposure to visuo-oculomotility only, the trigger features of visual units are similar to those of the free-moving kittens (column III). However, when all visuo-motor interactions are suppressed during exposure under Flaxedil, almost none of the units recorded displays any orientational selectivity (column IV). These results point out the crucial role of the ocular motility in the process of the restoration of the specific properties of cells in the kitten's visual cortex (Buisseret et al., 1978).

Discussion

These experiments are in agreement with those of Hubel and Wiesel (1963) in showing the presence of specific neurones in very young visually inexperienced kittens. However, for specificity to be maintained and to continue to develop, the kitten must be allowed active visual experience by its third week of life.

In very young kittens, under 3 weeks of age, there is clear evidence that a large proportion of neurones respond preferentially to horizontal and vertical orientations than to oblique stimuli. These results are in agreement with those obtained by Leventhal and Hirsch (1975). Moreover, these horizontal and vertical detectors are monocularly activated by the contralateral eye; they are stable as long as they remain monocularly activated.

The hypothesis that visual cortical neurones may develop in two groups, one "labile" group composed mainly of binocular cells and one "stable" group with particular receptive field organization (monocular cells responding preferentially to horizontal or vertical orientations), is emphasized by the results of other experiments (Buisseret et al., 1978; Gary Bobo et al., 1978). When normally reared kittens are settled in the dark room at the age of 4 or 6 weeks, the proportion of specific cells decreases with the duration of binocular deprivation but the specific cells which are recorded last have the same general features as the specific neurones recorded in kittens less than 3 weeks of age. The same results are obtained when DR kittens are returned to the dark room after a 6-hour light exposure.

Altogether, the results presented in this report appear to confirm that two groups of visual cortical units, with different characteristics of ocular dominance and orientation preference, have a different developmental history with and without visual experience.

References

Barlow, H. B., and J. D. Pettigrew (1971). Lack of specificity of neurones in the visual cortex of young kittens. J. Physiol. (Lond.) 218:98:-100P.

Blakemore, C., and R. C. Van Sluyters (1975). Innate and environmental factors in the development of the kitten's visual cortex. J. Physiol. (Lond.) 248:663-716.

Buisseret, P., and M. Imbert (1976). Visual cortical cells: their developmental properties in normal and dark-reared kittens. J. Physiol. (Lond.) 255:511-525.

Buisseret, P., E. Gary Bobo, and M. Imbert (1978). Ocular motility and recovery of orientational properties of visual cortical neurons in dark-reared kittens. Nature 272:816-817.

Fregnac, Y., and M. Imbert (1978). Early development of visual cortical cells in normal and dark-reared kittens: relationship between orientation selectivity and ocular dominance. J. Physiol. (Lond.) 278:27-44.

Fregnac, Y., P. Buisseret, E. Bienenstock, E. Gary Bobo, and M. Imbert (1978). Persistence of orientation-selective cells in the primary visual cortex of dark-reared kittens enucleated unilaterally at birth. C. R. Acad. Sci. (Paris), Serie D, 149-151.

Gary Bobo, E., and P. Buisseret (1978). Reversal of the physiological effect of dark rearing upon orientational properties of kitten's visual cortical cells. Neurosc. Letters S390.

Hubel, D. H., and T. N. Wiesel (1962). Receptive fields binocular interaction and functional architecture in the cat's visual cortex. J. Physiol. (Lond.) 160:106-154.

Hubel, D. H., and T. N. Wiesel (1963). Receptive fields of cells in striate cortex of very young, visually inexperienced kittens. J. Neurophysiol. 26:994-1002.

Imbert, M., and P. Buisseret (1975). Receptive field characteristics and plastic properties of visual cortical cells in kittens reared with or without visual experience. Exp. Brain Res. 22:25-36.

Leventhal, A. G., and H. V. B. Hirsch (1975). Cortical effect of early exposure to diagonal lines. Science (N.Y.) 190:902-904.

Pettigrew, J. D. (1974). The effect of visual experience on the development of stimulus specificity by kittens' cortical neurones. J. Physiol. (Lond.) 237:49-74.

Sherk, H., and M. P. Stryker (1976). Quantitative study of cortical orientation selectivity in visually inexperienced kitten. J. Neurophysiol. 39:63-70.

Singer, W., and F. Tretter (1976). Unusually large receptive fields with restricted visual experience. Brain Res. 26: 171-184.

Kinetics of the Development of Orientation Selectivity in the Primary Visual Cortex of Normally and Dark-reared Kittens

YVES FREGNAC

Laboratoire de Neurophysiologie
Collège de France
Paris, France

Abstract A kinetic model of orientation tuning is inferred from quantitative analysis of extracellular recordings in the primary visual cortex of normally and dark-reared kittens.

Seven hundred twelve visual cells were classified into three functional groups: (a) nonspecific cells; (b) immature cells which are not as orientation-selective as (c) specific cells. Power regression and covariance analysis show that the critical period begins before 19 days and that the kinetics of the immature pool are the same in both rearing conditions.

A catenary process of development of orientation selectivity is proposed, the immature compartment being a transit pool between nonspecific and specific cells. Two sequential stages occur: (a) the realisation of an intrinsic program of maturation by which cortical specificity appears at eye opening and increases independently of visual experience; (b) a phase of "epigenesis," beginning at 19 days during which functional modification depends on visuomotor experience in a nonlinear way.

In order to predict the effects of dark-rearing, normal rearing, and restricted visual experience, two assumptions are made: (a) the kinetics are first-order, time-dependent; (b) during delayed visuomotor experience the exchange coefficients take the value they would have had at this age, if the animal had been reared normally since birth.

This model suggests that visuomotor experience during the critical period allows the *expression* of a maturation process which may have been masked up to this point by the absence of vision or eye movements.

After the pioneering work of Hubel and Wiesel (1963), it is now accepted that orientation selectivity is an inborn feature of some, but perhaps not all, visual cortical cells (Blakemore and Van Sluyters, 1975; Buisseret and Imbert, 1976;

51

Sherk and Stryker, 1976; Fregnac and Imbert, 1978). Nevertheless, there is a lack of quantitative data describing the whole process of ontogeny of selectivity of orientation. More precisely, little is known about the *kinetics* of orientation tuning in the presence or total absence of visual experience. Is it just a pre-established specificity which, if not verified by a normal visual input, disappears with disuse, or does visual experience guide the maturation of the visual cortex?

The three-state analysis, based on previous extracellular recordings in the primary visual cortex of normally reared and dark-reared kittens (Fregnac and Imbert, 1978), presented here, attempts to answer the following three questions:

1. What is the degree of functional specialisation of the visual cortex at eye opening, i.e., around seven or eight days?

2. At which stage of postnatal development does visual experience play a role in the maintenance or the maturation of specific properties in the visual cortex? What is the precise date of the beginning of the so-called "critical period" (Hubel and Wiesel, 1970)?

3. How far is it possible to separate endogenous and exogenous factors in the maturation of orientation selectivity?

A kinetic catenary model of orientation tuning is inferred from the experimental regressions and its predictions are applied to describe the functional effects of delayed visual experience after dark-rearing.

Methods

Seven hundred twelve visual cells have been recorded in the primary visual cortex of 22 normally reared (NR) and 21 dark-reared (DR) anaesthetized and curarized kittens, aged between 9 and 50 days. All the receptive fields tested were situated within 10° of area centralis. For details of the recording procedure, see Fregnac and Imbert (1978).

Recording sensitivities The first crucial point is to define a method suitable for the comparison of data obtained in kittens of different ages and different rearing conditions: in the adult cat it has been observed that using the given recording procedures, an average of 50 visual cells are identified in an 8-hour experiment and this is thus designated as a "standard level." In kittens a distinct level of confidence is associated with each experiment and an "experimental sensitivity" index, varying with postnatal age, is defined for each rearing condition. The two indices, shown in Fig. 1(A), are obtained by tracing the upper envelopes of the two clusters of points, given by the age as the abscissa and the number of visual cells recorded in eight hours as the ordinate, according to each rearing condition.

Tuning classification Neurones of the primary visual cortex which were visually activated were classified into three functional types, according to a previous classification introduced by Imbert and Buisseret (1975).

A: Nonspecific cells, with a large circular receptive field, responding equally well to moving spots and bars.

I: Immature cells, responding better to bars than to spots and exhibiting an orientation selectivity wider than 60°.

S: Specific cells, simple or complex cells, as selective as those found in the adult cat.

Typical tuning curves, recorded in visually deprived kittens, are shown in Fig. 1(B).

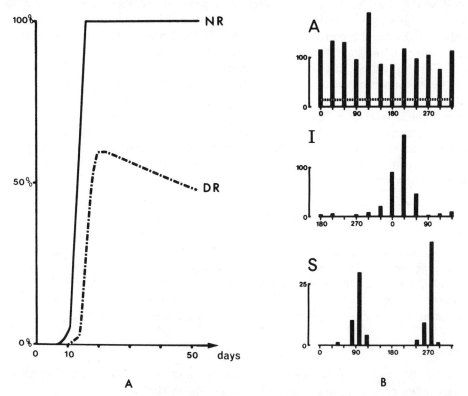

FIGURE 1 Recording sensitivities and tuning classification. (A) Recording sensitivity index in normally reared kittens (NR: solid line) and dark-reared kittens (DR: dotted line) with postnatal ages. (B) Orientation tuning curves and cell classification. From top to bottom: A, nonspecific cells; I, immature cells; S, specific cells. Each bar indicates the mean number of spikes elicited by the sweep of an oriented stimulus across the receptive field, in a direction orthogonal to its orientation. The data were obtained using a completely automated program of random stimulation and averaged over five runs. The mean level of spontaneous activity is represented by a dotted line.

Power analysis In order to achieve a trend analysis of the functional properties of cortical neurones with time, two assumptions are made:

1. The probability of recording A-, I-, or S-type neurones does not depend on the cortical locus of recording, so long as the sample concerns the projection area of 10° around area centralis ("homogeneity hypothesis").

2. The kinetics of transformation of nonvisual into visual cells, and vice versa, are ignored ("constant pool hypothesis").

A power analysis is performed, relating the percentage of each functional type among the total number of visual cells, with the postnatal age, for each rearing condition (NR and DR). Two models are used to define the minimal order of the polynom of the independent variable, i.e., postnatal age, sufficient to describe the kinetics of the different dependent variables, i.e., A, I, or S. The "simultaneous model" tests the contribution of each power of the independent variable in the correlation. The "hierarchical model" tests the increase of the explained variance which is obtained by increasing the order of the polynom used for the regression (Cohen and Cohen, 1975).

The best polynomial trends according to least mean square criteria are parabolic functions (y) which are plotted on a logarithmic abscissa (t) $(y = a_1 + a_2 \cdot \log t + a_3 (\log t)^2)$.

Two methods were used to calculate the trends: a *nonponderate* method, where the same statistical weight is given to each kitten; and a *ponderate* method, where the statistical weight of one data point is proportional to the number of visual cells recorded in this kitten.

Covariance analysis For each functional type A, I, or S, a covariance analysis is performed between data obtained in NR and DR kittens under a moving limit of age: two groups of kittens corresponding to the two rearing conditions are formed and ordered with increasing age. A common upper age limit is arbitrarily chosen. The Fisher (F) factor which compares the unexplained intragroup variance to the unexplained inter-group variance is calculated (Lison, 1958). The upper limit is then changed, new groups are formed, and the calculation of the F factor for another number of degrees of freedom goes on.

Results

Nonspecific effect of dark-rearing The comparison of the two recording sensitivity index curves, represented in Fig. 1(a), shows that:

1. The experimental sensitivities increase rapidly for each rearing condition around the second week of postnatal age.

2. However, the gain in the number of visual cells recorded (in eight hours) in DR kittens is delayed by four days, compared with that observed in NR kittens.

3. The recording sensitivity in NR kittens is 1.5 times better than in DR kittens.

This suggests that dark-rearing exerts a nonspecific effect by delaying the kinetics of the specialisation of cortical neurones to become visually sensitive neurones.

Kinetics of orientation tuning The left part of Fig. 2 illustrates the maturation process of orientation tuning observed in normally reared kittens. The ponderate and nonponderate trends give similar regressions. The proportion of specific cells (S) increases monotonously from eye opening and at around 50 days the properties of visual cortical cells are as specific as in the adult cat in terms of orientation specificity.

As shown in the right part of Fig. 2, a striking despecification effect is observed in the absence of visual experience. The proportion of nonspecific cells decreases and then increases with age. The kinetics of the immature population and the initial state of specification seem to be the same as in NR kittens. A final nonstable equilibrium is reached at around 6 weeks of age. A marked difference is seen between the ponderate and nonponderate trends, but this is linked to the fact that it was more difficult to record visual cells in young DR kittens than in older ones.

The initial conditions may be extrapolated from data obtained in NR kittens alone or from both NR and DR kittens less than 3 weeks old. The inferences lead to the same conclusion: at eye opening the visual cortical cells, which were up to this point nonspecific, begin their maturation process.

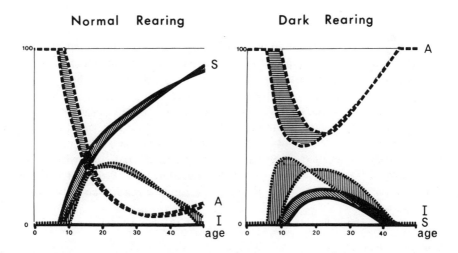

FIGURE 2 Kinetics of orientation tuning. The best polynomial trends given by the ponderate and nonponderate methods are plotted with postnatal age (in days) for each functional type (A, I, and S) in each rearing condition. The shaded areas indicate the deviations between the two methods.

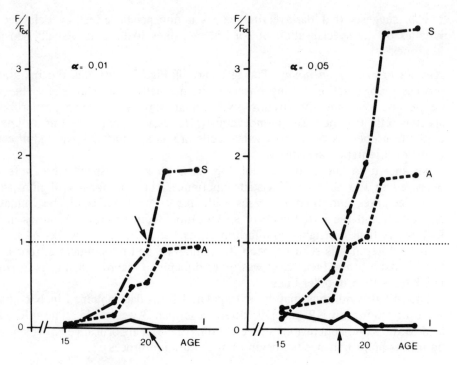

FIGURE 3 Beginning of the critical period. The Fisher factors given by the covariance analysis (see Methods) and normalized for two levels of confidence ($\alpha = 0.01$, $\alpha = 0.05$) are plotted with the upper limit of age used for the group formation for each functional type (A, I, and S). The arrows indicate the estimates of the beginning of the critical period.

Beginning of the critical period An estimate of the beginning of the period of sensibility to binocular deprivation is the earliest time at which visual experience significantly affects the kinetics of one of the three functional types.

The F factors normalized for a given level of confidence, and calculated separately for each class of visual neurones (F_A, F_I, F_S) are plotted with the upper limit of age used for the group formation. When greater than 1, these F factors indicate that the kinetics are dependent on visual experience.

From this covariance analysis it is concluded that (a) with a level of confidence of 0.05, the critical period begins at around 19 days; (b) the kinetics of the immature population are the same in the two rearing conditions.

Models

Kinetic system According to the experimental inferences given above, two sequential stages occur in the development of orientation selectivity: first, realisation of an intrinsic program of maturation by which cortical specificity appears at eye opening and increases independently of visual experience; second, a phase of epigenesis beginning at around the 19th day of postnatal life and during which functional modification depends on visual experience.

A catenary process of maturation between three pools of neurones is proposed, the immature compartment (I) being a transit pool between nonspecific (A) and specific (S) compartments.

In this model k_1 is called the "gate factor" and k_2 the "specialisation factor." The kinetics are assumed to be first-order and given by the following set of equations:

$$\begin{cases} \dfrac{dA}{dt} = -k_1 A + k_{-1} I \\[2mm] \dfrac{dI}{dt} = k_1 A - (k_2 + k_{-1}) I + k_{-2} S \\[2mm] \dfrac{dS}{dt} = k_2 - k_{-2} S \end{cases} \qquad (1)$$

with the initial conditions calculated from Fig. 2:

$$A\,(_0-) = A_0; I\,(_0-) = I_0; S\,(_0-) = S_0.$$

The exchange coefficients are identified, so that the predictions of the model fit equally well with data obtained from NR and DR kittens (Fregnac and Imbert, 1978) and from kittens having a brief visuomotor experience after dark rearing (Buisseret et al., 1978; Buisseret and Gary-Bobo, 1978).

Constant exchange coefficient model In the case in which the exchange coefficients are supposed to be independent of time, the analytic resolution of the matricial first-order differential equation is achieved by Laplace transform techniques and is described elsewhere (Fregnac, 1978).

The initial and final states of occupation of the three compartments are known from the experimental regressions shown in Fig. 2. According to the theorem of the final value, the emptiness of certain compartments at infinite time implies that certain exchange coefficients, according to the rearing condition, have a zero value. The steady-state equations are given by:

$$\begin{cases} A\,(\infty) = \dfrac{k_{-1} k_{-2} C}{k_1 k_2 + k_{-2}(k_1 + k_{-1})} \\[3mm] I\,(\infty) = \dfrac{k_1 k_{-2} C}{k_1 k_2 + k_{-2}(k_1 + k_{-1})} \\[3mm] S\,(\infty) = \dfrac{k_1 k_2 C}{k_1 k_2 + k_{-2}(k_1 + k_{-1})} \end{cases}$$

with $C = A_0 + I_0 + S_0$.

The simplest model, using only two parameters, is a BANG-BANG model in which the k_1 and k_2 factors and the k_{-1} and k_{-2} factors cannot be different from zero simultaneously.

The dependency of the parameters with visual experience is illustrated in Fig. 4(A,B).

While the predictions of the BANG-BANG model fit adequately with data obtained in NR and DR kittens, they do not reproduce the functional modifications observed in kittens having a brief visual experience after six weeks of visual deprivation. Buisseret et al. (1978) observed the filling up of the immature compartment after only a few hours of visuomotor experience.

Without going into details of the calculation (comparison of the response of the system to pulse (Dirac) and step (Heavyside) functions), the failure of the constant coefficient model might be explained in the following way: the final nonstable equilibrium that was reached after 40 days of visual deprivation, is, in fact, in terms of orientation specificity, the same as the initial state just before eye opening. A brief visual experience, delayed during the critical period, would give an increase in specificity which is comparable to that observed during the first days following eye opening, since the exchange coefficients are assumed to be independent of time. Thus, according to the BANG-BANG model, a few hours of visuomotor experience following six weeks of dark rearing cannot lead to a significant specification of visual cortical cells.

The contradiction between predicted and experimentally observed values of the three functional compartments suggests that the kinetics of maturation of tuning observed in consequence to a brief visual experience after dark rearing are not the delayed repetition of the maturation process observed in NR kittens or very young DR kittens. Physiological support for this argument is given by the orientation preference distribution of orientation-selective neurones in kittens younger than 3 weeks of age, which shows a significant bias towards an over-representation of horizontal and vertical orientations, whereas no bias is found towards any orientation preference in 6-week-old DR kittens having a brief delayed visual experience.

Expression-Extinction model In order to take into account nonlinearities observed during the critical period, the exchange coefficients k_1 and k_2 are assumed to be time-dependent and the kinetics of the immature compartment are shown, in this case, to be ruled by a Volterra equation of the second species (Fregnac, 1978):

$$I_{(t)} = f(t) + \int_t^0 K(t - x) I(x) \, dx.$$

The evolution in time of the exchange coefficients k_1 and k_2 is a "potpourri" of different data, simulating the development of the cortical neuropile for the ascending part (Cragg, 1974) and having the same decay as that found for the sensibility to monocular deprivation (Hubel and Wiesel, 1970; Blakemore and Van Sluyters, 1974).

The conditions verified by the four exchange coefficients are summarized in Fig. 4(C,D).

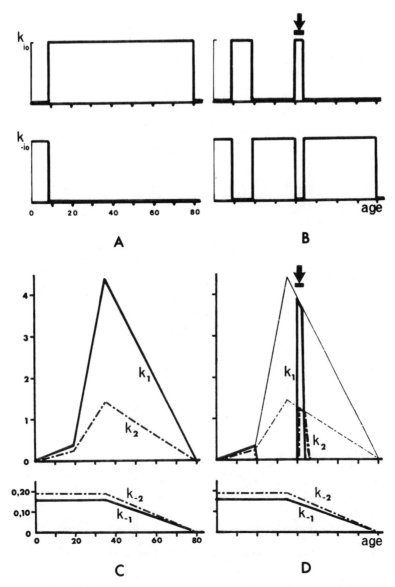

FIGURE 4 Models of development of orientation selectivity. Bang-bang model: A = normal rearing; B = dark rearing and restricted visual experience, indicated by the black arrow. The upper graphs represent, in each rearing condition, the evolutions of k_1 and k_2 with postnatal age given in days (d). The factors k_1 and k_2 just differ by a magnification factor: $k_{2(t)} = 0.8k_{-1(t)}$ and $k_{10} = 0.12(d^{-1})$. The lower graphs concern k_{-1} and k_{-2} with $k_{-2(t)} = 0.8k_{-1(t)}$ and $k_{-20} = 0.12(d^{-1})$. Expression-Extinction model: C = normal rearing; D = dark rearing and restricted visual experience, indicated by the black arrow. The exchange coefficients given in days^{-1} are plotted with age (in days). The values taken by k_{-1} and k_{-2} do not depend on rearing condition. The factors k_1 and k_2 are not affected by visual experience until 19 days. See text for details.

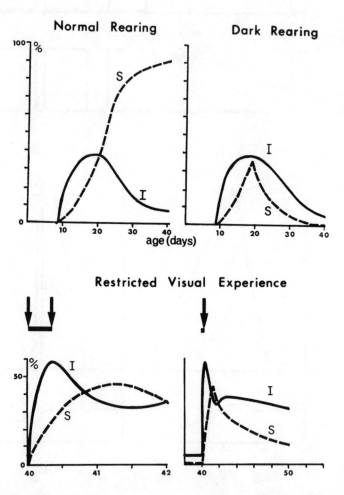

FIGURE 5 Prediction of the Expression-Extinction model. Immature proportion (I: solid line) and specific proportion (S: dotted line) are plotted with postnatal ages (in days) for different rearing conditions. Nonspecific proportion (A) is given by A = 100 - (I + S).

The coefficients k_{-1} and k_{-2}, which are responsible for the despecification process, act like a "forgetting mechanism," slowly varying with time. The coefficients k_1 and k_2, which are responsible for the nonlinear effects of specification described during the critical period, increase until 5 weeks of age and then decrease if the animal is reared normally (Fig. 4(C)). As shown in Fig. 4(D), in the case where the animal is reared in total darkness, k_1 and k_2 take a zero value at the beginning of the critical period. If a brief visual experience is given during the critical period, the model supposes that the exchange coefficients take the value they would have if the animal was reared normally from the beginning. When the animal is put back into total darkness, the "gate factor" goes instantly to zero, while the "specialisation factor" reaches zero with an extinction time constant. These two Expression-Extinction hypotheses are sufficient to reproduce experimental data obtained in such kittens having a restricted visual experience.

Figure 5 shows the results of the numerical simulation using a Runge-Kutta technique of the fourth order. The model predicts that after six weeks of dark rearing, leading to an almost complete loss of orientation-selective neurones (I and S), the immature compartment reaches its maximum at the end of eight hours of visual experience and that the specific compartment continues to fill up 12 hours after the animal is put back in the dark.

Discussion

In summary, this three-state analysis, as simple and arbitrary as it may appear, gives quantitative information on the kinetics of development of orientation selectivity.

At eye opening the visual cortical cells, which were up to this point nonspecific, begin their specification process independently of visual experience. The beginning of the critical period is shown to occur before 19 days and from this age, functional modification depends on visuomotor experience in a nonlinear way.

A model of development of orientation tuning is proposed on the assumptions that (a) the kinetics are first-order, time-dependent; (b) visuomotor experience allows the expression of a maturation process which may have been masked up to this point by the absence of vision or eye movements.

A comparison can be drawn between predictions of the model and experimental data obtained by Buisseret et al. (1978) in kittens reared in darkness for several weeks, then having a few hours of visual experience and put back in the dark. The model offers an alternative to a so-called "consolidation" hypothesis according to which return to the dark for a short period may facilitate the formation of orientation selectivity, consequently, to a brief visual experience. The increase of specification, observed a few hours after the end of the period of normal vision, might be the consequence of differences in the time constants associated with each exchange coefficient and interpreted as the effect of the extinction phase following the expression of the specification phase, rather than the result of an active consolidation process.

References

Blakemore, C., and R. C. Van Sluyters (1974). Reversal of the physiological effects of monocular deprivation in kittens: further evidence for a sensitive period. J. Physiol. (Lond.) 237:195-216.

Blakemore, C., and R. C. Van Sluyters (1975). Innate and environmental factors in the development of the kitten's visual cortex. J. Physiol. (Lond.) 248:663-716.

Buisseret, P., and E. Gary-Bobo (1978). In dark-reared kittens, is the restoration of the orientational properties observed after 6 hours of light exposure definitely stable? Neuroscience Letters S:387 (abstract).

Buisseret, P., E. Gary-Bobo, and M. Imbert (1978). Ocular motility and recovery of orientational properties of visual cortical neurones in dark-reared kittens. Nature 272:816-817.

Buisseret, P., and M. Imbert (1976). Visual cortical cells: their developmental properties in normal and dark-reared kittens. J. Physiol. (Lond.) 255:511-525.

Cohen, J., and P. Cohen (1975). Applied multiple regression/correlation analysis for the behavioral sciences. John Wiley & Sons, New York.

Cragg, B. G. (1974). Plasticity of synapses. Br. Med. Bull. 30:141-144.

Fregnac, Y. (1978). Cinétique de dévelopment du cortex visuel primaire chez le chat. Thèse d'Etat en Biologie Humaine. Paris V. 220 pp.

Fregnac, Y., and M. Imbert (1978). Early development of visual cortical cells in normal and dark-reared kittens: relationship between orientation selectivity and ocular dominance. J. Physiol. (Lond.) 278:27-44.

Hubel, D. H., and T. N. Wiesel (1963). Receptive fields of cells in striate cortex of very young visually inexperienced kittens. J. Neurophysiol. 26:994-1002.

Hubel, D. H., and T. N. Wiesel (1970). The period of susceptibility to the physiological effects of unilateral eye closure in kittens. J. Physiol. (Lond.) 206:419-436.

Imbert, M., and P. Buisseret (1975). Receptive-field characteristics and plastic properties of visual cortical cells in kittens reared with or without visual experience. Exp. Brain Res. 22:25-36.

Lison, L. (1958). Statistique appliquée à la biologie expérimentale. Ed. Gauthier Villars.

Sherk, H., and M. P. Stryker (1976). Quantitative study of cortical orientation selectivity in visually inexperienced kitten. J. Neurophysiol. 39:63-70.

Optokinetic Nystagmus and Single-cell Responses in the Nucleus Tractus Opticus After Early Monocular Deprivation in the Cat

K.-P HOFFMANN

Abteilung für Vergleichende Neurobiologie
Universität Ulm
Ulm, Germany

Abstract The normal cat's optokinetic nystagmus (OKN) can be elicited in the two horizontal directions even if one eye is stimulated alone. After early monocular deprivation OKN is only elicited by patterns moving from temporal to nasal if either the deprived or the non-deprived eye is stimulated alone. A specific class of neurons in the nucleus of the optic tract (NOT) in the pretectum of normal cats can be identified by the following criteria: (1) large area patterns rich in contour are the most effective visual stimuli; (2) all neurons in the left NOT are strongly excited by movements from right to left and all neurons in the right NOT by movements from left to right; (3) effective stimulus velocities are within a broad range of less than 0.1 deg/sec to greater than 50 deg/sec; (4) latency differences to electrical stimulation of the optic chiasma and optic tract indicate slow conducting (W-cell) retinal input to these NOT neurons; (5) all neurons could be antidromically activated from the inferior olive; (6) the neurons recorded in the NOT were either binocularly driven preferring the same direction in visual space in each eye (40%) or were activated only through the contralateral eye (60%).

After six months of early monocular deprivation the properties of NOT neurons listed above remained largely unaltered except that neurons in each NOT could only be influenced from the contralateral eye irrespective of whether it was the deprived one or not. This is considered to be the explanation for the asymmetry of OKN after early visual deprivation.

A normal cat shows a symmetric optokinetic nystagmus (OKN) to horizontal movements from left to right or from right to left even when one eye is stimulated alone. This monocular symmetric response can be rendered asymmetric by lesions of visual cortex (Wood, Spear, and Braun, 1973) or by early monocular visual deprivation due to lid closure (van Hof-van Duin, 1978). Only stimulus movements from left to right through the left eye and from right to left through the right eye are effective in eliciting the slow phase of OKN (see Fig. 5). Obviously, the visual cortex plays a major role in setting up the normal symmetric response. Where and how is the cortical information outflow integrated in the subcortical visuomotor pathway feeding into reflexes which serve to stabilize the overall retinal image?

Recent studies have described a class of cells in the nucleus of the optic tract (NOT) of the rabbit (Collewijn, 1975) and the cat (Hoffmann and Schoppmann, 1975; Hoffmann, Behrend, and Schoppmann, 1976) as having properties which make it seem very likely that they are the essential visual afferents in this pathway. These properties are:

1. Patterns rich in contour and covering a large area of the visual field are more effective stimuli than single spots or bars for these cells.

2. All cells of the NOT in the left brain hemisphere preferred stimulus movements from right to left in visual space and all cells in the right NOT preferred left to right movements (see Fig. 2). Stimuli moving in the direction opposite to the preferred one often inhibited the high spontaneous activity (20-50 spikes/sec; see Fig. 2).

3. Optimal stimulus velocities in the preferred direction were within a range of 1-10 deg/sec and, in most cases, the cells responded strongly to even slower (< 0.1 deg/sec) motion of the stimulus pattern.

4. In the normal cat more than one-third of these cells can be driven from both eyes and preferred the same direction in visual space for both eyes (see Fig. 3).

5. Latency differences after electrical stimulation of the optic chiasma and optic tract were compatible with direct W-fibre input from the contralateral retina to all these cells.

6. All cells with the properties listed above could be antidromically stimulated from the ipsilateral inferior olive. The dorsal cap of the inferior olive has been shown anatomically to receive NOT projections and in turn to project to the vestibulo-cerebellum (Takeda and Maekawa, 1976).

7. Experiments in our laboratory have shown that these cells in the NOT receive input from visual areas 17 and 18 (Schoppmann, in preparation), but that direction specificity is not lost after decortication.

In what way does early monocular deprivation alter the properties of these cells and can these alterations account for the specific changes in OKN found in deprived animals? Experiments were carried out on four adult cats which had one eye closed by a lid suture over the first year after birth. Methods for recording from units in the NOT and for visual stimulation were the same as described previously (Hoffmann and Schoppmann, 1975; Schoppmann and Hoffmann, 1976). Cats were initially anesthetized with Trapanal (i.v. injection of 20 mg), immobilized with Flaxedil, and artificially respirated with nitrous oxide and oxygen (70%:30%). Large random noise patterns (Julesz-patterns) projected onto a tangent screen by a slide projector via a double mirror system served as visual stimuli. These patterns were moved along a circular path without changing their orientation. The cell's responses were analyzed by on-line computation of average response histograms which then were displayed in polar coordinates (Fig. 2). One pair of electrodes was stereotactically inserted into the inferior olive ipsilateral to the NOT from which recordings were to be made. The placement of the stimulating electrodes was checked by recording the climbing fibre field potential in the cerebellar cortex. A second pair of stimulating electrodes was placed in the optic chiasma (OX) and a third in the optic tract (OT). NOT cells were located stereotactically and recorded with Insl-X varnished tungsten electrodes. All stimulating sites in the inferior olive and recording sites in the NOT were verified histologically (Fig. 1). The results from one experiment in which the deprived eye was opened at the beginning of recording will be described in detail. In this cat two successful penetrations were made in each NOT and 17 units were recorded and analyzed contralateral and 13 units ipsilateral to the deprived eye (Fig. 1). All these units showed the

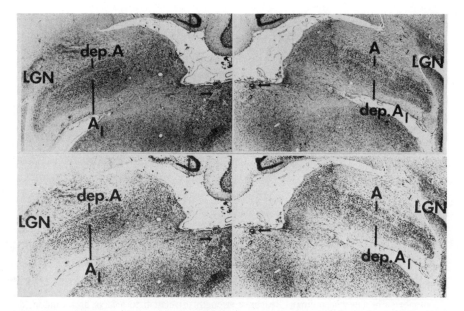

FIGURE 1 Frontal sections through the pretectum and dorsal lateral geniculate nucleus of a monocularly deprived cat. LGN: dorsal lateral geniculate nucleus; NOT: nucleus of the optic tract; MGB: medial geniculate body; dep. A: lamina A contralateral to the deprived eye; dep. A_1: lamina A_1 ipsilateral to the deprived eye. Arrows indicate lesion in each NOT. It can be clearly seen that in layer A on the left and in layer A_1 on the right that cell shrinkage has occurred proving the effectiveness of visual deprivation of the right eye. The influence of this deprivation on the units recorded in the marked penetrations through the NOT is described in the text.

properties listed above except that they could only be driven by the contralateral eye, i.e., through the deprived eye as well as through the non-deprived eye only units in the contralateral NOT could be stimulated and all showed the typical direction-specific response. Figure 2 shows the pooled responses for all units in the NOT contralateral to the deprived eye and stimulated through the deprived eye in the upper left polar histogram as well as when stimulated through the non-deprived eye in the upper right histogram. In the lower left of Fig. 2 the pooled responses for all units stimulated through the contralateral, now non-deprived, eye are presented and in the lower right the responses elicited in the same units through the ipsilateral, now deprived, eye are presented. There is a clear direction-specific response when the contralateral eye is stimulated, but no influence on the spontaneous activity when the ipsilateral eye is stimulated. The direction tuning curve is little more specific for the cells driven by the non-deprived eye.

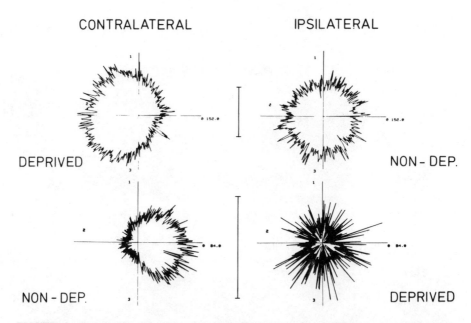

FIGURE 2 Peristimulus time histograms displaying in polar coordinates the responses to visual stimulation of all neurons recorded in the two nuclei of the optic tract (NOT) in one monocularly deprived animal. Response strength in relation to a large area random dot pattern moving on a circular path (see methods) is given for the various directions by the vector from the origin to the curve. Upper left: pooled responses of 17 units in the left NOT to stimulation of the contralateral deprived eye; upper right: pooled responses of the same units to stimulation of the ipsilateral non-deprived eye. This activity is not different from the spontaneous activity; lower left: pooled responses of 13 units in the right NOT to stimulation of the contralateral non-deprived eye; lower right: responses of the same 13 units to stimulation of the ipsilateral deprived eye. Again, this activity is not different from the spontaneous activity. Vertical calibration bars: 100 spikes/sec.

These results were confirmed by the other three animals in which four penetrations were made contralateral to the deprived eye (two animals) and three penetrations ipsilateral to the deprived eye (one animal). The ocular dominance distribution for all units is given in Fig. 3 (A,B). For comparison the data from 28 penetrations in 21 normal animals are presented in Fig. 3(C). Only one out of 52 units was found in the monocularly deprived animals to show an influence from both eyes compared with 39 out of 95 in the normal animals. These differences between deprived and normal animals are statistically highly significant (Yates corrected χ^2-test) even if the data are broken up into the two groups contralateral and ipsilateral to the deprived eye as presented in Fig. 3(A,B).

How can we be sure that we recorded from the same type of NOT-cells in the normal and in the deprived animals? Every recording site was checked histologically and the lesions contralateral and ipsilateral to the deprived eye are at exactly corresponding locations in the two hemispheres (Fig. 1). All 12 lesions in the pretectum of the four deprived animals correspond to the locations of the 38 lesions in the 21 normal animals. Additional support for the identity of the normal and the two deprived populations is given by the data from electrical stimulation of the optic chiasma and the inferior olive (see Fig. 4). After optic chiasma stimulation the orthodromic latency range and mean are the same in normal NOT and in the NOT ipsilateral or contralateral to the deprived eye (Fig. 4(A,C)). Antidromic activation from the inferior olive was found in all NOT-cells with the response properties given above. In fact, antidromic invasion from the inferior olive was used in addition to visual stimulation to search for these NOT units. Again, antidromic latency range and mean are identical in the normal and in the two populations from the deprived animals (Fig. 4(B,D)). Contralateral to the deprived eye, however (Fig. 4(D)), 10 out of 34 units were identified by antidromic stimulation as NOT-cells, but showed neither a response to optic chiasma stimulation nor to visual stimulation. All units contralateral to the non-deprived eye (Fig. 4(B)) which were driven antidromically from the inferior olive could also be orthodromically activated from the optic chiasma and visually stimulated.

In summary there are three lines of strong evidence for the identity of the NOT-cell population analyzed in the normal and in the monocularly deprived animals: (1) location in the pretectum; (2) orthodromic latencies after optic chiasma stimulation and antidromic responses to electrical stimulation of the inferior olive; (3) qualitatively identical response properties excluding binocular convergence.

Why, then, is binocular convergence lost in both NOTs after deprivation? Only hypothetical models can be suggested on the basis of the data available at present (Fig. 5). With respect to the NOT we have to consider three output channels leaving each retina. The first originates from W-cells. These axons cross in the optic chiasma and terminate directly on the contralateral NOT cells. The second and third originate most probably from other ganglion cell populations and project via the lateral geniculate nuclei to the contralateral visual cortex as well as to the ipsilateral visual cortex. In the cortex the pathways from

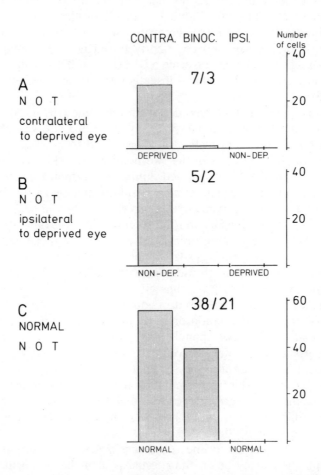

FIGURE 3 Frequency histograms of ocular dominance distribution of neurons recorded in the nucleus of the optic tract (NOT). Contralateral (A) and ipsilateral (B) to the deprived eye or in the normal animal's NOT (C). From left to right the columns in the histograms represent units driven exclusively by the contralateral eye, units driven from both eyes, units driven exclusively from the ipsilateral eye. Numbers above the histogram indicate number of penetrations/ number of animals.

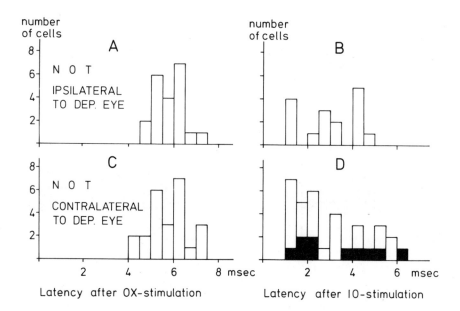

FIGURE 4 Frequency distribution of latencies measured for neurons of the nucleus of the optic tract after electrical stimulation of the optic chiasma (A, C) and inferior olive (B, D). *Y*-axis: number of cells; *X*-axis: latencies measured in msec. (A), (B) neurons recorded in the NOT ipsilateral to the deprived eye; (C), (D) neurons recorded contralateral to the deprived eye. Black boxes in (D) represent those neurons which could be antidromically identified from the inferior olive but could not be stimulated from the optic chiasma or visually.

the two eyes converge and binocularly excitable corticofugal fibres reach the NOT. The cells in the NOT can thus be activated from the contralateral eye by the crossing direct retino-pretectal W-axons or from both eyes via the visual cortex. The output from the NOT to the inferior olive is binocular but dominated by the contralateral eye (Fig. 3), and thus is a combination of the two input types.

In the monocularly deprived cat (Fig. 5(B)) we have to consider the two eyes and the two NOTs separately. The deprived eye projects directly to the NOT and this connection is not disrupted by deprivation. The deprived eye's influence through the cortex is, of course, disrupted by the mechanism of binocular competition at the cortical synapse. The non-deprived eye also projects directly to the contralateral NOT and its influence through the visual cortex is unaffected or even strengthened by the monocular deprivation. The NOT ipsilateral to the deprived eye should receive direct plus cortical visual input only from the contralateral non-deprived eye and all cells should be monocular (see Fig. 3(B)).

The other NOT should receive input directly from the contralateral deprived eye and input from the ipsilateral non-deprived eye via the visual cortex. Fig. 3(A) shows, however, that cells in this NOT also can only be activated from

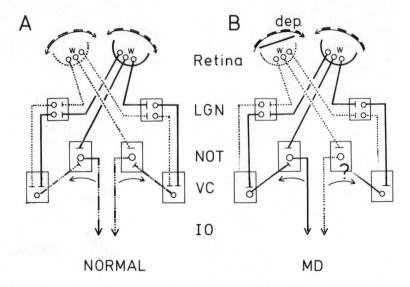

FIGURE 5 Summary schematic diagrams showing the afferents to the nucleus of the optic tract (NOT) in the normal cat (on the left) and monocularly deprived cat (on the right). LGN: dorsal lateral geniculate nucleus; NOT: nucleus of the optic tract; VC: visual cortex; IO: inferior olive; MD: monocularly deprived cat; dep.: deprived eye; W: retinal ganglion cell of the W-type. These diagrams are hypothetical explanations for much of the data from this and related studies. The afferents from the left eye (dotted lines) and from the right eye (continuous lines) converge in the visual cortex and each NOT in the normal cat receives a binocularly balanced input via the visual cortex (lines interrupted by dots) in addition to the purely contralateral direct retino-pretectal W-projection. This direct retino-pretectal projection is unaffected by monocular deprivation, but the connection of the deprived eye to NOT via the visual cortex is disrupted at the cortical synapse. Also, the connection of the non-deprived eye to the ipsilateral NOT is not functioning due to a miswiring or disruption at the NOT (question mark). In the normal as well as in the MD-cat all neurons of the left NOT prefer movement of large area random dot patterns from right to left and all neurons in the right NOT movements from left to right, as indicated by the arrows below each NOT. For further explanations see text.

the contralateral eye. We have to assume that the cortico-pretectal pathway dominated by the ipsilateral non-deprived eye does not develop its proper function in the NOT contralateral to the deprived eye. One possible explanation for this could be the relative late development of the cortico-fugal influence on the midbrain (Stein, Labos, and Kruger, 1973; Norton, 1974). It may be necessary for the integration of cortical influence in a highly direction-specific way that the activity of the target NOT cells is modulated by visual stimulation. This can never happen because the direct retinal influence from the contralateral eye, probably present at birth, is obstructed by the deprivation. The NOT cells contralateral to the deprived eye always display their monotonous high spontaneous activity and the cortical information is not integrated at all or unspecifically. Stimulus movement presented to the non-deprived ipsilateral eye does not even alter the spontaneous activity in an unspecific way so that the second possibility seems less likely.

It is interesting to relate this model to the OKN observed in the normal and the deprived animal. In the normal cat the two eyes project into both NOTs and can thus activate the appropriate direction-specific output system. OKN can be elicited through each eye in both horizontal directions, temporo-nasally through the contralateral and naso-temporal through the ipsilateral NOT. At four weeks after birth only an asymmetric temporo-nasal OKN is present. The symmetric OKN develops during the next four weeks (van Hof-van Duin, 1978). If one eye is closed from birth, the other eye never develops the symmetric OKN because the cortico-pretectal information from this eye cannot be compared to and integrated in the direction-specific responses of the cells in the NOT contralateral to the closed eye. After eye opening only a presumably genetically determined asymmetric OKN can be elicited through the deprived eye because it has lost its functional connections in the visual cortex. After monocular deprivation OKN can be elicited through each eye only via the contralateral NOT in the temporo-nasal direction. The connections to the ipsilateral NOT eliciting an OKN in response to monocular naso-temporal stimulus movement are disrupted.

Acknowledgements I should like to thank Dr. A. Schoppmann and C. M. Morrone for their help during the experiments and data analysis, Ms. R. Barthel and Ms. L. Oechsle for technical and secretarial help. The work was supported by DFG-grants Ho 450/6 and 7.

References

Collewijn, H. (1975). Direction-selective units in the rabbit's nucleus of the optic tract. Brain Res. 100:489-508.

Hoffmann, K.-P., K. Behrend, and A. Schoppmann (1976). A direct afferent visual pathway from the nucleus of the optic tract to the inferior olive in the cat. Brain Res. 115:150-153.

Hoffmann, K.-P., and A. Schoppmann (1975). Retinal input to direction-selective cells in the nucleus tractus opticus of the cat. Brain Res. 99:359-366.

Norton, Thomas T. (1974). Receptive-field properties of superior colliculus cells and development of visual behavior in kittens. J. Neurophysiol. 37:674-690.

Schoppmann, A., and K.-P. Hoffmann (1976). Continuous mapping of direction-selectivity in the cat's visual cortex. Neuroscience Letters 2:177-181.

Stein, Barry E., E. Labos, and L. Kruger (1973). Determinants of response latency in neurons of superior colliculus in kittens. J. Neurophysiol. 36:680-689.

Takeda, T., and K. Maekawa (1976). The origin of the pretecto-olivary tract. A study using the horseradish peroxidase method. Brain Res. 117:319-325.

van Hof-van Duin, J. (1978). Direction-preference of optokinetic responses in monocularly tested normal kittens and light-deprived cats. Arch. ital. Biol. 116:471-477.

Wood, C. C., P. D. Spear, and J. J. Braun (1973). Direction-specific deficits in horizontal optokinetic nystagmus following removal of visual cortex in the cat. Brain Res. 60:231-237.

Neuronal Activity in the Afferent Visual System and Monocular Pattern Deprivation

O.-J. GRÜSSER

Department of Physiology
Freie Universität
Berlin, Germany

Summary

In a paper entitled "The loss of a specific cell type from dorsal lateral geniculate nucleus in visually deprived cats," Sherman, Hoffmann, and Stone (1972) described the probability of recording from Y-cells in layer A or A_1 of the cat LGN being reduced for the respective layer connected with the deprived eye (monocular deprivation). Several years ago, while recording from single neurons in the visual cortex or the optic radiation (OR) of monocularly deprived animals, we found that monocular deprivation did not change the relative distribution of latency class I (Y) or latency class II (X) neurons in the neuronal samples recorded from the OR. Because these findings seemed to contradict the data published by Sherman, Hoffmann, and Stone, a more extensive study was performed in a larger population of neurons in five adult cats which had undergone monocular deprivation since the second week of life (Eysel, Grüsser, and Hoffmann, 1978, 1979). In two animals, some of the recordings were obtained before the eyelids of the deprived eye were opened. Single neurons were recorded in the optic tract (OT), the lateral geniculate body (pars dorsalis, layer A and A_1; LGN) and the OR by means of tungsten microelectrodes (pentobarbital anaesthesia). The neurons were classified as Y- or X-neurons according to their visual response properties to light on-off, grid patterns moved through the RF, flicker responses and receptive field sizes relative to the position of the RF in the visual field or as class I or class II neurons by their latencies evoked by electrical optic tract stimulation.

1. Visual classification (Y/X) and latency classification (I/II) of OT-, LGN-, or OR-neurons were compared with each other and yielded a very close positive correlation (> 95%): Y-neurons = class I, X-neurons = class II.

2. Recording from the LGN confirmed the results of Sherman, Hoffmann, and Stone; monocular visual deprivation led to a reduction in probability of recording Y (class I) neurons (Figs. 1(C), (D), (E)).

3. There was, however, no significant difference in the frequency of class I/II neurons or Y/X neurons when the samples of monocularly deprived neurons and normal neurons recorded from the OR were compared to each other (Figs. 1(A), (B)). It seems possible to conclude that the reduction in the probability of recording the action potentials of Y (class I) cells from the deprived LGN laminae might be due to the effect of prolonged pattern deprivation on the *growth of Y-cells* as compared with the deprivation effects on the growth of the X-cells. Because of the lesser size difference between both cell populations, a shift in the recording probability might have been the consequence.

4. *Sum potentials* were recorded from the OR of both sides; the optic nerve of the deprived and the normal eye were stimulated electrically by intraorbital electrodes. The evoked potentials were computer-averaged. The sizes of the evoked potential waves r_1 and r_2 elicited by *supramaximal* electrical stimuli of the deprived eye optic nerve were smaller than the amplitudes of the r_1 and r_2 waves elicited by stimulation of the normal optic nerve (Fig. 2). The relationship between the amplitudes of r_1 and r_2 waves, however, was not significantly altered in the evoked potentials elicited by electrical stimulation of the deprived eye versus the normal eye optic nerve. This finding supports the interpretation that monocular deprivation does not selectively interrupt the signal transmission through the Y-channel of the deprived afferent visual system. Monocular deprivation might, however, have an effect on the overall number of active LGN cells. This would explain the reduced probability of recording from deprived OR-neurons as compared to normal OR Y- *and* X-neurons.

5. Measurements of the spontaneous activity of OT-neurons, LGN-cells, or OR-neurons recorded while the eyelids of the monocularly deprived eye were still closed exhibited no reduction in the spontaneous activity as compared to the data obtained in corresponding neurons activated by the normal eye. The average activity elicited by light stimulation of the normal eye was also not higher than the spontaneous activity of neurons in the deprived retina (eyelids still closed). The pattern deprivation effects found morphologically and physiologically in the afferent visual pathway and the visual cortex are, therefore, not induced by reduction in the average maintained activity of neurons activated by the deprived eye. These findings were confirmed in a recent experiment performed with U. Schreiter in LGN-neurons of a monocularly deprived *squirrel monkey.*

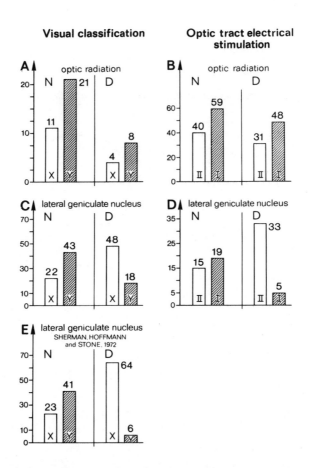

FIGURE 1 Distribution of Y/X or class I/II neurons in different neuron samples recorded from the optic radiation or lateral geniculate nucleus; layers A and A_1. N = activation by the normal retina, D = activation by the pattern-deprived retina (from Eysel et al., 1978).

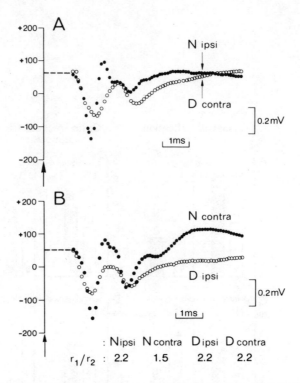

FIGURE 2 Computer-averaged evoked potentials recorded from the right and the left optic radiation. Responses to supramaximal electrical stimulation of the right and the left optic nerve. N = stimulation of the normal eye optic nerve, D = stimulation of the deprived eye optic nerve. 1 per sec stimuli, 0.2 msec duration, 48 responses averaged. Shock artifacts excluded. (From Eysel et al., 1978, 1979.)

There is, however, a distinct change in the temporal properties of the respective input impulse trains. High-frequency discharge bursts and longer inhibitory periods are absent in the input to a pattern-deprived neuron while the eyelids are still closed. Such impulse patterns are normally elicited by moving contrast patterns seen with the normal eye.

6. The activity pattern of monocularly driven neurons (OT, LGN, OR) elicited by moving visual black-white contours, stationary flicker stimuli and electrical stimulation of the OT at different stimulus frequencies (1-300 stimuli/sec) was compared to the response pattern of corresponding normal neurons. With the exception of a few LGN neurons recorded from the monocularly deprived laminae which exhibit sluggish responses to visual stimulation, no differences between normal and monocularly deprived neurons were found. We did not measure, however, the capacity of the neurons for spatial resolution.

References

Eysel, U. Th., O.-J. Grüsser, and K.-P. Hoffmann (1978). The effect of monocular pattern deprivation on the signal transmission by neurons of the cat lateral geniculate body. Arch. Ital. Biol. 116:427-443.

Eysel, U. Th., O.-J. Grüsser, and K.-P. Hoffmann (1979). Monocular deprivation and the signal transmission by X- and Y-neurons of the cat lateral geniculate nucleus. Exp. Brain Res. 34:521-539.

Sherman, S. M., K.-P. Hoffmann, and J. Stone (1972). The loss of a specific cell type from dorsal lateral geniculate nucleus in visually deprived cats. J. Neurophysiol. 35:532-541.

Development of the Lateral Geniculate Nucleus in Cats Raised with Monocular Eyelid Suture

S. MURRAY SHERMAN

Department of Physiology
University of Virginia School of Medicine
Charlottesville, Virginia USA

Since the pioneering work of Wiesel and Hubel (1963a,b; 1965), neurobiologists have appreciated the kitten's central visual pathways as an elegant model system for studies of the role of the postnatal environment in neural development. A particularly useful approach has been a comparison of the geniculo-cortical pathways in normally reared cats with those in cats raised with monocular eyelid closure. This paper concentrates on the developmental abnormalities seen in the lateral geniculate nucleus of such monocularly deprived cats. Although most studies of visually deprived cats have focused upon striate cortex, we have emphasized the lateral geniculate nucleus, because an understanding of cortical abnormalities requires a fairly complete description of the status of its geniculate inputs.

A simple version of the cat's retino-geniculo-cortical pathways is shown in Fig. 1. The dorsal two geniculate laminae, A and A1, provide a reasonably matched representation of each eye, and nearly all of our data are derived from this laminar pair. The ventral C complex includes laminae C, C1, and C2 (Guillery, 1970), and virtually nothing is known regarding the postnatal development of these laminae. Mainly for these reasons, this paper is further limited to a consideration of deprivation effects in laminae A and A1. However, before these are considered, it is useful to divide the lateral geniculate nucleus further into binocular and monocular segments and X- and Y-cells.

Binocular and Monocular Segments

Definition Each geniculate neuron has a small receptive field limited in visual space, and neighboring neurons tend to map neighboring spatial coordinates. As a consequence, an orderly, fairly precise point-to-point map of visual space

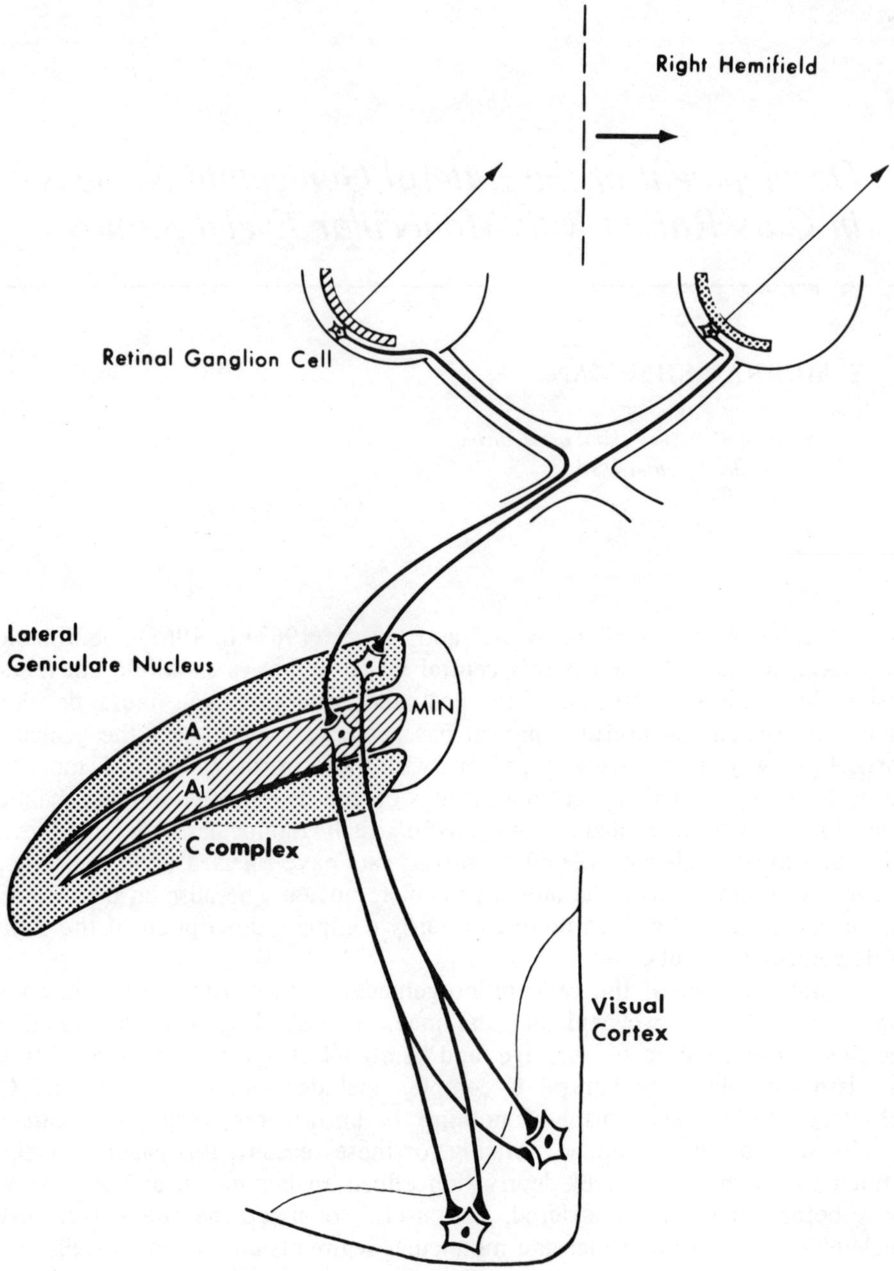

FIGURE 1 Retino-geniculo-cortical pathways in the cat (see text for details). The lateral geniculate nucleus is diagrammed in a coronal plane. The C complex refers to ventral laminae (C, C1, and C2), which receive input from retina and project to cortex, although these pathways are not drawn. Likewise, the medial interlaminar nucleus (MIN) contains cells which receive retinal input and project to cortex.

exists in the lateral geniculate nucleus (Sanderson, 1971): lateral (or medial) displacements in the nucleus map more peripheral (or central) visual space in the contralateral hemifield, and the medial edge of the nucleus represents the vertical meridian of visual field; rostral (or caudal) displacements map more inferior (or superior) visual space. The maps in laminae A and A1 are in register such that lines perpendicular to the laminae represent the same general area of visual space. As a result of this mapping, the binocular segment is that part of the nucleus which maps the central visual field seen by both eyes (roughly 45° to either side of the vertical meridian; Sherman, 1973). This includes all of lamina A1 and the corresponding portion (i.e., the medial three-fourths) of lamina A. The monocular segment is that part which maps the extreme peripheral crescent of visual field which can be viewed only by one eye (roughly 45-90° ipsilateral to that eye; Sherman, 1973). This is represented in the lateral one-fourth of lamina A which extends beyond lamina A1.

Binocular competition vs. deprivation per se Guillery and Stelzner (1970) first made use of this division into binocular and monocular segments in their histological studies of monocularly deprived cats. They confirmed and extended an earlier observation of Wiesel and Hubel (1963a). That is, Guillery and Stelzner (1970) reported that although cells in deprived laminae (i.e., those receiving direct retinal afferents from the sutured eye) were abnormally small, this effect was limited to the binocular segment of the nucleus. The deprived monocular segment of lamina A had cells of normal size which were indistinguishable from those in the nondeprived monocular segment on the other side (however, see Hickey, Spear, and Kratz, 1977).

The significance of this differential effect of monocular suture on the binocular and monocular segments of the nucleus is outlined in Fig. 2. The concept represented is that at least two different mechanisms can operate to produce the deprivation effects. We refer to one as "binocular competition" and the other as "deprivation *per se*," and they are described more fully below (see also, Sherman, Guillery, Kaas, and Sanderson, 1974).

The idea that a competitive mechanism is involved originated with Wiesel and Hubel (1965) and was elaborated by Guillery and coworkers (Guillery and Stelzner, 1970; Guillery, 1972; Sherman, Hoffmann, and Stone, 1972; Sherman, 1973; Sherman et al., 1974; Sherman, Wilson, and Guillery, 1975; Wilson and Sherman, 1977). Wiesel and Hubel (1965) suggested that during early postnatal development, pathways from each eye compete with one another for dominance of central connections. The actual site of this competition remains unknown, and for illustration purposes only, Fig. 2 is drawn as if the competition occurs between sets of geniculocortical synapses related to each eye. During development, these synapses proliferate in strength, number, or both, and they compete for total control of the cortical cell. If the visual environment is normal (Fig. 2, left), neither set of synapses related to one or the other eye has a competitive advantage conferred upon it, a balance is struck, and normal, binocular cortical neurons emerge (Hubel and Wiesel, 1962). However, if one

eye is sutured (i.e., the right eye in Fig. 2 (right), so that lamina A in the drawing is deprived), an advantage is somehow conferred upon the development of
nondeprived geniculocortical connections. The advantage may be related to
higher peak firing rates, more synchronous firing, etc., but in fact, we have no
evidence as yet to suggest why nondeprived cells should be given an advantage.
In any case, because of this advantage during competitive development, the
nondeprived eye gains essentially total control over the cortical neurons, as is
the case in monocularly deprived cats (Wiesel and Hubel, 1963b, 1965; Wilson
and Sherman, 1977). Notice, however, that cells in the deprived monocular
segment, by definition, cannot suffer the deleterious consequences of developing at a competitive disadvantage. Although they are deprived just as much as
their counterparts in the binocular segment, they can form many stable

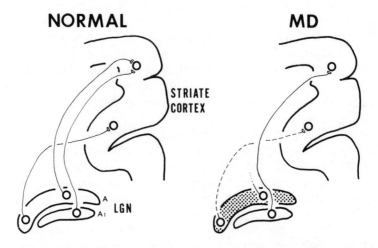

FIGURE 2 Diagram to illustrate developmental mechanisms of binocular competition and
deprivation *per se* in monocularly deprived cats. The deprived example (MD) is shown as if
the right eye and thus, lamina A, were deprived. It is suggested that during postnatal development, competition occurs among geniculo-cortical synapses for control of cortical neurons.
During normal rearing (left), no competitive advantage is present, a balance is struck, and
binocular cortical cells emerge. During monocular deprivation (right), the deprived cells are
somehow placed at a competitive disadvantage such that the nondeprived eye develops nearly
complete dominance over cortical cells. The deprived cells in the monocular segment of lamina A cannot by definition be placed at a competitive disadvantage, so they develop and/or
maintain at least some cortical connections. To the extent that the deprived monocular segment cells are completely normal, whereas those in the binocular segment are not, a developmental mechanism of binocular competition is indicated. To the extent that equal deficits are
seen throughout deprived lamina A, a noncompetitive developmental mechanism of deprivation *per se* is suggested. A combination of these two developmental processes is also possible.

geniculocortical connections simply because they are not fighting with a superior foe for these synaptic sites. Therefore, to the extent that the deprived
monocular segment develops much more normally than does the deprived binocular segment, support for a mechanism of binocular competition is indicated.
This is one explanation for the histological observations of Guillery and
Stelzner (1970).

The concept of the noncompetitive mechanism of deprivation *per se* is much
simpler. By this mechanism, the development of central pathways is determined solely by the quality of the visual environment experienced by each eye;
and interocular interactions, such as binocular competition, play no role. As
applied to Fig. 2, this mechanism would require that deprived cells develop
equal abnormalities in the binocular and monocular segments.

From the consideration above, it should be clear that an important comparison to be made in these studies of monocularly deprived cats is between the
deprived binocular and monocular segments. When deprivation-induced
deficits are apparent, three possible conclusions can be drawn from such comparisons: (1) if the deprived monocular segment develops completely normally
while the binocular segment does not, a mechanism of binocular competition
alone can parsimoniously account for the results; (2) if the deprived monocular
and binocular segments develop equal deficits, competitive mechanisms are not
indicated, and deprivation *per se* can account for the results; and (3) if deficits
are seen both in monocular and binocular segments, but the monocular segment deficits are less severe, a combination of binocular competition and
deprivation *per se* is indicated. Two other points can be made. First, if only
the binocular segment is studied, one cannot easily distinguish between effects
due to binocular competition and those due to deprivation *per se* (cf. Sherman
et al., 1974). Second, the deprived monocular segment may be the only place
where deprivation *per se* without competition influences can be studied.

X- and Y-cells

The other important division of the cat's retino-geniculo-cortical pathways
stems from the classical optic tract study of Enroth-Cugell and Robson (1966).
They defined two distinct populations of retinal ganglion cells as X (linear spatial summation) and Y (nonlinear spatial summation).* Since then, numerous
laboratories have concentrated on this distinction, and the scope of this literature is much too broad to cover in the present paper (for reviews, see Rowe
and Stone, 1977; Rodieck, 1979). X- and Y-cells have been described in the

---

*Recently, a third cell type (W) has been described among retinal ganglion cells. Some of these
cells project through geniculate neurons in the C complex to cortex (Wilson and Stone, 1975; Wilson, Rowe, and Stone, 1976). W-cells differ in many ways from X- and Y-cells, but relatively little
is known of them in normal cats and virtually nothing is known of their properties following early
visual deprivation. For this reason, plus the fact that nearly all of our analysis has been limited to
laminae A and A1, which lies outside the W-cell pathway, W-cells are not considered further in this
paper.

lateral geniculate nucleus with nearly identical properties to their retinal counterparts (Cleland, Dubin, and Levick, 1971; Hoffmann, Stone, and Sherman,
1972; Shapley and Hochstein, 1975). We now know that X- and Y-cells differ
among many electrophysiological characteristics. Compared to X-cells, Y-cells
possess: faster conducting axons, less linear spatial summation in the receptive
field, larger fields, more phasic responses to standing contrasts, slightly better
sensitivity to temporal changes, slightly poorer sensitivity to high spatial frequencies, and much greater sensitivity to low spatial frequencies (Cleland et al.,
1971; Hoffmann et al., 1972; Shapley and Hochstein, 1975; Hochstein and
Shapley, 1976a,b; Lehmkuhle, Kratz, Mangel, and Sherman, 1979a).

Although it seems clear that X- and Y-cells represent two parallel, fairly
independent pathways from retina to cortex (Cleland et al., 1971; Hoffmann et
al., 1972), the significance of X- and Y-cells for cortical processing remains
unclear and somewhat controversial. In a strong departure from the Hubel and
Wiesel (1962) "serial processing" hypothesis (Fig. 3A), whereby a single chain
of cells from geniculate to cortical simple cell to cortical complex cell, etc., processed visual information, Stone and coworkers (Hoffmann and Stone, 1971;
Stone and Dreher, 1973) suggested a hypothesis of "parallel processing" (Fig.
3B), whereby two independent cell chains—X-cells through cortical simple cells
and Y-cells through cortical complex cells—processed different aspects of the
visual scene in parallel.

Whatever the functional significance of the division of retinal and geniculate
neurons into X- and Y-cells (see also below), it seemed reasonable to investigate the possibility that these two systems were differentially affected by
deprivation in an analogous fashion to the differences seen between binocular
and monocular segments. The importance of this possibility was underscored
recently by the observations of Daniels, Pettigrew, and Norman (1978) who
concluded that kitten geniculate X-cells normally attain maturity earlier than do
Y-cells, and that Y-cells are thus more susceptible to environmental deficiencies
during the "critical period" (Hubel and Wiesel, 1970) of early postnatal development.

Effects of Monocular Deprivation upon Y-cells

Sherman et al. (1972) reported that genicular Y-cells seemed much more
affected by early lid suture than did X-cells. Deprived X-cells generally seemed
normal both in numbers and response properties, although a subtle abnormality
is described in the next section. Figure 4 represents a redrawing of Fig. 2 from
Sherman et al. (1972) with added data points and limited to data from laminae
A and A1. This shows that, with our recording techniques in monocularly
deprived cats, few normal Y-cells were encountered throughout the binocular
segment, whereas normal numbers were seen in the monocular segment.
Furthermore, the receptive field properties of the encountered deprived Y-cells
(mostly in the monocular segment) were completely normal (Sherman et al.,
1972; Lehmkuhle et al., 1979b). These properties included response rate, field

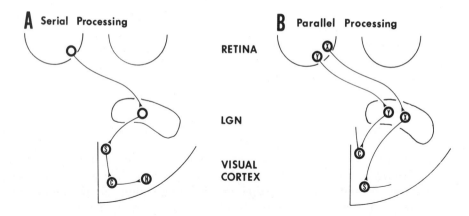

FIGURE 3 Hypotheses of serial and parallel processing. A: Wiring diagram for serial processing hypothesis of Hubel and Wiesel (1962). They suggested a single hierarchy of neurons from retina through cortex for visual processing. In this scheme, a fairly homogeneous population of geniculate cells feeds onto the first-order, or simple (S) cortical cell, then to the complex (C) and hypercomplex (H) cells. Note that this hypothesis was proposed before knowledge of X- and Y-cells. B: Wiring diagram for parallel processing scheme suggested by Stone and coworkers (Hoffmann and Stone, 1971; Stone, 1972; Stone and Dreher, 1973) and incorporating the concept of X- and Y-cells. At least two fairly independent pathways from retina through cortex which process different aspects of the visual scene in parallel are suggested. Retinal X-cells project to geniculate X-cells which project to cortical simple (S) cells. Retinal Y-cells project to geniculate Y-cells which project to cortical complex (C) cells. These diagrams represent simplified versions of hypotheses and should not be treated literally.

size, temporal and spatial contrast sensitivity (see also below), and area response functions. This pattern of few normal Y-cells in the deprived binocular segment and many perfectly normal Y-cells in the deprived monocular segment corresponds closely to the histological observations of Guillery and Stelzner (1970) and suggests a mechanism of binocular competition.

The interpretation of these results is not straightforward since uncontrolled electrode sampling biases are possible. Anatomical correlates, described below, help somewhat in our understanding of these results. One suggested by Eysel, Grüsser, and Hoffmann (1978) is that these results merely reflect a changed electrode sampling artifact caused by relatively selective shrinkage of Y-cells (see also, LeVay and Ferster, 1977; Garey and Blakemore, 1977). Even if this is the sole explanation, it supports the general notion that geniculate Y-cells are more affected by early lid suture than are the X-cells. Furthermore, whatever the reason for our failure to record deprived Y-cells, soma size alone cannot be the general explanation. This point is made most clearly in studies of cats reared in total darkness (Kratz, Sherman, and Kalil, 1979). In these animals, we found very few geniculate Y-cells (Fig. 4), yet the soma size distribution among laminae A and A1 neurons was completely normal in the same cats from which Y-cells went unrecorded. These cells clearly did not show the lack of growth seen in monocularly deprived cats.* This suggests that Y-cell "losses" need not be correlated with changed soma size. Indeed, we have recently obtained evidence that deprived laminae contain abnormal cells with poor or no visual responsiveness that might represent the "missing" Y-cells (Kratz, Webb, and Sherman, 1978b; unpublished observations; and see Norton, Casagrande, and Sherman, 1977 for similar observations in monocularly sutured tree shrews).

In any case, there have been two types of anatomical studies which correlate with, but cannot yet explain, the physiological absence of recordable Y-cells from deprived laminae, which was described above. First, Garey and Blakemore (1977) and Lin and Sherman (1978) tried to isolate geniculate Y-cells for anatomical study by capitalizing on the observation (Stone and Dreher, 1973) that geniculate X-cells project only to area 17, while the Y-cells project both to areas 17 and 18. Horseradish peroxidase was injected into area 18 of monocularly deprived cats to label only a Y-cell population, and it was found that, in deprived laminae, labeled cells were smaller (Garey and Blakemore, 1977), much rarer, and more poorly stained (Lin and Sherman, 1978) than they were in nondeprived laminae. Area 17 injections provided relatively little asymmetry in labeling of deprived and nondeprived laminae, presumably because of the many fairly normal X-cells labeled in deprived laminae.

Second, LeVay and Ferster (1977) suggested a histological marker to distinguish between X- and Y-cells in the cat's lateral geniculate nucleus. They found that some cells had a curious cytoplasmic structure—a "cytoplasmic laminar body" (CLB)—while others did not. Based upon several lines of converging but indirect evidence, they concluded that cells with CLBs were X-cells. Larger cells without CLBs would thus be Y-cells; and the few smaller cells without CLBs, interneurons. Furthermore, they correlated these cell types with Golgi studies and concluded that Y-cells were Guillery's (1966) class 1 (large soma, extensive, cruciate dendritic arbor with few appendages or spines), X-cells were class 2 (intermediate soma size, curved dendrites with grape-like structures appended at dendritic branch points), and interneurons were class 3 (small soma, fine tortuous dendritic arbor with numerous stalked appendages of variable morphology). Although Guillery (1966) reported that 40% of his sample was intermediate or nonclassifiable, LeVay and Ferster (1977) do not mention such cells, so it is not clear whether these cells contain CLBs. LeVay and Ferster (1977) then applied their CLB classification to one monocularly deprived cat and concluded that, compared to deprived X-cells, deprived Y-cells were both fewer in number and considerably more shrunken (see also, Kalil and Worden, 1978).

---

*This raises another perplexing question. That is, what environmental factors and/or mechanisms control cell size? On the one hand, monocular suture retards cell growth in deprived laminae (Wiesel and Hubel, 1963a; Guillery and Stelzner, 1970; Hickey et al., 1977), whereas binocular suture or total dark rearing has little or no effect on cell size (Guillery, 1973; Hickey et al., 1977; Kalil, 1978; Kratz et al., 1978). Perhaps competitive mechanisms control cell size, so that only during appropriately unbalanced environmental conditions between the eyes will significant abnormalities in geniculate cell size develop.

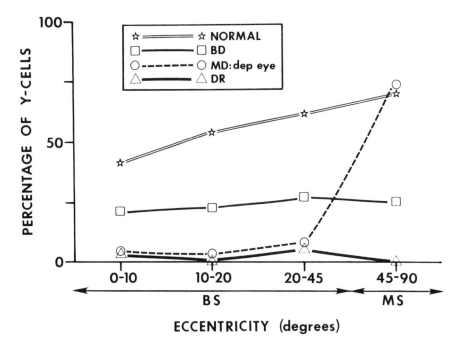

FIGURE 4 Percentage of Y-cells recorded in laminae A and A1 as a function of receptive field eccentricity from the area centralis. The few interneurons and unclassified relay cells (< 5%) are excluded, and the ordinate represents the percent fraction of Y-cells recorded among the total X- and Y-cell sample. The abscissa is broken into four eccentricity groups, and the binocular and monocular segments (BS and MS) are indicated. Each point on the graph for normal and monocularly deprived (MD) cats represents a sample of 62 to 212 cells. The data from nondeprived laminae of monocularly deprived cats (not shown) are indistinguishable from the normal data. Also shown are data from binocularly sutured (squares) and dark-reared (triangles) cats.

Such an anatomical means of identifying X- and Y-cells, based on a presumably nonselective histological method for locating CLBs, is of obvious potential importance for studying differential deprivation effects. For this reason, we have initiated a study to evaluate critically the LeVay and Ferster (1977) hypothesis by obtaining structure/function correlates at the single cell level for geniculate neurons in normal cats. Our method is to record intracellularly from these neurons with a fine micropipette filled with horseradish peroxidase, identify the cell as X or Y with conventional electrophysiological tests, and then iontophorese peroxidase into the cell for later morphological study. The filled cells present a Golgi-like appearance that allows ready identification into the classes described by Guillery (1966) and used by LeVay and Ferster (1977). To date, we have made such a correlation for ten Y-cells and eight X-cells (Friedlander, Lin, and Sherman, 1979). Of the Y-cells, six were class 1, two were class 2, and two were intermediate or could not be classified. Of the X-cells, one was class 3, and the remaining seven were intermediate between classes 2 and 3, varying from nearly complete class 3 morphology to mostly class 2 structure. We thus tentatively conclude from a small sample that the LeVay and Ferster (1977) hypothesis requires some modification. Class 1 cells seem to be Y-cells, but cells with class 2 characteristics can also be Y-cells. X-cells seem to occupy the structural ground between classes 2 and 3. To the extent that class 2 cells seem to be Y-cells, and this was the only neuron class significantly affected by monocular eyelid suture (LeVay and Ferster, 1977), the anatomy again suggests that geniculate Y-cells are much more affected by early eyelid suture than are geniculate X-cells.

Effects of Monocular Deprivation upon Geniculate X-cells

Until recently, we were unable to detect any obvious effect of lid suture upon X-cell development. In deprived laminae, these cells were encountered in normal numbers and possessed fairly normal response properties (Sherman et al., 1972). Also, none of the anatomical studies suggested significant structural abnormalities for deprived X-cells (LeVay and Ferster, 1977; Garey and Blakemore, 1977; Lin and Sherman, 1978).

However, the recent literature suggested that more sensitive receptive field methods might uncover subtle deficits for deprived X-cells. For instance, Ikeda, Tremain, and Einon (1978) report that geniculate X-cells in cats raised with artificial esotropia have abnormally poor spatial acuity (defined as the highest spatial frequency to which the cell responds). Similarly, Maffei and Fiorentini (1976) and Hoffmann and Sireteneau (1977) reported poorer spatial acuity for deprived geniculate cells, but neither report distinguished between X- and Y-cells. We reinvestigated this question of geniculate X-cell normality in monocularly deprived cats by obtaining for these cells spatial and temporal contrast sensitivity functions to counterphased, sine-wave gratings (Lehmkuhle et al., 1978, 1979b). That is, we measured the grating contrast necessary to evoke a threshold neuronal response as spatial frequency (cycles/degree) and/or temporal frequency (cycles/sec counterphase rate) was varied. We found that, compared to nondeprived or normal geniculate X-cells, deprived X-cells had normal temporal sensitivity and normal sensitivity to lower spatial frequencies but were relatively insensitive to the higher spatial frequencies. Consequently, their spatial acuity was consistently reduced to the point that, on average, a normal X-cell could respond to a grating twice as fine as one that would excite a deprived X-cell (see Table 1). An important additional point also evident from Table 1 is the observation that deprived X-cells in the monocular segment were just as affected as were those in the binocular segment.

This last point is in stark contradistinction to the deprivation abnormalities described for Y-cells (compare Fig. 4 with Table 1) and suggests that different mechanisms are involved. Whereas some form of binocular competition effectively accounts completely for the Y-cell pattern of deficits, deprivation *per se* seems the simplest explanation for the X-cell pattern, since the abnormalities are equal for deprived binocular and monocular segments.

TABLE 1 Spatial resolution (highest spatial frequency sine-wave grating at 0.6 contrast and 2 cycles/sec counterphase rate to which the cell responds) for geniculate X-cells in deprived and nondeprived laminae A or A1 of monocularly deprived cats; data from Lehmkuhle et al. (1979b). These values (number of cells and mean ± standard error) are indicated for each of five eccentricity groups (receptive field eccentricity from the area centralis), including the monocular segment (> 45°), plus the total of all cells. No deprived X-cells were studied with a receptive field eccentricity between 20° and 45°. The reduction in resolution for each group is also shown and is fairly constant with eccentricity. This reduction is calculated as 100% [1 - (deprived resolution)/(nondeprived resolution)].

	0°-5°	5°-10°	10°-15°	15°-20°	> 45°	Total
Nondeprived						
N	32	29	8	9	11	89
Mean ± S.E.	2.8 ± 0.2	2.6 ± 0.2	2.5 ± 0.3	1.9 ± 0.2	1.2 ± 0.1	2.4 ± 0.2
Deprived						
N	5	17	9	7	17	55
Mean ± S.E.	1.5 ± 0.3	1.2 ± 0.1	1.1 ± 0.2	1.3 ± 0.2	0.6 ± 0.1	1.0 ± 0.2
Reduction	47%	54%	56%	33%	50%	58%

Further Evidence for Binocular Competition

The evidence presented above for binocular competition as a developmental mechanism is based upon differences between the reactions of the binocular and monocular segments to early monocular deprivation. The underlying assumption has been that these developmental differences are due to the binocular/monocular distinction between these segments. However, there are other differences that seem unrelated to this distinction. For instance, compared to centrally represented portions of the visual field (i.e., binocular segment), the peripherally represented portions (i.e., monocular segment) tend to have cells with less selective receptive field properties, and thus their development may be less sensitive to environmental irregularities. Also, differences in geniculocortical pathways between these areas have been suggested. Tusa, Rosenquist, and Palmer (1979) report that, whereas cortical area 17 includes a complete representation of the visual field, the area 18 map essentially covers only the binocular segment. Perhaps only the geniculocortical pathways to area 18, which involve Y-cells but not X-cells (Stone and Dreher, 1973), are affected by early lid suture, and this would not require a competitive mechanism.

Guillery (1972) designed an elegant experiment to demonstrate that the developmental differences between the binocular and monocular segments are due to the binocular/monocular distinction—and thus binocular competition—rather than other factors suggested above. He created a centrally located "critical segment" or "artificial monocular segment" by placing a neonatal retinal lesion centrally in the open eye at the time the other eye was sutured. Figure 5 summarizes the results obtained with this preparation which now includes two monocular segments for the deprived eye: the natural one related to extreme nasal retina and the artificial one related to central retina homonymous to the

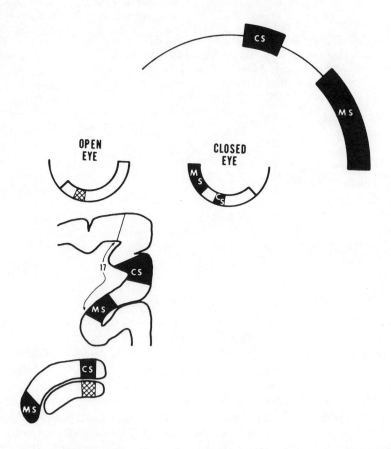

FIGURE 5 Summary of results from critical segment or artificial monocular segment preparation described by Guillery (1972) and studied also by Sherman et al. (1974, 1975). At the time the right eye is neonatally closed, a small lesion is placed in the left retina. This creates two monocular segments relative to the deprived eye: a natural one (MS) and an artificial one (critical segment, CS). Both segments develop in the same way (see text), and this supports the concept of binocular competition during development.

open eye's lesion. Note that the artificial monocular segment occupies regions in central pathways which, without the lesion, would have developed as binocular segment.

With this preparation, Guillery (1972) showed that in deprived geniculate laminae, cells were of normal size only in the natural and artificial monocular segments. Sherman et al. (1974) then showed that while using the deprived eye, such a cat could visually orient to targets placed only in the natural or artificial monocular segments; also, only in the natural and artificial monocular segments of striate cortex did the deprived eye influence significant numbers of neurons. Finally, Sherman et al. (1975) reported that in the deprived laminae, only the natural and artificial monocular segments contained significant numbers of recordable Y-cells. The pattern of results illustrated in Fig. 5 indicates that the differential response of the binocular and monocular segments to early monocular lid suture is due to some form of binocular competition. While these studies clearly implicate such a developmental mechanism, we still know virtually nothing about the details of the mechanism or even its central site of action.

However, one additional speculation can be made based upon the observation that the geniculate Y-cells, but not X-cells, seem to develop by way of a mechanism of binocular competition. Recently, Ferster and LeVay (1978) suggested that axons from X-cells in layer IVc of cat striate cortex arborize within a single ocular dominance column. The Y-cells, on the other hand, seem to possess axons which ramify across many ocular dominance columns in layer IVab. X-cells may not show binocular competition simply because their projections from one geniculate lamina are not in a position to interact with those from another lamina. Y-cells alone may be in a position to compete binocularly along the lines suggested in Fig. 2, simply because of their more extensive axonal arborizations which permit interactions among axons and terminals from different geniculate laminae.

Effects of Monocular Deprivation upon Retinal Ganglion Cells

Theoretically, it is possible to account for deprivation defects in geniculate cells on the basis of similar defects in their retinal inputs. That is, Y-cells could be missing from the deprived retina, and retinal X-cells in the closed eye could develop poor spatial acuity. Any mechanism requiring interocular interactions (i.e., binocular competition for Y-cells) must almost certainly occur central to the retina, and an earlier study (Sherman and Stone, 1973) reported unchanged proportions of X- and Y-cells in the deprived retina. On the other hand, the noncompetitive deprivation *per se* mechanism implicated for X-cells could well have a retinal origin. We reinvestigated retinal ganglion cells in monocularly deprived cats by recording from optic tract, and we found no evidence for abnormalities in the spatial or temporal contrast sensitivity functions for deprived X- or Y-cells (Kratz, Mangel, Lehmkuhle, and Sherman, 1979). Thus, the retina seems to develop fairly normally despite the lid suture, and the defects described above have a more central origin.

If the data for X-cells have been correctly interpreted, this raises a difficult conceptual problem. Why should deprived geniculate X-cells display spatial deficits for only higher frequencies if their presumed retinal inputs have normal sensitivity throughout the spatial frequency domain? The population of X-cells (and Y-cells) shows considerable scatter in terms of the sensitivity to high spatial frequencies or spatial acuity, and it may be that only the units with poorer spatial acuity sampled in the optic tract make or maintain effective connections in deprived geniculate laminae.

Summary and Conclusions

Patterns of X- and Y-cell effects   It seems clear that geniculate cells do not develop normally during monocular lid suture, and that the consequences and underlying mechanisms of these deprivation effects are quite different for X- and Y-cells. These differences probably depend to some extent on the finding that when they enter the "critical period," X-cells have completed more of their development than have Y-cells (Daniels et al., 1978). Both physiological and anatomical evidence suggests that in the deprived, binocular segment, Y-cells are much more profoundly affected by lid suture than are X-cells. On the other hand, Y-cells seem completely normal in the deprived monocular segment, whereas X-cells are not. This suggests the very different deprivation mechanisms of binocular competition for Y-cells and deprivation *per se* for X-cells.

Functional implications   In order to understand these results in a functional or clinical framework, we must first know what the significance of the X- and Y-cell division is for normal cats. Unfortunately, we have only intuitive speculations that can be addressed to this critical point. The most common suggestion (cf. Ikeda and Wright, 1972, 1975) is that X-cells are most concerned with the analysis of spatial patterns; and Y-cells with temporal patterns. However, our recent contrast sensitivity studies (Lehmkuhle et al., 1979a) suggested fairly small differences between these cell groups in terms of sensitivity to high spatial or temporal frequencies (X-cells were slightly more sensitive than were Y-cells to the former, Y-cells more than were X-cells to the latter). These data do not support a differential role for X- and Y-cells based upon spatial and temporal processing. The most dramatic difference in sensitivity between X- and Y-cells occurred in response to low spatial frequencies. To such stimuli, X-cells are fairly insensitive, whereas Y-cells are quite sensitive.

We have thus suggested a different functional dichotomy based upon the psychophysical observations that low spatial frequencies in a visual scene carry the basic form information, whereas the high frequencies add detail (Kabrisky, Tallman, Day, and Radoy, 1970; Ginsberg, Carl, Kabrisky, Hall, and Gill, 1976; Hess and Garner, 1977; Hess and Woo, 1978). Because of their unique sensitivity to these important low spatial frequencies, Y-cells are probably important to basic spatial analysis. X-cells, because of their better acuity and spatial phase dependency (Hochstein and Shapley, 1979a; Lehmkuhle et al., 1979a), probably add detail, such as better acuity, perhaps stereopsis, etc. (for a more complete discussion of this suggestion, see Lehmkuhle et al., 1979a,b). Without Y-cells, spatial vision might be at best rudimentary, but if only X-cells were affected, reasonable spatial vision might still be possible, since low spatial frequency analysis is possible. In support of the latter consequence of the suggestion, Berkley and Sprague (1978) found that nearly total lesions of area 17, which destroy the X pathways but leave many or most of the geniculocortical projections of Y-cells intact (Stone and Dreher, 1973; Gilbert and Kelly, 1975; Kratz et al., 1977a), produce a cat with excellent spatial vision and only a 20% loss of spatial acuity.

These hypotheses might also explain some of the variability reported in clinical studies of amblyopia of central origin (cf. Hess and Woo, 1978; and many others). If X- and Y-cells are both affected, as in a lid sutured cat, the amblyopia might be maximal. If only the X-cells are affected at higher spatial frequencies, as seems to be the case in cats raised with esotropia or anisometropia (Ikeda and Wright, 1976; Ikeda and Tremain, 1978), the amblyopia would be much less severe and affect only the acuity level for fine details. Finally, the fact that Y-cells are very sensitive to low spatial frequencies could explain why lid suture, which attenuates all spatial frequencies, prevents their normal development, whereas anisometropia, which essentially attenuates only higher spatial frequencies, permits their normal development. X-cells, which are somewhat more sensitive to higher spatial frequencies develop abnormally under any deprivation condition, such as lid suture or anisometropia, which attenuates these frequencies.

Acknowledgements The research described in this paper was supported by USPHS Grant EY01565, NSF Grant BNS77-06785, and a grant from The A. P. Sloan Foundation. The author also received support from a USPHS RCDA EY00020.

References

Berkley, M. A., and Sprague, J. M. (1978). Behavioral analysis of the geniculocortical system in form vision. In: Frontiers in Visual Science. S. J. Cool and E. L. Smith (eds.). Springer-Verlag, New York.

Cleland, B. G., Dubin, M. W., and Levick, W. R. (1971). Sustained and transient neurons in the cat's retina and lateral geniculate nucleus. J. Physiol. 217:473-496.

Daniels, J. D., Pettigrew, J. D., and Norman, J. L. (1978). Development of single neuron responses in kitten's lateral geniculate nucleus. J. Neurophysiol. 41:1373-1393.

Enroth-Cugell, C., and Robson, J. G. (1966). The contrast sensitivity of retinal ganglion cells of the cat. J. Physiol. 187:517-552.

Eysel, U. Th., Grüsser, O.-J., and Hoffmann, K.-P. (1978). The effect of monocular pattern deprivation on the signal transmission by neurons of the cat lateral geniculate body. Arch. Ital. Biol. 116:427-443.

Ferster, D., and LeVay, S. (1978). The axonal arborizations of lateral geniculate neurons in the striate cortex of the cat. J. Comp. Neurol. 182:923-944.

Friedlander, J. J., Lin, C.-S., and Sherman, S. M. (1979). Structure of physiologically identified X- and Y-cells in the cat's lateral geniculate nucleus. Science (in press).

Garey, L. J., and Blakemore, C. (1977). The effects of monocular deprivation on different neuronal classes in the lateral geniculate nucleus of the cat. Exp. Brain Res. 28:259-278.

Gilbert, C. C., and Kelly, J. P. (1975). The projections of cells in different layers of the cat's visual cortex. J. Comp. Neurol. 163:81-106.

Ginsburg, A. P., Carl, J. W., Kabrisky, M., Hall, C. F., and Gill, P. A. (1976). Psychological aspects of a model for the classification of visual images. In: Advances in Cybernetics and Systems, vol. III. J. Rose (ed.). Gordon and Breach Science Publishers, Ltd., London.

Guillery, R. W. (1966). A study of Golgi preparations from the dorsal lateral geniculate nucleus of the adult cat. J. Comp. Neurol. 128:21-50.

Guillery, R. W. (1970). The laminar distribution of retinal fibers in the dorsal lateral geniculate nucleus of the cat: a new interpretation. J. Comp. Neurol. 138:339-368.

Guillery, R. W. (1972). Binocular competition in the control of geniculate cell growth. J. Comp. Neurol. 144:177-230.

Guillery, R. W. (1973). The effect of lid suture upon the growth of cells in the dorsal lateral geniculate nucleus of kittens. J. Comp. Neurol. 148:417-422.

Guillery, R. W., and Stelzner, D. J. (1970). The differential effects of unilateral lid closure upon the monocular and binocular segments of the dorsal lateral geniculate nucleus of the cat. J. Comp. Neurol. 139:413-422.

Hess, R. F., and Garner, L. R. (1977). The effects of corneal edema on visual function. Invest. Ophthal. & Vis. Sci. 16:5-13.

Hess, R., and Woo, G. (1978). Vision through cataracts. Invest. Ophthal. & Vis. Sci. 17:428-435.

Hickey, T. L., Spear, P. D., and Kratz, K. E. (1977). Quantitative studies of cell size in the cat's lateral geniculate nucleus following visual deprivation. J. Comp. Neurol. 172:265-282.

Hochstein, S., and Shapley, R. M. (1976a). Quantitative analysis of retinal ganglion cell classifications. J. Physiol. 262:237-264.

Hochstein, S., and Shapley, R. M. (1976b). Linear and nonlinear spatial subunits in Y cat retinal ganglion cells. J. Physiol. 262: 265-284.

Hoffmann, K.-P., and Sireteanu, R. (1977). Interlaminar differences in the effects of early and late monocular deprivation on the visual acuity of cells in the lateral geniculate nucleus of the cat. Neuroscience Letters 5:171-175.

Hoffmann, K.-P., and Stone, J. (1971). Conduction velocity of afferents to cat visual cortex: a correlation with cortical receptive field properties. Brain Res. 32:460-466.

Hoffmann, K.-P., Stone, J., and Sherman, S. M. (1972). Relay of receptive field properties in dorsal lateral geniculate nucleus of the cat. J. Neurophysiol. 35:518-531.

Hubel, D. H., and Wiesel, T. N. (1962). Receptive fields, binocular interaction, and functional architecture in the cat's visual cortex. J. Physiol. 160:106-154.

Hubel, D. H., and Wiesel, T. N. (1970). The period of susceptibility of the physiological effects of unilateral eye closure in kittens. J. Physiol. 206:419-436.

Ikeda, H., and Tremain, K. E. (1978). Amblyopia resulting from penalisation: neurophysiological studies of kittens reared with atropinisation of one or both eyes. Brit. J. of Ophthal. 62:21-28.

Ikeda, H., Tremain, K. E., and Einon, G. (1978). Loss of spatial resolution of lateral geniculate nucleus neurones in kittens raised with convergent squint produced at different stages in development. Exp. Brain Res. 31:207-220.

Ikeda, H., and Wright, M. J. (1972). Receptive field organization of sustained and transient retinal ganglion cells which subserve different functional roles. J. Physiol. 227:769-800.

Ikeda, H., and Wright, M. J. (1975). Spatial and temporal properties of "sustained" and "transient" neurones in area 17 of the cat's visual cortex. Exp. Brain Res. 22:363-383.

Ikeda, H., and Wright, M. J. (1976). Properties of LGN cells in kittens reared with convergent squint: a neurophysiological demonstration of amblyopia. Exp. Brain Res. 25:63-77.

Kabrisky, M., Tallman, O., Day, C. M., and Radoy, C. M. (1970). A theory of pattern perception based on laminar physiology. In: Contemporary Problems in Perception. A. T. Welford and L. Houssiadas (eds.). Taylor & Francis, Ltd., London.

Kalil, Ronald (1978). Dark rearing in the cat: effects on visuomotor behavior and cell growth in the dorsal lateral geniculate nucleus. J. Comp. Neurol. 178:451-468.

Kalil, Ronald, and Worden, Ian (1978). Cytoplasmic laminated bodies in the lateral geniculate nucleus of normal and dark-reared cats. J. Comp. Neurol. 178:469-486.

Kratz, K. E., Mangel, S. C., Lehmkuhle, S., and Sherman, S. M. (1979). Retinal X- and Y-cells in monocularly lid-sutured cats: normality of spatial and temporal properties. Submitted for publication.

Kratz, K. E., Sherman, S. M., and Kalil, R. (1979). Lateral geniculate nucleus in dark-reared cats: loss of Y-cells without changes in cell size. Science 203:1353-1355.

Kratz, K. E., Webb, S. V., and Sherman, S. M. (1978a). Studies of the cat's medial interlaminar nucleus: a subdivision of the dorsal lateral geniculate nucleus. J. Comp. Neurol. 181: 601-614.

Kratz, K. E., Webb, S. V., and Sherman, S. M. (1978b). Effects of early monocular lid suture upon neurons in the cat's medial interlaminar nucleus. J. Comp. Neurol. 181:615-625.

Lehmkuhle, Stephen W., Kratz, Kenneth E., Mangel, Stuart C., and Sherman, S. Murray (1978). An effect of early monocular lid suture upon the development of X-cells in the cat's lateral geniculate nucleus. Brain Res. 157:346-350.

Lehmkuhle, S., Kratz, K. E., Mangel, S. C., and Sherman, S. M. (1979a). Spatial and temporal sensitivity of X- and Y-cells in the dorsal lateral geniculate nucleus of the cat. Submitted for publication.

Lehmkuhle, S., Kratz, K. E., Mangel, S. C., and Sherman, S. M. (1979b). The effects of early monocular lid suture on spatial and temporal sensitivity of neurons in the dorsal lateral geniculate nucleus of the cat. Submitted for publication.

LeVay, S., and Ferster, D. (1977). Relay cell classes in the lateral geniculate nucleus of the cat and the effects of visual deprivation. J. Comp. Neurol. 172:563-584.

Lin, C.-S., and Sherman, S. M. (1978). Effects of early monocular eyelid suture upon development of relay cell classes in the cat's lateral geniculate nucleus. J. Comp. Neurol. 181:809-831.

Maffei, L., and Fiorentini, A. (1976). Monocular deprivation in kittens impairs the spatial resolution of geniculate neurones. Nature 264:754-755.

Norton, Thomas T., Casagrande, Vivien A., and Sherman, S. Murray (1977). Loss of Y-cells in the lateral geniculate nucleus of monocularly deprived tree shrews. Science 197:784-786.

Rodieck, R. W. (1979). Visual pathways. Ann. Rev. Neurosci. 2:193-225.

Rowe, M. H., and Stone, J. (1977). Naming of neurons. Classification and naming of cat retinal ganglion cells. Brain, Behav., & Evol. 14:185-216.

Sanderson, K. J. (1971). The projection of the visual field to the lateral geniculate and medial interlaminar nuclei in the cat. J. Comp. Neurol. 143:101-118.

Shapley, R., and Hochstein, S. (1975). Visual spatial summation in two classes of geniculate cells. Nature 156:411-413.

Sherman, S. M. (1973). Visual field defects in monocularly and binocularly deprived cats. Brain Res. 49:25-45.

Sherman, S. M., Guillery, R. W., Kaas, J. H., and Sanderson, K. J. (1974). Behavioral, electrophysiological, and morphological studies of binocular competition in the development of the geniculo-cortical pathways of cats. J. Comp. Neurol. 158:1-18.

Sherman, S. M., Hoffmann, K.-P., and Stone, J. (1972). Loss of a specific cell type from dorsal lateral geniculate nucleus in visually deprived cats. J. Neurophysiol. 35:532-541.

Sherman, S. M., and Stone, J. (1973). Physiological normality of the retina in visually deprived cats. Brain Res. 60:224-230.

Sherman, S. M., Wilson, J. R., and Guillery, R. W. (1975). Evidence that binocular competition affects the postnatal development of Y-cells in the cat's lateral geniculate nucleus. Brain Res. 100:441-444.

Stone, J. (1972). Morphology and physiology of the geniculocortical synapse in the cat. The question of parallel input to the striate cortex. Invest. Ophthal. 11:338:344.

Stone, J., and Dreher, B. (1973). Projection of X- and Y-cells of cat's lateral geniculate nucleus to areas 17 and 18 of visual cortex. J. Neurophysiol. 36:551-567.

Tusa, R., Rosenquist, A. C., and Palmer, L. A. (1979). Retinotopic organization of areas 18 and 19 in the cat. J. Comp. Neurol. (in press).

Wiesel, T. N., and Hubel, D. H. (1963a). Effects of visual deprivation on morphology and physiology of cells in the cat's lateral geniculate body. J. Neurophysiol. 26:978-993.

Wiesel, T. N., and Hubel, D. H. (1963b). Single-cell responses in striate cortex of kittens deprived of vision in one eye. J. Neurophysiol. 26:1003-1017.

Wiesel, T. N., and Hubel, D. H. (1965). Comparison of the effects of unilateral and bilateral eye closure on cortical responses in kittens. J. Neurophysiol. 28:1029-1040.

Wilson, J. R., and Sherman, S. M. (1977). Differential effects of early monocular deprivation on binocular and monocular segments of cat striate cortex. J. Neurophysiol. 40:891-903.

Wilson, P. D., Rowe, M. H., and Stone, J. (1976). Properties of relay cells in cat's lateral geniculate nucleus: a comparison of W-cells with X- and Y-cells. J. Neurophysiol. 39:1193-1209.

Wilson, P. D., and Stone, J. (1975). Evidence of W-cell input to the cat's visual cortex via the C laminae of the lateral geniculate nucleus. Brain Res. 92:472-478.

The Consequence of a "Consolidation" Period Following Brief Monocular Deprivation in Kittens

R. D. FREEMAN
School of Optometry
University of California
Berkeley, California USA

Abstract Brief periods of limited visual exposure, such as monocular deprivation, have been reported to cause changes in striate cortex that are accentuated by a delay prior to physiological study. This notion, called "consolidation" in previous work, has been tested in the present study. Four-week-old normally reared kittens were monocularly occluded for brief periods (8 or 24 hours). Extracellular study of striate cortex was undertaken either immediately after occlusion or following a period during which the animals were kept in darkness. Substantial monocular deprivation effects were found in both groups, but the most pronounced changes were observed for the kittens studied immediately after exposure. Therefore, no "consolidation" has been observed. On the contrary, it appears that the period spent in darkness actually diminished the consequences of the monocular deprivation.

Introduction

Monocular deprivation of relatively short duration can have marked effects at the level of striate cortex in kittens (Hubel and Wiesel, 1970; Olson and Freeman, 1975). Recently, periods of a day or less have been reported to result in reduced numbers of binocular cells (Movshon and Dursteler, 1977). Thus, the consequences of unilateral occlusion develop rapidly and it may be useful, in the search for underlying mechanisms, to examine other short-term processes of the brain, such as learning and memory.

A concept from this latter field, "consolidation," has been applied in several studies to visual deprivation (Pettigrew and Garey, 1974; Peck and Blakemore, 1975; Buisseret, Gary-Bobo, and Imbert, 1978). The basic idea is that memory is a two-stage process in which an event is initially unstable and then becomes "consolidated" into its final form (Agranoff, 1974). This term has been applied

99

in investigations in which dark-reared kittens received limited visual exposures and then were studied physiologically either immediately or after a delay. Effects were reported to be most pronounced following an intervening period between exposure and study, and this result was attributed to a process during which the limited visual experience was "consolidated" (Pettigrew and Garey, 1974; Peck and Blakemore, 1975). This interpretation of the term "consolidation" is that the effects of exposure are accentuated during the delay before physiological study. In the work on memory, the term refers to a conversion process whereby a labile trace develops into a permanent form (Agranoff, 1976). Therefore, the original notion of "consolidation" has not been addressed by the visual deprivation experiments. Moreover, the evidence for the accentuated effects during the period between exposure and physiological testing is not strong.

It is not clear why the effects of limited exposure should become more pronounced during this intervening period that an animal spends binocularly deprived. On the contrary, in the case of monocular exposure, presumed competition between afferent pathways from each eye is halted during this delay period. It might be predicted, therefore, that some binocular connections might be re-established and the consequences of the monocular exposure would be reduced.

The experiments presented here address this question by assessment of the effect of a delay period between monocular deprivation and physiological study. Two groups of kittens were tested. Both were reared normally for four weeks and then monocularly occluded for brief periods. One group was studied physiologically immediately after the unilateral exposure and the other was recorded after an intervening session during which the animals were kept in darkness. Effects were most pronounced for the first group, and therefore, the results are in conflict with a "consolidation" notion.

Methods

Normally reared kittens were monocularly occluded with very large opaque contact lenses on postnatal day 29. This is near the peak of the vulnerability period to the effects of monocular deprivation (Hubel and Wiesel, 1970). During the period of occlusion, the animals were kept awake as much as possible to offset their tendency to sleep a great deal at this age. Occluder lenses were removed after 24 hours and the kittens were then prepared for physiological study or they were placed in a darkroom for 48 hours prior to recording.

Before inducing anaesthesia, atropine and dexamethasone were given. Animals were anaesthetized with Fluothane while a vein was cannulated. Anaesthesia was continued with Brevital during placement of a tracheal tube and removal of a small area of skull bone and dura slightly anterior to the lambda suture next to the midline. Kittens were then positioned to face a translucent screen, 57 cm from the eyes, on which visual stimuli could be presented. After a loading dose of Flaxedil was given, animals were artificially ventilated and expired CO_2, body temperature, EEG, and EKG were monitored. Action potentials, isolated from individual cells with tungsten-in-glass microelectrodes, were amplified, displayed, and fed into audio monitors. These procedures are described in more detail elsewhere (Freeman, 1978).

Visually unresponsive units were noted, all responsive cells were studied, and receptive fields were plotted. Ocular dominance ratings based on the standard scale of 1 to 7 (Hubel and Wiesel, 1962) were assigned after careful evaluation of the relative response strengths elicited by optimal stimuli presented to each eye. Groups 1 or 7 designate monocular categories with response solely through the eye contralateral or ipsilateral, respectively, to the hemisphere in which the electrode was placed. The other groups signify binocular cells dominated by the contralateral (2 or 3) or ipsilateral (5 or 6) eye or equally by both eyes (4).

One important aspect of the procedure should be noted. Nearly all kittens were exposed and studied physiologically in pairs, each in a different recording laboratory. This allowed independent assessment of each condition before results were compared, which reduced very much the possible effects of experimenter bias.

Results

Ocular dominance data for control animals are given in Fig. 1. Results for a normally reared adult cat (a) show that most cells are binocular (84% in this case) as known from previous work (Hubel and Wiesel, 1962). The cortical cells of 4-week-old kittens are also largely binocular and findings for a typical animal are given in (b). If monocular deprivation is initiated at 4 weeks and allowed to continue for ten days, the open eye gains control over nearly all cortical cells. Data illustrating this are shown in (c). The normally experienced eye, ipsilateral to the recording hemisphere, exclusively dominates all but one cell. Although information on response strength is not included here, very few cells in any of these preparations were unresponsive. Most responded well to optimal stimuli.

Data for the experimental animals are shown in Fig. 2. Each histogram is a summary of data for several cats. It should be noted that variability between cats of a given group is small so that the combination of data is statistically acceptable (χ^2 statistic). A summary ocular dominance histogram for four kittens studied immediately after 24 hours of monocular deprivation is shown in (a). There are two clear effects. First, nearly all cells are monocular. Second, the nonoccluded eye dominated most cells so that the contralateral eye, which in all cases was the one deprived, controlled only 27% of the total. In addition, most units were quite responsive to optimal stimuli and, as shown in the histogram, very few nonresponsive cells were encountered. Results for the four animals placed in a darkroom for two days following the monocular exposure are shown in the summary crosshatched histogram of (b). Although the

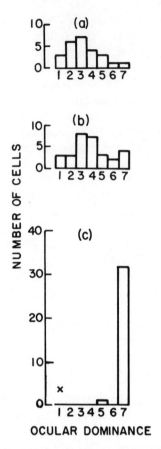

FIGURE 1 Ocular dominance histograms are shown for control animals. The numbers on the abscissae represent subjectively determined classifications of the degree to which the two eyes influence a given cortical cell. (Binocular neurons are designated by groups 2 through 6, and monocular cells by groups 1 and 7. Groups 1 through 3 (or 5 through 7) represent cells dominated by the eye contralateral (or ipsilateral, respectively) to the hemisphere from which electrode recordings are made.) In (a), data for a normally reared adult cat are shown, indicating a high degree of binocularity. Data in (b), for a normally reared 4-week-old kitten, also show that most cells are binocular. The kitten for which data are given in (c) underwent ten days of monocular occlusion beginning on postnatal day 29. The deprived eye (represented by an "X") was contralateral to the hemisphere in which the electrode was placed. Except for one cell, all units are activated exclusively by the normally experienced eye.

number of binocular cells is reduced compared to normal, there are substantially more than in the first group (binocular cells comprise 14% in the first group (a) and 47% in the second (b)). Data for the group for which recording was delayed also show a dominance of ipsilateral cells (64%). This is less than the proportion found in the group recorded immediately after monocular exposure (73%), but the difference is not significant. One additional noticeable

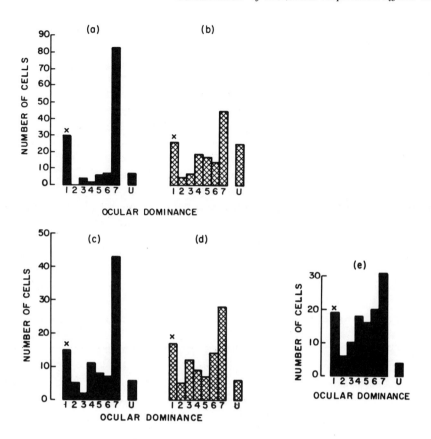

FIGURE 2 Ocular dominance data are shown for experimental kittens, with groups designated as in the previous figure. Visually unresponsive cells are indicated by the category "U". After four weeks of normal postnatal rearing, 18 kittens were unilaterally occluded with large opaque contact lenses worn in their left eyes. Eight animals were occluded for 24 hours and then half were studied physiologically immediately after occlusion (a), while the other half spent 48 hours in darkness before recording (b). Similarly, six kittens were occluded for eight hours and half were recorded immediately (c), while the other half were studied after a delay of eight hours (d). Data for four kittens occluded for four hours and then recorded are shown in (e). The "X" in each graph indicates the occluded eye, which, in all cases, is contralateral to the hemisphere that contained the electrode.

difference between the two groups may be observed. Nonresponsive units constitute only 5% of the total in the first group, but for the second group, 16% were unresponsive cells. Moreover, the time spent in darkness appeared to cause a degenerative effect since cells seemed to be less responsive than normal and exhibited more habituation and spontaneous activity. These characteristics gave one the impression that responses were relatively variable. This suggestion, that a short period of darkness may initiate changes at the cortical level, is of considerable interest but it requires confirmation using quantitative methods.

The findings illustrated in Fig. 2(a) and (b) are clearly at odds with a notion of consolidation. Instead, they support the possibility that some functional recovery of binocular connections occurs during the intervening period between monocular occlusion and extracellular study. However, one may draw this conclusion only for a given time set. It could be argued that a different time course for both the deprivation and delay periods is necessary to demonstrate a consolidation effect. Since one day of monocular occlusion produced a very marked effect, it is possible that the consequences of a shorter period of deprivation would be heightened during a subsequent delay prior to recording.

Therefore, a second set of kittens was studied. As before, they were normally reared, monocularly occluded on postnatal day 29, and studied either immediately after exposure or following a delay during which they were kept in a darkroom. Only for this set, the occlusion and delay periods were each eight hours. Since the exposures were very short, a special procedure was instituted to insure that the animals were alert during most of the period. A clear plexiglass container into which the kittens were placed was rotated very slowly (about 1 cycle/minute) for 15 minutes every 20 minutes. Kittens were exposed in pairs and seemed comfortable but alert inside the enclosure.

Surprisingly, both groups of kittens had considerably reduced numbers of binocular cells, as shown in Fig. 2(c) and (d). For three kittens recorded immediately after the occlusion, 36% of the total were binocular (c), while for three animals studied after the eight-hour delay period, 51% were binocular (d). Moreover, the nonoccluded eye controlled the majority of cells in each case (70% and 58%, respectively, in (c) and (d)). Thus, the pattern of the effects obtained is exactly like that found for the first set of kittens, only less marked. Once again, it appears that binocular recovery rather than fixation of a monocular effect occurs during the delay between exposure and study.

It is rather unexpected that only eight hours of monocular deprivation can have a substantial effect on ocular dominance. To explore the minimal time required to cause changes in binocular connections, a shorter period of monocular occlusion, four hours, was studied in another group of kittens. Animals were reared and exposed as with the previous groups and cortical study was undertaken immediately after the exposures. Ocular dominance data for these four kittens (Fig. 2(e)) show clear abnormalities, although they are not nearly as prominent as those for the previously described animals. Of 120 cells, 58% were binocular and 63% were dominated by the ipsilateral eye (groups 5-7). These figures compare with 77% and 42%, respectively, for the normal control. The differences are statistically significant (χ^2 test). This result shows that at the peak of the sensitive period, functional changes in normal binocular connections occur within four hours of disruption of those pathways. Since these alterations take place during a period shorter than that required to obtain a sample of cortical cells, it is natural to wonder if changes in ocular dominance occur during the acute physiological experiment. Therefore, separate analyses were made for all kittens in this study, of ocular dominance data for cortical cells recorded during the first and second halves of the experimental sessions. For the experimental groups whose data are shown in Fig. 2 (a, b, c, and e), there is a higher proportion of binocular cells in the second as compared to the first halves of the experiments. The differences, however, are all small and not statistically significant for three of the four groups. For the animals occluded monocularly for four hours (e), significance is marginal ($\chi^2 = 3.53$, d.f. $= 1$, significance level $= 0.06$). No systematic differences were found in proportions of ipsilateral cells. It appears, then, that there might be slight recovery of binocular connections during the course of the experiment, but if so, it is extremely limited.

Summary results for all experimental conditions are given in Fig. 3. On the ordinate, proportions of binocular cells (solid symbols and lines) and ipsilateral cells (open symbols and dashed lines) are indicated for each rearing condition (abscissa). As can be seen, the proportions of binocular cells decrease monotonically, while those of ipsilateral cells increase, with increasing durations of monocular deprivation. For comparison, proportions of binocular and ipsilateral cells are also shown for the groups studied physiologically after a delay period following monocular occlusion. In each case, these data show that the deprivation effects are relatively smaller.

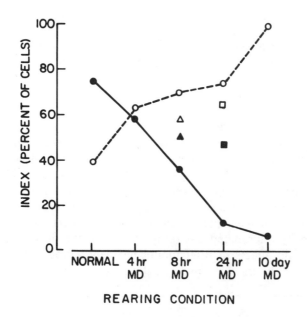

FIGURE 3 Summary findings for all experimental conditions (abscissa) are shown as a function of two indices. The first gives the proportion of binocular cells for the collected data of each of the seven conditions (binocularity index: filled symbols, solid lines). Similarly, the second gives the corresponding proportions of ipsilateral cells (ipsilateral index: unfilled symbols, dashed lines). For comparison, results for the animals not recorded immediately after unilateral deprivation are also shown. Conditions of 8-hour occlusion plus 8-hour darkness (triangles) and 24-hour occlusion plus 48-hour darkness (squares) are indicated.

Discussion

There are two central conclusions that may be drawn from the work reported here. First, very short periods of monocular occlusion are capable of disrupting normal binocular connections in a 4-week-old kitten. Second, this effect does not seem to be "consolidated" by a delay between exposure and physiological study. On the contrary, there is apparently some degree of functional recovery that occurs during a period of darkness prior to recording.

As little as four hours of monocular deprivation can cause dysfunction of binocular connections. This is a unique result since, in former investigations, brief periods of unilateral occlusion were preceded by bilateral deprivation (Peck and Blakemore, 1975; Schechter and Murphy, 1976), a procedure which addresses the possibility for monocular *recovery* rather than *deprivation*. Although competitive processes may be operant in both cases, the actual mechanism could be quite different. In the only parallel study (Movshon and Dursteler, 1977), effects were not as pronounced as those reported here. This may be attributed to two factors. First, the procedures used in the present case to insure that kittens were awake during most of the exposure periods means that their monocular experience may have been substantially greater than that for the animals of the previous study. Second, the two-day period of darkness used in the former work, as shown by the current results, seems to allow some functional recovery of binocular connections.

It is intuitively understandable that there is a recovery from, rather than an accentuation of, effects of monocular occlusion during the delay before cortical study. This is because the competitive advantage of activity transmitted via afferent pathways from the open eye no longer obtains. Is it possible that a different result would be found if, instead of a period of darkness, animals spent the post-monocular delay session while exposed to normal binocular input? In this case, it seems that even more recovery of binocular function would be found. To determine this, two kittens were reared normally for four weeks and then monocularly exposed for 24 hours followed by 48 hours of normal binocular vision. Results showed that these animals had a normal complement of binocular cells (Freeman, unpublished). Therefore, this finding suggests an extensive recovery pattern of binocular connections when kittens are kept in a normally illuminated environment during the delay before recording. Recovery from the effects of brief monocular deprivation appears to be nearly complete within 48 hours of normal visual exposure.

Acknowledgements This work was done in collaboration with C. R. Olson. Support was provided by grant EY01175 from the National Eye Institute, U.S. Public Health Service, National Institutes of Health, and by Research Career Development Award EY00092 from the same agency.

References

Agranoff, W. (1974). Biochemical concomitants of the storage of behavioral information. In: Biochemistry of Sensory Functions. L. Jaenicke (ed.). Springer Verlag, Berlin, pp. 597-623.

Agranoff, W. (1976). Learning and memory: approaches to correlating behavioral and biochemical events. In: Basic Neurochemistry, 2d ed. Siegel, Albers, and Agranoff (eds.). Little, Brown, and Co., Boston, pp. 765-784.

Buisseret, P., E. Gary-Bobo, and M. Imbert (1978). Ocular motility and recovery of orientational properties of visual cortical neurones in dark-reared kittens. Nature 272:816-817.

Freeman, R. D. (1978). Restricted visuomotor coordination during development in kittens: striate cortex and behavior. Exp. Brain Res. 33:51-63.

Hubel, D. H., and T. N. Wiesel (1962). Receptive fields, binocular interaction, and functional architecture in the cat's visual cortex. J. Physiol. 160:106-154.

Hubel, D. H., and T. N. Wiesel (1970). The period of susceptibility to the physiological effects of unilateral eye closure in kittens. J. Physiol. 206:419-436.

Movshon, J. A., and M. R. Dursteler (1977). Effects of brief periods of unilateral eye closure on the kitten's visual system. J. Neurophysiol. 40:1255-1265.

Olson, C. R., and R. D. Freeman (1975). Progressive changes in kitten striate cortex during monocular vision. J. Neurophysiol. 38:26-32.

Peck, C. K., and C. Blakemore (1975). Modification of single neurons in the kitten's visual cortex after brief periods of monocular visual experience. Exp. Brain Res. 22:57:68.

Pettigrew, J. D., and L. J. Garey (1974). Selective modification of single neuron properties in the visual cortex of kittens. Brain Res. 66:160-164.

Schechter, P. B., and E. H. Murphy (1976). Brief monocular visual experience and kitten cortical binocularity. Brain Res. 109:165-168.

Competitive Interactions in Postnatal Development of the Kitten's Visual System

M. CYNADER

Department of Psychology
Dalhousie University
Halifax, Nova Scotia, Canada

Abstract Recent evidence indicates that many consequences of visual deprivation cannot be accounted for simply on the basis of disuse. Rather, accumulating evidence indicates that competition between different inputs may underly many deprivation effects. Monocular deprivation provides a striking example of a competitive interaction during development. If one eye is sutured shut during early life, the developmental consequences for the sutured eye at the level of the lateral geniculate body and visual cortex are much more severe if the other eye is allowed vision than they are if the other eye is also sutured. Two possible mechanisms for this effect have been proposed. According to one view, the principal competitive interactions occur at the lateral geniculate body, and cortical changes are secondary. Alternatively, the primary changes may occur at the cortical level with lateral geniculate alterations a secondary consequence of cortical changes. By rearing kittens with one eye viewing normally, while the other eye views through a cylindrical lens, we have been able to make binocular competition effects depend on the orientation preferences of cortical cells. Since orientation selectivity first occurs at the cortical level, the data indicate that the primary locus of the competition between the two eyes occurs at the orientation-selective cortical cell rather than at the lateral geniculate body.

Competitive interactions may also account for some of the effects of visual deprivation on the cat superior colliculus. In normal cats, single cells in the deeper layers of the colliculus receive auditory and somatosensory inputs, as well as visual afferents. In dark-reared cats, however, visual responses are markedly depressed relative to auditory and tactile responses. These data indicate that inputs from the different sense modalities compete during postnatal development in a fashion analogous to that which occurs with monocular deprivation at the cortical level. Since the visual input is at a competitive disadvantage relative to the auditory and tactile inputs, it loses the ability to influence the multimodal cells of the deep colliculus.

109

A central concept in postnatal development has been that of the critical period. This term underscores numerous observations which show that competitive interactions during postnatal development occur only during a certain period in the organism's early development. We have found that the critical period for monocular deprivation can be prolonged, *apparently indefinitely* by rearing cats in total darkness and then suturing one eyelid shut when the animals are brought into the light. The extent and rapidity of these changes in cortical binocular connectivity in the previously dark-reared animal indicate that the visual system retains residual plasticity in these cats even if dark-rearing is prolonged well beyond the duration of the naturally occurring critical period.

It is now well established that the postnatal development of the visual system can be altered markedly by rearing animals under unusual environmental conditions. The effects of suboptimal visual exposure have been extensively documented and it has been shown that some of the consequences of deprivation may be the result of disuse of visual pathways (Hubel and Wiesel, 1963; Wiesel and Hubel, 1965; Imbert and Buisseret, 1975). However, the effects of preventing vision in one eye during early development appear to involve, in addition, competitive interactions between the two eyes. When Wiesel and Hubel (1965) reared kittens with both eyelids sutured, they found that cortical cells responded less vigorously to visual stimuli and that the incidence of orientation-selective cells was reduced. If, however, only one eye was sutured shut, while the other eye was allowed normal vision during early development, the cortical consequences for the sutured eye were much more severe (Wiesel and Hubel, 1963a, 1965). Stimuli that were presented through the sutured eye failed to influence the vast majority of cortical cells. Instead, the nonsutured eye became the sole effective route for visual stimuli. Subsequently, physiological and anatomical changes were found in cells of the lateral geniculate nucleus connected to the sutured eye (Wiesel and Hubel, 1963b; Sherman, Hoffmann and Stone, 1972; Sherman, this volume). Since the effects of *monocular* eyelid suture were so much more severe for the deprived eye than those of *binocular* eyelid suture, it was evident that the alterations in the lateral geniculate body and visual cortex were largely due to a competition between inputs from the deprived and normal eyes. The synaptic mechanisms underlying such competitive effects remain unclear, but we have tried to get some insights into these mechanisms by determining the location and timing of competitive interactions during postnatal development.

Locus of Binocular Competition

The two principal hypotheses concerning the primary location of the binocular competition occurring after monocular deprivation were put forward by Guillery (1972, 1973). One hypothesis states that cellular changes in the lateral geniculate body are a secondary consequence of a competitive interaction among geniculate terminals at the visual cortex. Here the primary event would be the loss of functional connections from the deprived eye with the cortical cell. Alternatively, an intrageniculate competition mediated through inhibition between the layers of the geniculate may result in a shrinkage of cells connected to the deprived eye. The loss of functional connections with the cortical units would then be a secondary consequence of the intrageniculate competition. We have dissociated these hypotheses by rearing kittens with one eye viewing through a negative (axis vertical) 12-diopter cylindrical lens. This lens allows clear vision of horizontal contours but defocuses the image progressively more as the stimulus orientation approaches vertical. As such, it simulates the clinical condition of astigmatism (Freeman, Mitchell, and Millodot, 1972; Mitchell, Freeman, Millodot, and Haegerstrom, 1973).

Three kittens were reared from birth to 25 days old in the dark. They were then given visual exposure wearing goggles which forced one eye to view through the cylindrical lens while the other viewed normally, for 4-6 hrs per day for a total of 80-120 hrs. They were then returned to the dark until they were at least 3 months old at which time single-cell responses were examined in the visual cortex. Our procedures for recording responses from single cortical cells are described elsewhere (Cynader and Berman, 1972; Cynader, Berman, and Hein, 1976). Kittens were initially anesthetized with intravenous sodium thiopental and cortical units were recorded extracellularly with glass-coated platinum-iridium microelectrodes. We used the sampling methods of Stryker and Sherk (1975), sampling units at intervals of approximately 100 μm in order to minimize the bias of our recorded population. A blind procedure in which the experimenter knew neither the axis of the cylindrical lens nor which eye wore the negative lens, was used throughout the recording sessions. If the competition between the two eyes occurs at the orientation-selective cortical cell, the effect of the lens should be to give the normal eye a competitive advantage at cells preferring vertical stimuli and no competitive advantage at cells preferring horizontal stimuli. The "cortical" mechanism proposed by Guillery (1972) thus predicts that the effect of the deprivation would be maximal at vertically oriented cells and minimal at horizontally oriented cells. If the competition occurs at a more peripheral level, where orientation selectivity among the cells is not present, the "geniculate" mechanism predicts that the deprivation should reduce the effectiveness of the "astigmatic" eye equally across cortical units of all orientations.

The distribution of ocular dominance for 292 units that were encountered in these kittens is shown in Fig. 2. As can be seen, almost twice as many units can be influenced through the normal eye as through the astigmatic eye. The distribution of ocular dominance is broken down according to the preferred orientation of the cortical units in the lower parts of Fig. 1. It is clear that the magnitude of the deprivation effect varies depending on the orientation preference of the cortical cell. It is marked for units preferring orientations within 30° of vertical, weaker for diagonally oriented units, and absent for units preferring horizontally oriented (±30°) stimuli. The results thus confirm the cortical hypothesis outlined earlier; they show that binocular competition occurs at the orientation-selective cortical cell and that this competition seems to be sufficient to account for all ocular dominance changes consequent to monocular image blur.

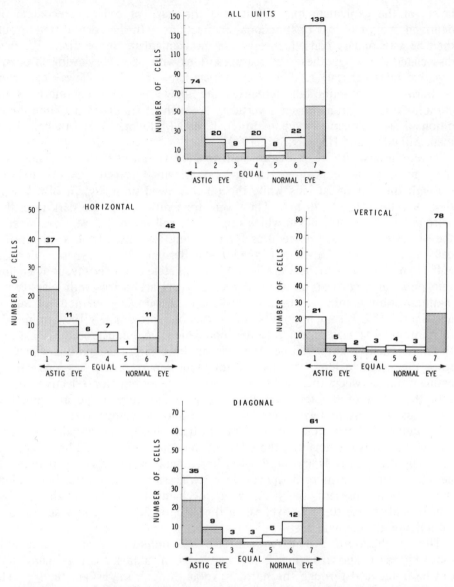

FIGURE 1 The distribution of ocular dominance for units in the visual cortex of cats reared with one eye viewing through a cylindrical lens and the other eye viewing normally. The numbers from 1 to 7 represent a trend from the "astigmatic" eye to the normal eye. Cells in group 1 are driven exclusively by the astigmatic eye; cells in group 4 are driven equally by both eyes; cells in group 7 exclusively by the normal eye. The top part of the figure shows the distribution of ocular dominance for all cells in these cats. The lower parts show the distribution of ocular dominance as a function of the orientation of the unit. The vertical and horizontal distributions represent cells which prefer orientations within 30° of vertical and horizontal. The "diagonal" distribution represents cells preferring stimulus orientations within 15° of either diagonal. The hatched parts of the distribution represent cells in the hemisphere contralateral to the astigmatic eye.

The total number of units that were encountered which prefer a particular orientation is shown for each eye separately in Fig. 2. The "astigmatic" eye drives far fewer units which prefer vertically oriented stimuli than units which prefer horizontal. The histograms of Fig. 2 show that the effect is similar in both hemispheres. One might imagine that this is a simple consequence of this eye having less exposure to focused stimuli with vertical orientations (Hirsch and Spinelli, 1970; Freeman and Pettigrew, 1973). Such an explanation cannot, however, account for the striking bias in the distribution of preferred orientations which is evident in the *normally viewing* eye. In this eye, 55% more cells prefer stimuli oriented within 30° of vertical than prefer stimuli within 30° of horizontal. This is at first glance surprising, since the normal eye has, after all, viewed all orientations with equal frequency and clarity. An examination of the histograms of Fig. 2 reveals that the distributions in the two eyes are complementary. The trough near vertical for the astigmatic eye is balanced by a corresponding bulge in the normal eye. Combining distributions for the two eyes, the total number of cortical units that is encountered preferring vertical (±30°) and horizontal (±30°) stimuli is nearly identical. The overall distribution of orientation-selective neurons in the visual cortex has remained unaltered but the normal eye seems to have made compensatory, orientation-selective inroads into the cortical territory normally occupied by the astigmatic eye. The orientation-specific expansion of territory by one eye provides further evidence for the "cortical" hypothesis outlined earlier.

Intermodal Competition

The concept of competition during development may have much wider applicability than we have suspected. Whenever two or more fiber pathways converge on a population of postsynaptic target cells, these input pathways may either share input to the postsynaptic elements or one pathway may have an opportunity to increase its input to the recipient cells if it is placed at a competitive advantage relative to the other inputs. We have recently encountered another example of what appears to be a competitive developmental interaction when we studied the effects of visual deprivation on the cat superior colliculus.

In normally reared cats, cells in the superficial layers of the superior colliculus are driven exclusively by visual stimuli (Sterling and Wickelgren, 1969; Berman and Cynader, 1972). As the electrode is advanced into the intermediate layers of the colliculus, however, responses to auditory and somatic stimuli can be observed as well. Individual units may be influenced in various two-out-of-three combinations or by stimuli from all three sense modalities.

When we studied the responses of cells in the superior colliculus of dark-reared cats, we observed changes in the response characteristics of collicular cells as a consequence of this deprivation. In the superficial layers, we found the same sort of changes that others had previously described, namely, a loss of direction selectivity and a shift of ocular dominance in favor of the contralateral eye (Wickelgren and Sterling, 1969). However, brisk visual responses could still be elicited from nearly all units that were encountered in the superficial

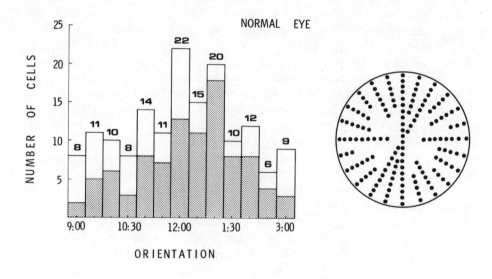

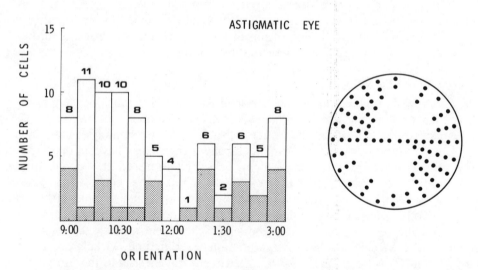

FIGURE 2 The top and bottom parts of the figure illustrate the distribution of preferred orientations for cortical units driven by the normal and astigmatic eyes, respectively. On the right are polar plots in which each cell represents a single dot. The length of the dotted line along a given orientation is thus proportional to the number of units that are encountered which have this preferred orientation. On the left- hand side, the same data are presented in histogram form and are further subdivided by cortical hemisphere. The height of the bar at any orientation is proportional to the number of units preferring that orientation. The hatched part of the distribution is derived from the hemisphere contralateral to the astigmatic eye. The horizontal cells of the polar plots are divided equally between 9:00 and 3:00 on the associated histograms.

collicular layers. In the intermediate and deep layers of the colliculus, where multimodal responses normally predominate, we found that responses to visual stimuli were profoundly depressed or, in other cases, abolished entirely, while responses to auditory and somatic stimuli were relatively normal.

Figure 3 illustrates the consequences of dark-rearing on visual responses in the superior colliculus. On the left is a reconstruction of an electrode track through the superior colliculus of a normal cat. The right-hand side of the figure represents a similar penetration through the colliculus of a dark-reared animal. It is evident from Fig. 3 that the consequences of visual deprivation are not simply to render all units in the colliculus unresponsive to visual stimuli. Only in the intermediate and deep layers of the colliculus, where inputs from other sense modalities normally mingle with those of vision, do we observe a depression of visual responsivity. We interpret the selective loss of visual responsivity in the intermediate and deep collicular layers in terms of a competitive interaction which is similar to that which occurs at the visual cortex following monocular deprivation. At the cortical level, the protagonists of the competition represent inputs from the left and right eyes. In the deeper layers of the colliculus, however, the competing inputs represent the different sense modalities. Since vision is at a competitive disadvantage relative to the other senses in the dark-reared cat, its afferents lose the ability to influence cells in these collicular layers just as inputs from the deprived eye lose their ability to influence cortical cells after one eye is sutured. The observation that cells in the superficial layers retain their visual inputs while those in the deep layers lose them may be due to the finding that vision forms the exclusive input to the superficial collicular layers. Hence the opportunity for competition between the different senses, and the loss of visual responsivity, occurs only in the deeper layers of the colliculus.

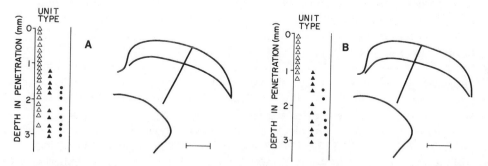

FIGURE 3 Schematic reconstructions of electrode penetrations through the superior colliculus of (A) a normally reared cat and (B) a dark-reared animal. In each case, the electrode track has been angled 20° medialward in order to avoid damage to the overlying visual cortex. As the electrode enters the colliculus it first encounters visual units (open triangles), then units with auditory (filled circles) or tactile (filled triangles) input. In the dark-reared cat, visual units are rarely encountered below 1.2 mm from the surface, whereas they are not uncommon in the normally reared cat.

The Critical Period

An important aspect of the cortical competition which occurs when one eye is sutured is its timing during development. Previous workers have shown that monocular deprivation is effective only if it is instituted early in the kitten's life (Hubel and Wiesel, 1970; Blakemore and Van Sluyters, 1974; Movshon, 1976). The period of maximal vulnerability occurs during the fourth postnatal week when only a few days of monocular suture can severely disrupt cortical binocular connectivity. The effects of monocular deprivation become progressively less marked as the animal becomes older and no effects are observed if one eyelid is sutured in adult cats. From these and other experiments has emerged the concept of the *critical period,* a period early in development during which the organism is maximally vulnerable to environmental manipulation. Previous studies have indicated that the critical period extends from three weeks to three- four months postnatally in the cat (Hubel and Wiesel, 1970; Blakemore and Van Sluyters, 1974).

In our experiments in dark-reared cats we have found that it is possible to *extend the critical period for competition between inputs from the two eyes.* In previous experiments on dark-reared cats, we found (Cynader et al., 1976), as had other workers (Hubel and Wiesel, 1963a; Wiesel and Hubel, 1965; Imbert and Buisseret, 1975), that cortical responses to visual stimuli were less vigorous and that orientation selectivity was less pronounced. These effects were, however, not irreversible. If the animals were given a period of normal visual exposure subsequent to prolonged (12 months starting at birth) deprivation, cortical responsivity was much improved and the incidence of orientation-selective cortical units was increased markedly. Our ability to alter cortical properties with normal exposure following such prolonged deprivation suggested that it might be possible to extend the critical period for competitive interactions beyond the naturally occurring critical period. Accordingly, we raised cats in the dark until they were 4 months old and then brought them into the light and sutured one eyelid shut. The animals were then allowed normal visual exposure for an additional one or two months. When we recorded from cortical cells of these kittens, we found evidence for marked changes in cortical binocular connectivity. As shown in Fig. 4 (L.H.S.) the large majority of cortical cells that were encountered responded to stimuli presented through the nonsutured eye and only a few cells could be influenced through the deprived eye. These data appeared to provide clear evidence for a prolongation of the critical period in dark-reared cats. Our control experiments were, however, somewhat disappointing. Figure 4 (R.H.S.) shows that effects of monocular deprivation can still be observed even in 4-month-old normally reared cats. These experiments indicate that the critical period in normal cats is more prolonged than had previously been thought (Hubel and Wiesel, 1970; Blakemore and Van Sluyters, 1974). While the effects of monocular deprivation commencing at 4 months of age are much more marked in dark-reared cats than in normal cats, one cannot conclude unequivocally from these data that we have succeeded in extending the period of susceptibility for competitive binocular interactions. To obtain

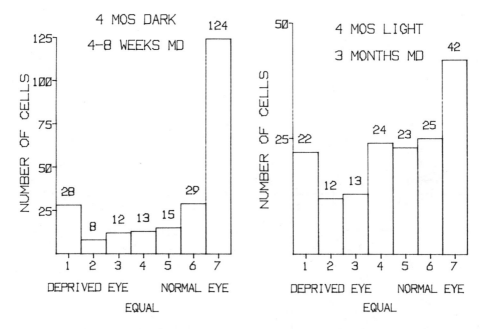

FIGURE 4 On the left is the distribution of ocular dominance for units in the visual cortex of cats reared in the dark for four months and then allowed four to eight weeks of monocular visual exposure. On the right is the comparable distribution for units of cats reared in the light for four months and then monocularly sutured for an additional three- month period. The relative excitatory strength of the two eyes is classified into the seven groups of Hubel and Wiesel (1962). Cells in group 1 are monocularly driven by the contralateral eye and those of group 7 exclusively by the ipsilateral eye. Cells in all other groups are binocularly driven with a spectrum of dominance from those driven much more strongly through the contralateral eye (group 2) to those driven much more strongly through the ipsilateral eye. All recordings in these cats were made in the hemisphere contralateral to the sutured eye so that units dominated by that eye are found in ocular dominance groups 1-3.

definitive evidence on this point, we reared kittens in the dark for still longer periods of time and then brought them into the light and sutured one eyelid for an additional three-month period. The results for these longer deprivations and the associated control data are shown in Fig. 5. The data of Fig. 5 show that strong monocular deprivation effects can be obtained following six, eight, or ten months of dark-rearing. In all cases, most cortical cells can be driven only through the nonsutured eye. The control data in the lower part of Fig. 5 show that no strong trend toward the opened eye was observed in the cats that were allowed eight or ten months of normal vision before the monocular deprivation was instituted. The similarities among the effects of monocular deprivation instituted following different periods of dark-rearing and the difference with normal cats indicate that the animal's age is not the sole determinant of his susceptibility to monocular deprivation. Rather, it appears that the type of experience which the animal has had and the associated state of cortical maturity determines the susceptibility to later monocular deprivation. These data thus

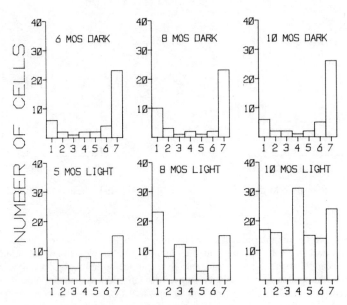

FIGURE 5 The effects of three months of monocular deprivation starting at different ages on the distribution of ocular dominance in dark-reared and light-reared animals. Conventions as in Fig. 4.

show that it is possible to extend the period of susceptibility to monocular deprivation far beyond the normal duration of the critical period.

Cats that have been reared in the dark for four months also express prolonged immaturity in the speed with which their cortical binocular connectivity may be modified. Figure 6 shows the effects of one or two weeks of monocular deprivation instituted when kittens were taken out of the darkroom at 4 months of age. Clear effects of monocular deprivation are seen after only one week and by two weeks the effect is not very different from that obtained with much longer periods of monocular deprivation (see Fig. 4). If one compares the speed of these competitive changes with those occurring in normally reared kittens of different ages following monocular deprivation (Movshon, 1976), the rates of change of ocular dominance are comparable to those observed in a 5- or 6-week-old normally reared kitten. They are much higher than those which occur when deprivation is instituted in normal kittens at later ages. In this sense, the "cortical age" of these 4-month-old dark-reared cats may be only 5 or 6 weeks.

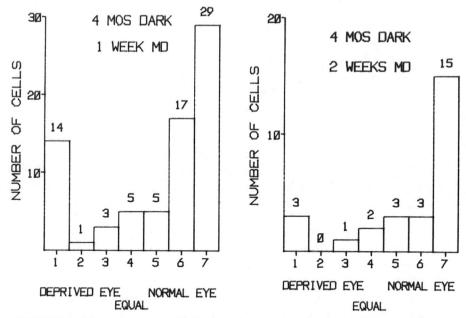

FIGURE 6 Ocular dominance distributions from cortical cells in two kittens which were reared in darkness from birth until 4 months of age and then allowed one or two weeks of monocular visual exposure. Conventions as in Fig. 4.

Conclusion

Taken together, the results that have been described in this paper indicate that competition between afferent fiber populations plays an important role in post-natal development. With monocular deprivation this competition occurs at the visual cortex between lateral geniculate terminals representing input from the two eyes. In dark-reared cats a similar competition appears to take place between inputs from the different sense modalities at the level of the superior colliculus. Under normal circumstances these competitive interactions are age-dependent, but it is possible to demonstrate such interactions even in much older animals if the system is maintained in an immature state by prolonged deprivation.

Acknowledgements This research was supported by Grant No. MT-5201 from M.R.C. of Canada and Grant No. A9939 from N.R.C. of Canada.

References

Berman, N., and M. Cynader (1972). Comparison of receptive- field organization of the superior colliculus of Siamese and normal cats. J. Physiol. 224:363-389.

Blakemore, C., and R. C. Van Sluyters (1974). Reversal of the physiological effects of monocular deprivation in kittens; further evidence for a sensitive period. J. Physiol. 237:195-216.

Cynader, M., and N. Berman (1972). Receptive-field organization of monkey superior colliculus. J. Neurophysiol. 35:187-201.

Cynader, M., N. Berman, and A. Hein (1976). Recovery of function in cat visual cortex following prolonged deprivation. Exp. Brain Res. 25:139-156.

Cynader, M., and D. E. Mitchell (1977). Monocular astigmatism effects on kitten visual cortex development. Nature 270: 177-178.

Freeman, R. D., and J. D. Pettigrew (1972). Alteration of visual cortex from environmental asymmetries. Nature 246: 359-360.

Freeman, R. D., D. E. Mitchell, and M. Millodot (1972). A neural effect of partial visual deprivation in humans. Science 175:1384-1386.

Guillery, R. W. (1972). Binocular competition in the control of geniculate cell growth. J. Comp. Neurol. 144:117-127.

Guillery, R. W. (1973). The effect of lid suture upon the growth of cells in the dorsal lateral geniculate nucleus of kittens. J. Comp. Neurol. 148:417-422.

Hirsch, H. V. B., and D. N. Spinelli (1970). Visual experience modifies distribution of horizontally and vertically oriented receptive fields in cats. Science 168:869-871.

Hubel, D. H., and T. N. Wiesel (1962). Receptive fields, binocular interaction and functional architecture in the cat's visual cortex. J. Physiol. 160:106-154.

Hubel, D. H., and T. N. Wiesel (1963). Receptive fields of cells in striate cortex of very young, visually inexperienced kittens. J. Neurophysiol. 26:994-1002.

Imbert, M., and P. Buisseret (1975). Receptive field characteristics and plastic properties of visual cortical cells in kittens reared with or without visual experience. Exp. Brain Res. 22:25-36.

Mitchell, D. E., R. D. Freeman, M. Millodot, and G. Haegerstrom (1973). Meridional amblyopia: evidence for modification of the human visual system by early visual experience. Vis. Res. 13:535-558.

Movshon, J. A. (1976). Reversal of the physiological effects of monocular deprivation in the kitten's visual cortex. J. Physiol. 261:125-174.

Sherman, S. M. This volume.

Sherman, S. M., K. P. Hoffmann, and J. Stone (1972). Loss of a specific cell type from dorsal lateral geniculate nucleus in visually deprived cats. J. Neurophysiol. 35: 532-541.

Sterling, P., and B. G. Wickelgren (1969). Visual receptive fields in the superior colliculus of the cat. J. Neurophysiol. 32:1-15.

Stryker, M. P., and H. Sherk (1975). Modification of cortical orientation selectivity in the cat by restricted visual experience: a reexamination. Science 190:904-905.

Wickelgren, B. G., and P. Sterling (1969). Influence of visual cortex of receptive fields in the superior colliculus of the cat. J. Neurophysiol. 32:16-23.

Wiesel, T. N., and D. H. Hubel (1963a). Single-cell responses in the striate cortex of kittens deprived of vision in one eye. J. Neurophysiol. 26:1003-1017.

Wiesel, T. N., and D. H. Hubel (1963b). Effects of visual deprivation on morphology and physiology of cells in the cat's lateral geniculate body. J. Neurophysiol. 26: 978-993.

Wiesel, T. N., and D. H. Hubel (1965). Comparison of the effects of unilateral and bilateral eye closure on cortical unit responses in kittens. J. Neurophysiol. 28: 1029-1040.

Orientation-dependent Changes in Response Properties of Neurons in the Kitten's Visual Cortex

J. P. RAUSCHECKER

Max-Planck-Institut für Psychiatrie
Munich, Germany

Abstract Kittens were reared wearing goggles with cylindrical lenses, which restricted the visual experience of one eye to a very narrow range of orientations. The other eye was either occluded or allowed normal vision. Physiological changes caused by the selective exposure were assessed by means of single-cell recording from striate cortex.

In all cases the majority of neurons driven by the "cylinder eye" preferred the experienced orientation. When the second eye had been covered during the exposure, most units were dominated by the cylinder eye and had receptive field orientations in register with the orientation experienced by this eye. Neurons with other orientation preferences were shared between the two eyes.

When the second eye was allowed to view normally, the cylinder eye became strongly inferior. In this case many binocular cells were found, almost all of which preferred that orientation experienced by both eyes together. Neurons with other orientation preferences were dominated by the normal eye.

In a two-stage experiment, restricted vision of one eye (using the cylinder lens) followed normal experience of the other eye. Polar plots of preferred orientations for the two eyes show complementary distributions: the cylinder eye had selectively taken over neurons with corresponding receptive field orientations from the previously normal eye.

These experiments support the hypothesis that circuit changes in the visual cortex do not depend solely on asymmetries in the activation level of the afferents from the two eyes, but also on the response properties of the cortical target cells. Such a mechanism can account for both maintaining and specifying influences of visual experience on cortical response properties during early development.

Most workers in the field of developmental neurobiology of vision now agree that the visual cortex of a newborn, visually inexperienced kitten contains neurons with oriented receptive fields (Hubel and Wiesel, 1963; Barlow, 1975;

Blakemore and Van Sluyters, 1975; Imbert and Buisseret, 1975; Sherk and Stryker, 1976; Pettigrew, 1978). On the other hand, many neurons in such kittens are not fully specified. These neurons tend to respond less vigorously to light stimulation and often respond over a wider range of orientations (Hubel and Wiesel, 1963; Barlow and Pettigrew, 1971; Pettigrew, 1974; Blakemore and Van Sluyters, 1975; Imbert and Buisseret, 1975).

When the kittens are totally deprived of vision during development, the number of neurons that are unresponsive or non-specific for orientation increases even more (Wiesel and Hubel, 1965a; Blakemore and Van Sluyters, 1975; Singer and Tretter, 1976; Fregnac and Imbert, 1978). From this it is obvious that visual experience is necessary to maintain those response properties that have been prespecified genetically.

When the kittens grow up normally, on the other hand, the number of unresponsive and non-oriented cortical cells rapidly declines, while the vigor and the specificity of unit responses increase (Hubel and Wiesel, 1963; Pettigrew, 1974; Blakemore and Van Sluyters, 1975; Buisseret and Imbert, 1976). Such improvement has never been observed without visual experience, and therefore it seems obvious that light stimulation also plays an active part in specifying poorly determined cortical connections.

Thus, it appears that visual experience has both a maintaining and a specifying role. In trying to reduce these two aspects of visual development to a unitary concept, one of them has often been highly emphasized, while the other has been neglected. From theoretical arguments (Hebb, 1949; Stent, 1973) it can be derived, however, that a single mechanism does exist that could account for both effects of visual experience: the only prerequisite is that any changes of cortical circuitry do not depend solely on the afferent activity, but are gated by a matching operation between presynaptic activity and postsynaptic response properties (Singer, 1976; Singer, Rauschecker, and Werth, 1977; Rauschecker and Singer, 1978, 1979a).

The experiments discussed below are grouped into three paradigms. In each, the hypothesis above has been tested by combining two techniques that have been widely and independently used in studying plasticity of vision, namely, monocular deprivation (e.g., Wiesel and Hubel, 1963, 1965a) and selective visual experience restricted to one orientation (e.g., Blakemore and Cooper, 1970; Hirsch and Spinelli, 1970; Freeman and Pettigrew, 1973; Tretter, Cynader, and Singer, 1975). Only one eye received restricted experience, while the other eye either was totally deprived or was allowed to see normally. With this procedure no conflicting orientational input is offered to the two eyes, but the degree of asymmetry between the eyes is manipulated in such a way that only particular subsets of neurons are addressed, depending on their selectivity or preference for orientation.

Restriction in the orientation domain was performed by means of cylindrical lenses that were inserted into polyurethane helmets. Care was taken to align the optic axis of the lens with that of the eye. Cylindrical lenses possess high refractive power in one orientation (-25 D) and zero power in the orthogonal orientation. Thus, only those contours that are in register with the zero axis of the lens remain clearly visible; all orientations more than 10-15° off the zero axis become so blurred that their contrast effectively drops below the visual threshold, as determined by the contrast-sensitivity function (Campbell and Robson, 1968; Campbell, Maffei, and Piccolino, 1973). Because the orientation of the visible bar remains constant relative to retinal coordinates (regardless of head or body tilt), this method is most effective in restricting visual experience to a narrow range of oriented contours. At the same time, it is a comparatively natural way of controlling the visual input, because the environment itself remains unchanged and consists of real objects that can be moving or stationary. The kittens remain unrestrained, interacting with the visual world around them; eye and head movements are unimpaired. (For further details see Rauschecker and Singer, 1979b.)

For every kitten, exposure was started at the peak of the "critical period" (between 4 and 6 weeks of age) with no prior experience. Total exposure amounted to at least 100 hours distributed over about a two-week period with the average-daily experience being eight hours. For the remaining time the kittens were kept in the dark. After the end of exposure the kittens underwent the physiological experiment, using standard procedures (c.f. Singer and Tretter, 1976). The eyes of each kitten were checked carefully for any spherical and astigmatic aberrations with a Rodenstock refractometer. Except for some occasional hypermetropia, no abnormalities were found; in particular, there were no signs of astigmatism. Single-unit responses to light stimulation were then analyzed in area 17. In all kittens the average distances between cells encountered on the long oblique ($\alpha > 30°$) electrode penetrations were very similar to those in normal adult cats (about 80 μm). High impedance ($\sim$10 $M\Omega$) micropipettes (1.5 M K^+—citrate filled) were used; multi-unit responses were not included in the analysis. Great care was taken not to overlook any cells lacking spontaneous activity by frequently depolarizing the micropipette. Ocular dominance, preferred stimulus orientation, and orientational selectivity were determined from peri-stimulus time histograms using a computer-controlled set-up.

In the first paradigm, the visual experience of two kittens was restricted for 200 hours as follows: the left eye was occluded by the mask, while the right eye received restricted experience through the cylinder lens. This lens was oriented vertically in one kitten and horizontally in the other.

The effect of this selective visual experience on neurons with different orientation preferences is shown in Fig. 1. From the two diagrams at the top (Fig. 1A), it becomes evident that an ocular dominance shift has occurred, a shift which is orientation-dependent. The ocular dominance shift is confined to the experienced orientation and is not apparent in the orientation orthogonal to it. The differential effect of the exposure on the respective neuron populations is also reflected in the two lower diagrams (Fig. 1B): here, for the kitten with horizontal exposure, polar plots of preferred orientations are shown classified according to the units' ocular dominance. It can be seen that the main axes of

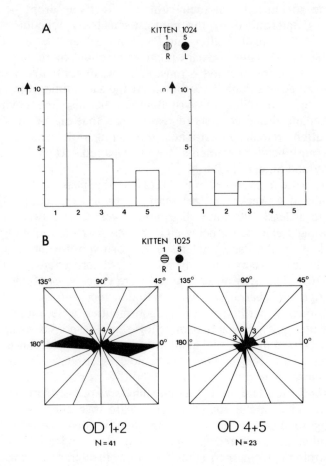

FIGURE 1 Results from the two kittens that were totally deprived of vision in the left eye and received visual experience in the right eye that was restricted to a very narrow range of orientations by means of a cylindrical lens. (A) Ocular dominance histograms as a function of preferred stimulus orientation for kitten 1024 (right eye vertical). Five ocular dominance classes are used: neurons in groups 1 and 5 are monocularly driven through the right or left eye, respectively; cells in groups 2-4 are binocular, those in groups 2 and 4 being dominated by the right or left eye, respectively; neurons in group 3 are about equally sensitive to stimulation through either eye. Left: neurons preferring 90°. Right: neurons preferring 0° stimulus orientation. (B) Polar diagrams of preferred orientations as a function of eye dominance for kitten 1025 (right eye horizontal). The radius at every angle (n = 22.5°) gives the number of neurons that responded best to that particular orientation (±11.25°) regardless of their tuning for orientation. Left: neurons exclusively driven or dominated by the right eye. Right: neurons exclusively driven or dominated by the left eye.

the two distributions are complementary to each other: while the "cylinder eye" clearly prefers horizontal stimulus orientations, the deprived eye shows a slight preference for vertical orientations.

These findings can be best interpreted if one considers the response properties of an inexperienced cortex, as mentioned in the introduction. Among those neurons that had already prespecified orientation preferences, only those that possessed receptive field orientations more or less in register with the orientation of the lens were stimulated adequately. Since only one eye was exposed, these neurons were in a situation equivalent to monocular deprivation. The connections from the open eye became consolidated, while those from the closed eye became disconnected due to competitive interactions between the afferents from the two eyes (Wiesel and Hubel, 1965a; Sherman et al., 1974; Cynader and Mitchell, 1977). On the other hand, neurons with built-in preferences for orientations very different from the experienced one were not stimulated adequately. In contrast to neurons with matching orientation preferences, they could not respond to the input pattern on the condition that they were reasonably well tuned for orientation. Nevertheless, these neurons should also become dominated by the open eye, if an imbalance in the afferent activity from the two eyes were sufficient to produce an ocular dominance shift. In fact, the afferent pathways from the two eyes for such cortical neurons *were* stimulated by different levels of light intensity and hence these pathways would convey asymmetric levels of neuronal activity. If, however, the postsynaptic neurons are required to *respond* to the afferent activity, then no ocular dominance shift should occur in these neurons. Recent results of an independent study (Singer, Rauschecker, and Werth, 1977) support the second alternative: if one eye is stimulated with diffuse light and the other eye is totally deprived of vision, the neurons remain symmetrically driven by both eyes; as in binocular deprivation, the only consequence is a general reduction in specificity.

So far, the interpretation of the experimental results is in line with the hypothesis that visual experience maintains or consolidates those response properties which are prespecified genetically. But we have yet to consider the neurons that were unspecific at the start of exposure.

If one would like to explain the whole process with just alterations of ocular dominance and deprivation effects, one has to assume that at least some neurons originally responding to orientations other than the experienced one have become unselective for orientation, or even unresponsive to light stimulation (Stryker and Sherk, 1975; Stryker, Sherk, Leventhal, and Hirsch, 1978). In a "Gedankenexperiment," one can consider the strongest case for this hypothesis as follows: if one assumes that all neurons that were found to be non-oriented or visually unresponsive *originally* preferred orientations that were not experienced during the second stage of exposure, one can add them to those neurons that actually did show this preference. In this case, any apparent bias for the experienced orientation should disappear if a selective theory is sufficient to explain all findings. Indeed, a considerable number of non-specific neurons was found: 16% of all neurons fully analyzed were unresponsive and 10% of the responsive neurons were non-oriented in this paradigm. However, these numbers are still too small to account for all deprivation effects to be expected from a purely selective theory. In fact, no bias-free, geometrically round, polar distribution is obtained even when adding all non-oriented and unresponsive units to those in the non-experienced orientations (Fig. 2). Since the recording density (number of units per mm penetration) was the same as for normal adult cats and since precautions were taken to preclude any recording bias (see above), it has to be concluded that the exposure stimulus has "instructed" orientation selectivity, at least in some of the non- or poorly selective neurons. This again could be achieved most elegantly by a mechanism that picks out those sets of afferents that were active at the same time when the cortical cell responded. Synaptic contacts from these afferents are strengthened, whereas afferents that are inactive while the postsynaptic element is firing get weakened. Thus, a quite selective part of the originally unspecific responsivity spectrum remains, which corresponds to the experienced environment. This notion includes the possibility of slight changes in preferred orientation within the limits given by the original orientation tuning.

These predictions were further tested in the second paradigm. For three kittens, artificial astigmatism was imposed for 100 hours on one eye, while the other eye, at the same time, was allowed normal experience. In Fig. 3A the ocular dominance distributions of cortical neurons are displayed as a function of their preferred orientation. It can be seen that neurons with optimal orientations close to vertical—the meridian in which normal binocular vision was possible—show a normal ocular dominance distribution, most of the cells being

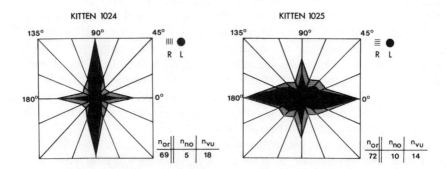

FIGURE 2 Polar diagrams of preferred orientations summarizing the effect of exposure in each of the two kittens from Figure 1. The hatched parts of the diagrams indicate the shape of the hypothetical graph which is obtained if neurons that are non-selective for orientation or unresponsive to light stimulation are added to the units in the non-experienced orientations (see text) following the algorithm stated below. The numbers of neurons included in each of the distributions are given on the right: $n(or)$ = number of units with a preference for a particular orientation regardless of their tuning for orientation; $n(no)$ = number of units with no orientation preference; $n(vu)$ = number of visually unresponsive units. Algorithm for outer graphs:

$$n_i = n(or)_i + n(no) + \frac{n(vu)}{16} * i,$$

with i being the angle difference between a particular orientation preference and the experienced orientation in multiples of 22.5°.

binocularly driven. All other neurons have become dominated by the eye with unrestricted experience. However, the number of neurons with vertical orientation preferences is greater than the number of all other orientation-selective neurons taken together. Therefore, an explanation of these results in terms of selective ocular dominance changes alone does not seem possible. Some neurons must have acquired the preference for vertical stimulus orientations as a consequence of the selective experience.

The same data are displayed in a different way in Fig. 3B: preferred orientations are plotted in polar diagrams as a function of ocular dominance. It becomes evident that the eye with normal exposure (corresponding to ocular dominance class 1) has taken over virtually all neurons preferring orientations that were not experienced by the other eye. On the other hand, all neurons that receive some input from the eye that had only vertical experience (ocular dominance classes 2-5) show a clear bias towards vertical. This is true even for neurons in ocular dominance class 2, where the "astigmatic" eye has only minor influence. Once more, all polar plots, taken together, indicate a clear bias towards the vertical orientation, a bias that cannot be equalized by including the non-oriented and unresponsive neurons. Coactivation of pre-existing binocular connections seems to be a very powerful stimulus both in terms of maintaining and specifying influence on response properties of visual cortical neurons. Binocular neurons that are not oriented or are poorly oriented prior to experience may become tuned to the orientation that was experienced by both eyes together; successful coactivation is used as the signal for their co-consolidation. From this it might be concluded that binocular interactions during cortical development are not only competitive, but also cooperative, in nature.

In order to preclude any kind of synergistic binocular interactions and to generate neurons with highly selective properties before orientation-specific exposure was started, the third experimental scheme was chosen as follows: beginning with 4 weeks of age, three kittens were allowed to see normally through the right eye for 100 hours distributed over nine days, while the left eye was sutured closed. A control experiment showed that by that time 88% of the neurons had become dominated by the open eye, three-fourths of them monocularly. Most neurons were found to have mature orientation tuning and the polar distribution of preferred orientations was round.

In the subsequent second stage of exposure (still within the critical period) the previously deprived eye was reopened and received 100 hours of visual experience through the cylinder lens, horizontal in two animals, vertical in the third one, while the eye with previously normal experience was occluded.

If, as following early monocular deprivation, the experimental situation is reversed between the two eyes, the ocular dominance distributions may also become reversed (Wiesel and Hubel, 1965b; Blakemore and Van Sluyters, 1974; Movshon, 1976). According to the hypothesis presented above, reversal should take place selectively only for those neurons that are stimulated adequately during the second stage, and all other neurons should remain dominated by the eye that was open in the first stage.

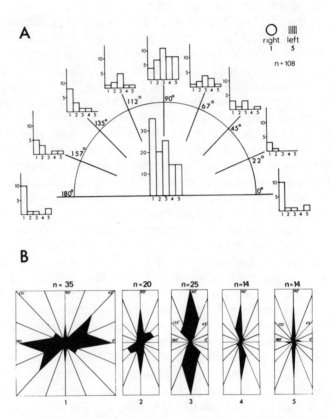

FIGURE 3 Results from the three kittens with normal experience in the right eye and simultaneous vertical exposure in the left eye. (A) Ocular dominance distributions as a function of preferred orientation (in steps of 22.5°). The diagram in the center is the conventional ocular dominance histogram of all neurons encountered in the three kittens. Ocular dominance classes as in Fig. 1(A). (B) Polar plots of preferred orientations as a function of the five ocular dominance classes (c.f. Fig. 1(B)).

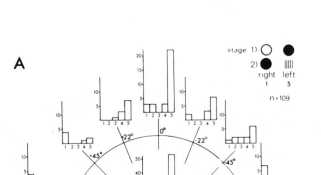

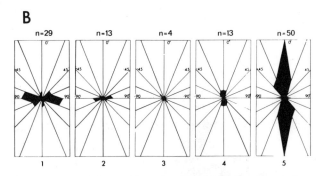

FIGURE 4 Results from the three kittens with unrestricted monocular experience in the right eye and subsequent restricted monocular experience in the left eye, as shown schematically in the top right corner. (A) and (B) as in Fig. 3. Since restriction was not done in the same orientation in the three kittens (two kittens saw horizontal, one vertical), the angles in the polar plots here represent the difference between the experienced orientation and the preferred orientation of the neurons.

Since the results were very similar in all three kittens, the pooled data are displayed in Fig. 4, using the same presentation as for the preceding paradigm. It can be seen clearly that only those neurons whose receptive field orientations corresponded to the orientation experienced during the second stage (Fig. 4A) were taken over selectively by the left eye. Neurons with differing orientation preferences remained dominated by the other, previously open, eye. Consequently, this eye is left with a complementary bias towards the orientation orthogonal to the one experienced by the other eye (Fig. 4B).

When the polar plot for all neurons in this third paradigm is examined (Fig. 5), there is still an overall bias towards the orientation that was experienced in the second stage, but it is somewhat weaker than in the preceding paradigms. After the first stage, only very few unspecific neurons that could have become "instructed" by the exposure stimulus of the second stage were present. Therefore, in this case the distribution bias is best interpreted as resulting from a loss of a specific class of neurons: cells tuned to orientations that were not experienced during the second stage were lacking a maintaining influence of visual experience and became partly unspecific.

In summary, although competitive changes in ocular dominance of a cortical neuron are the most common form of modification, it is clear that changes in the specificity of other parameters, e.g., preferred orientation, are possible in individual neurons without such parameters being coupled to ocular dominance. In particular, the full specification of orientational selectivity is brought about by visual experience.

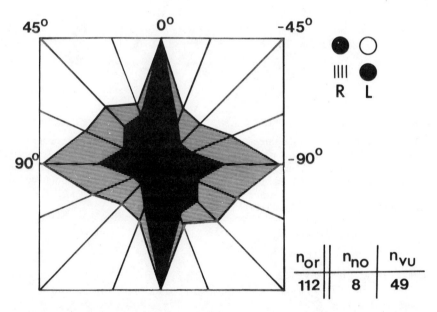

FIGURE 5 Polar diagram of preferred orientations for the three kittens of the third paradigm (c.f. Fig. 4). Outer graph: non-oriented and unresponsive neurons added as in Fig. 2.

Concerning the mechanisms underlying experience-dependent modification of the visual cortex, the present results suggest that such changes do not only depend on differences in the activation level of the afferents, but are guided by the response properties of the respective cortical cells. Adequate stimulation to which the cortical neuron can respond leads to selective consolidation or improvement of those afferents that were active at the same time when the postsynaptic neuron was firing, and to selective disruption of others that were inactive at this time. Competitive interactions between the afferents from the two eyes that lead to changes in ocular dominance would be a special case of this more general mechanism. The sharpening of orientation selectivity in previously non-oriented neurons according to the experienced environment is another case that could be explained by "postsynaptic resonance." In this case various sets of afferents encoding different orientations could be thought of as being in mutual competition. That set of afferents whose activity causes a postsynaptic response in the cortical cell will eventually determine the properties of the cortical cell. Furthermore, coactivation of certain afferents will lead to their co-consolidation, if they are effective in stimulating the postsynaptic neuron when activated together. Conversely, if none of the afferents is able to drive the cortical neuron, all become slightly impaired. Finally, this concept also shows the possible extent of cortical plasticity: response properties can only be changed or "imprinted" within the limits set by genetic programs. In particular, it should not be possible for a cortical neuron to change its preferred orientation directly once it has acquired full specificity.

Acknowledgements I wish to thank Drs. W. Singer, R. D. Freeman, B. Freeman, and U. Yinon for their helpful comments on the manuscript. This work was part of a Ph.D. thesis at the Faculty of Electrical Engineering, Technical University, Munich.

References

Barlow, H. B. (1975). Visual experience and cortical development. Nature 258:199-204.

Barlow, H. B., and J. D. Pettigrew (1971). Lack of specificity of neurones in the visual cortex of young kittens. J. Physiol. 218:98-100P.

Blakemore, C., and G. F. Cooper (1970). Development of the brain depends on the visual environment. Nature 228:477-478.

Blakemore, C., and R. C. Van Sluyters (1974). Reversal of the physiological effects of monocular deprivation in kittens: further evidence for a sensitive period. J. Physiol. 237:19-216.

Blakemore, C., and R. C. Van Sluyters (1975). Innate and environmental factors in the development of the kitten's visual cortex. J. Physiol. 248:663-716.

Buisseret, P., and M. Imbert (1976). Visual cortical cells: their developmental properties in normal and dark-reared kittens. J. Physiol. 255:511-525.

Campbell, F. W., and J. G. Robson (1968). Application of Fourier analysis to the visibility of gratings. J. Physiol. 197:551-566.

Campbell, F. W., L. Maffei, and M. Piccolino (1973). The contrast sensitivity of the cat. J. Physiol. 229:719-731.

Cynader, M., and D. E. Mitchell (1977). Monocular astigmatism effects on kitten visual cortex development. Nature 270:177-178.

Freeman, R. D., and J. D. Pettigrew (1973). Alteration of visual cortex from environmental asymmetries. Nature 246:359-360.

Fregnac, Y., and M. Imbert (1978). Early development of visual cortical cells in normal and dark-reared kittens: relationship between orientation selectivity and ocular dominance. J. Physiol. 278:27-44.

Hebb, D. O. (1949). The Organization of Behavior. New York, Wiley.

Hirsch, H. V. B., and D. N. Spinelli (1970). Visual experience modifies distribution of horizontally and vertically oriented receptive fields in cats. Science 168:869-871.

Hubel, D. H., and T. N. Wiesel (1963). Receptive fields of cells in striate cortex of very young, visually inexperienced kittens. J. Neurophysiol. 26:994-1002.

Imbert, M., and P. Buisseret (1975). Receptive field characteristics and plastic properties of visual cortical cells in kittens reared with or without visual experience. Exp. Brain Res. 22:25-36.

Movshon, J. A. (1976). Reversal of the physiological effects of monocular deprivation in the kitten's visual cortex. J. Physiol. 261:125-174.

Pettigrew, J. D. (1974). The effect of visual experience on the development of stimulus specificity by kitten cortical neurones. J. Physiol. 237:49-74.

Pettigrew, J. D. (1978). The paradox of the critical period for striate cortex. In: Neuronal Plasticity. C. W. Cotman (ed.). Raven Press, New York, pp. 311-330.

Rauschecker, J. P., and W. Singer (1978). Experience-dependent modification of response properties in striate cortex: instructive versus selective mechanisms. Neuroscience Letters, Suppl. 1, S395.

Rauschecker, J. P., and W. Singer (1978). Changes in the circuitry of the kitten's visual cortex are gated by postsynaptic activity. Nature (submitted).

Rauschecker, J. P., and W. Singer (1979b). Selective and instructive effects of early visual experience on the cat's cortex: the same neural mechanism. In preparation.

Sherk, H., and M. P. Stryker (1976). Quantitative study of cortical orientation selectivity in visually inexperienced kittens. J. Neurophysiol. 39:63-70.

Sherman, S. M., R. W. Guillery, J. H. Kaas, and K. J. Sanderson (1974). Behavioral, electrophysiological, and morphological studies of binocular competition in the development of the geniculo-cortical pathways of cats. J. comp. Neurol. 158:1-18.

Singer, W. (1976). Modification of orientation and direction selectivity of cortical cells in kittens with monocular vision. Brain Res. 118:460-468.

Singer, W., and F. Tretter (1976). Receptive-field properties and neuronal connectivity in striate and parastriate cortex of contour-deprived cats. J. Neurophysiol. 39:613-630.

Singer, W., J. P. Rauschecker, and R. Werth (1977). The effect of monocular exposure to temporal contrasts on ocular dominance in kittens. Brain Res. 134:568-572.

Stent, G. S. (1973). A physiological mechanism for Hebb's postulate of learning. Proc. Nat. Acad. Sci. (U.S.A.) 70:997-1001.

Stryker, M. P., and H. Sherk (1975). Modification of cortical orientation selectivity in the cat by restricted visual experience: a reexamination. Science 190:904-906.

Stryker, M. P., H. Sherk, A. G. Leventhal, and H. V. B. Hirsch (1978). Physiological consequences for the cat's visual cortex of effectively restricting early visual experience with oriented contours. J. Neurophysiol. 41:896-909.

Tretter, F., M. Cynader, and W. Singer (1975). Modification of direction selectivity of neurons in the visual cortex of kittens. Brain Res. 84:143-149.

Wiesel, T. N., and D. H. Hubel (1963). Single-cell responses in striate cortex of kittens deprived of vision in one eye. J. Neurophysiol. 26:1003-1017.

Wiesel, T. N., and D. H. Hubel (1965a). Comparison of the effects of unilateral and bilateral eye closure on cortical unit responses in kittens. J. Neurophysiol. 28:1029-1040.

Wiesel, T. N., and D. H. Hubel (1965b). Extent of recovery from the effects of visual deprivation in kittens. J. Neurophysiol. 28:1060-1072.

Evidence for a Central Control
of Developmental Plasticity
in the Striate Cortex of Kittens

W. SINGER
Max-Planck-Institut für Psychiatrie
Munich, West Germany

It is well established that the functional organization of the mammalian visual cortex can be influenced by manipulating early visual experience. In the striate cortex of visually inexperienced cats and of cats raised with undisturbed binocular vision, the large majority of neurons are binocular and can be driven equally well from either eye (Hubel and Wiesel, 1962). When, however, vision is restricted to one eye during a critical period of early development, the afferents from the deprived eye lose the ability to excite cortical cells; most neurons become monocular and excitable only from the experienced eye (Wiesel and Hubel, 1963). Disruption of binocularity occurs also when binocular fusion is prevented, as it occurs, e.g., after surgically induced strabismus. Again, most neurons become monocular, but they remain excitable either from the right or the left eye (Hubel and Wiesel, 1965). These changes in neuronal circuitry are commonly attributed to competitive interactions between the afferents from the two eyes at their common cortical target cells (Wiesel and Hubel, 1965; Guillery, 1972; Cynader et al., 1977). Asymmetries in the activity of converging pathways are thought to lead to competitive suppression of the less efficient connections.

More recently, evidence has been obtained suggesting that these changes in circuitry are not solely determined by asymmetries in the level of presynaptic activity (Singer et al., 1977). An additional requirement for the induction of competitive disconnection is that the common postsynaptic target cells actually respond to the more active input (Singer, 1977; Rauschecker and Singer, 1979; see also, Rauschecker, this volume).

This raised the question whether it is a sufficient or only a necessary condition for the induction of circuit changes that the pattern of afferent activity matches the response properties of cortical cells. It is conceivable that an additional prerequisite for the alteration of cortical circuitry is that the afferent sensory signals can be matched with more integral sensory-motor processes. The experiments described below were designed to test this hypothesis: their results indicate that activity-dependent changes in striate cortex circuitry are not solely dependent on local matching operations between the presynaptic activity in thalamic afferents and the postsynaptic responses of cortical target cells. They rather suggest that local changes in connectivity are gated by a central evaluation system that enables and/or disables adaptive changes according to the behavioural relevance of the sensory signals.

Dark-reared kittens were monocularly deprived by unilateral lid suture. At the same time the other open eye was surgically rotated within the orbit. This intervention leaves the spatio-temporal pattern of retinal responses to contours unaffected and consequently should not interfere with the ability of cortical cells to respond to afferent activity. But eye rotation disturbs the correspondence between retinal coordinates and other sensory and motor maps which leads to severe conflicts in visuomotor integration as manifested by abnormal visually guided behaviour. In this preparation it is thus possible to determine whether experience-dependent changes in cortical circuitry are gated only by local matching operations between pre- and postsynaptic responses or whether the adequacy of sensory signals is evaluated in a more complex behavioural context before sensory activity is enabled to alter local circuitry.

In five dark-reared kittens one eye was rotated and the other was closed by lid suture at age 28 days. Subsequently, the kittens lived together with their mothers in a normal animal house environment. For the surgical rotation of the eye the conjunctiva was dissected around the eyeball and the extraocular muscles were severed at their insertion to the bulbus. The eyes were then rotated and fixed in the new position by attaching one of the severed distal muscle tendons to the intraorbital conjunctive tissue. This operation resulted in a transient immobilization of the rotated eye. But after a few weeks the eye muscles had apparently reattached and small eye movements reappeared. The degree of rotation was determined from the orientation of retinal vessels on fundus pictures taken before and after the operation and immediately before the neurophysiological experiment. Three of the five kittens had their open eyes rotated by 180° and two by 90°. Recordings were taken from them when they were 6 months old. A number of controls were made to determine whether eye rotation had caused damage to the retina or the optic nerve but none of them provided evidence that such might have occurred (for details see Singer et al., 1979a).

To assess the effect of eye immobilization per se two control kittens were prepared: one eye was closed and in the other eye all extraocular muscles were severed and the four distal muscle tendons were sutured to the orbital fringe leaving the eye in its normal position. Preparation for single-unit recording was conventional and was described in detail elsewhere (Singer, 1977). Besides the usual RF parameters, the vigour of responses to optimally aligned stimuli was also evaluated and rated in five classes, "1" corresponding to a barely detectable response and "5" to the most vigorous reactions encountered in normal cats.

In all kittens with the open eye rotated, visually guided behaviour was highly abnormal. During the first weeks after rotation they displayed partially vigorous head nystagmus and circling behaviour. As more time elapsed, these behavioural patterns disappeared and a number of tests (forced jumping, obstacle avoidance, visual tracking, visually guided reaching and placing) revealed that kittens no longer used their open eye for visual orientation. Nevertheless, when alert they kept the rotated eye open.

This behavioural abnormality is reflected in the response properties of cortical neurons which differed markedly from cats that had undergone normal monocular deprivation (MD). The percentage of cells responding to light stimulation of either eye had markedly decreased. Of the 287 analyzed cells only 54% could be driven with light stimuli. For comparison, in MD cats of similar age (Fig. 1(A)), clear and vigorous responses could be obtained with hand-held light stimuli in 87% of the recorded neurons. Furthermore, the average quality of light responses of cells still responding was considerably worse than in cats with conventional monocular deprivation (Fig. 1(E)). The most striking finding was that the open eye had failed to induce the shift in ocular dominance that usually occurs with monocular deprivation (compare Figs. 1(A,B)). Of the 156 responsive cells the open rotated eye drove 114 neurons and the deprived eye 108 cells. On the average, the quality of responses elicitable from the open rotated and the deprived eye was equally poor (Fig. 1(E)). Although the percentage of binocular cells is markedly reduced when compared to normal cats, it is still surprisingly high (42%) with respect to the fact that these cats had only monocular visual experience.

A significant correlation was found between the neurons' ocular dominance and the vigorousness of their responses. Cells with weak responses had remained binocular or were still dominated by the closed eye. The few neurons (19% of responsive cells) with vigorous reactions to light (quality classes 4 and 5) were predominantly monocular (65%) and 75% of these cells were dominated by the open rotated eye (OD-classes 4 and 5).

The symmetrical impairment of the pathways from both eyes and the lack of a clear shift in ocular dominance towards the open eye is also in good agreement with LGN morphology. In contrast to monocularly deprived cats, it proved impossible to tell from inspection of Nissl stained LGN sections which laminae were connected to the deprived and the rotated eye. Cell shrinkage, if it had occurred at all, must have been similar in the layers connected to the rotated and the closed eye. To substantiate this finding the diameters of LGN cells were measured in Nissl stained sections according to the method described by Holländer and Vanegas (1977). Neither the individual data from the two LGNs of the three cats analyzed so far nor the pooled data shown in Fig. 2(A) revealed a significant difference between the cell size in the respective laminae. By contrast, a clear difference in cell diameter was found with identical methods in monocularly deprived control cats (Fig. 2(B)).

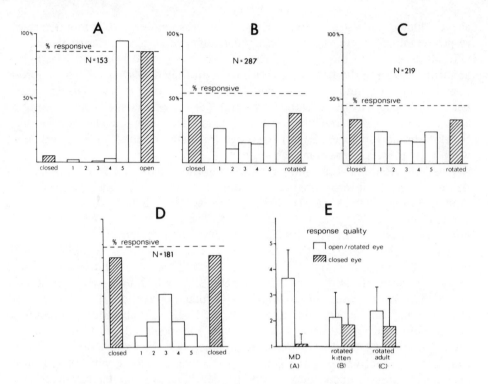

FIGURE 1 Ocular dominance (OD) distributions (blank columns), percentages of light reactive cells (interrupted horizontal lines), and percentages of neurons excitable from the right or the left eye (hatched columns) from striate cortex of monocularly deprived animals (A), experimental groups with one eye rotated (B,C), and binocularly deprived cats (D). The OD distributions are calculated as a percentage of the light-responsive cells; the percentages of cells excitable from either of the two eyes (hatched columns) are calculated from the sum of OD classes 1-4 and 2-5 and refer to the total sample of analyzed cells corresponding to the respective "N" in the graphs. (A) Data from two MDs (6 months old) that were dark reared until lid closure at age 28 days. (B) Pooled data from five kittens (6 months old) that were dark reared until age 28 days and then had one eye closed and the other rotated. (C) Pooled data from three adult cats with previously normal vision that were recorded six months after lid closure and eye rotation. (D) Pooled data from three binocularly deprived (lid suture) kittens (age 3 months). (E) Mean response quality of neurons in the experimental groups A, B, and C after stimulation of the open or rotated eye (blank columns) and the closed eye (hatched columns). The vertical bars indicate the standard deviation (from Singer et al., 1979a).

The results from the two kittens with the open eye immobilized but not rotated suggest that the effects described above are specific to the inversion of retinal coordinates. In both animals (survival time 5 weeks and 3 months) the ocular dominance was as skewed as with conventional monocular deprivation and the geniculate nuclei showed morphological changes typical of monocular deprivation.

These results clearly indicate that local correspondence between retinal signals and response properties of cortical target cells is not sufficient to induce

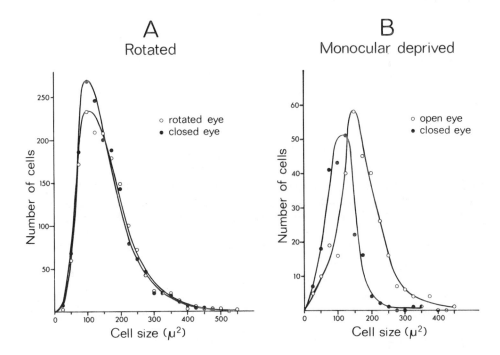

FIGURE 2 Distribution of cell size in the lateral geniculate of three kittens raised with one eye closed and the open eye rotated (A) and in a kitten raised with conventional monocular deprivation (B). To account for the normal difference in cell size between layers A and Al, the measurements from cells in layers A and A1 which are connected to the deprived eye were pooled (filled circles) and compared to the sum derived from cells in layers A and A1 connected to the open rotated eye (open circles). In contrast to the marked difference in cell size between deprived and nondeprived layers in a conventional MD cat (B), there is no significant difference between deprived and nondeprived laminae in cats with the open eye rotated. Data in (A) and (B) cannot be compared in terms of absolute cell size because the sections in (A) are from celloidin-embedded blocks, while those in (B) were cut from frozen blocks (from Singer et al., 1979a).

consolidation and disruption of afferent connections such as it occurs with conventional monocular deprivation. The inadequacy of retinal signals with respect to more integral sensory-motor processes has apparently prevented selective consolidation of afferents from the open eye and competitive suppression of input from the closed eye. Since the normal, non-rotated eye was permanently closed, the inadequacy of afferent retinal signals could have been detected only through comparison of retinal signals with other sensory maps or motor commands. This leads one to postulate additional, internally generated gating signals that control consolidation and disruption of local circuits as a function of a more integral evaluation of stimulus adequacy.

A similar conclusion can be drawn from recent experiments in cats with symmetrically induced squint. Maffei and Fiorentini (1976) and Maffei and Bisti (1976) have emphasized that cortical binocularity is not solely dependent on the congruency of the retinal signals from the two eyes but is controlled also by proprioceptive signals from extraocular muscles. These authors reported that cortical binocularity becomes disrupted also in the absence of visual experience when one eye is immobilized by section of the cranial nerves innervating the extraocular muscles. Their hypothesis that cortical binocularity is controlled to a critical extent by proprioceptive and not only by retinal signals has received further support from the finding that cortical binocularity is maintained in spite of misaligned visual axes when squint is induced by the section of the same muscles in both eyes (Maffei, 1978, and personal communication). From this observation it has been concluded that the main requirement for the maintenance of cortical binocularity is the symmetry of proprioceptive feedback signal rather than the correspondence of retinal images. The following experiments were designed to re-examine this hypothesis and to determine whether the maintenance of binocularity in squinting animals could be attributed to the same factors that had led to maintained binocularity in the kittens with one eye closed and the other rotated.

Five kittens were dark reared until the age of 28 days and then the optical axes of the two eyes were made to deviate by bilateral surgical intervention. In two kittens the recti laterales and in three kittens the recti mediales were severed on either side, resulting in symmetrical convergent, c.f. divergent, strabismus. Until the electrophysiological investigation (at which time they were at least 6 months old) the kittens were raised in a normal animal house environment. Several weeks after the operation the visually guided behaviour of these animals was examined and, in particular, their ability to fixate objects of interest. According to the techniques described by J. van Hof-van Duin (1976), visually guided orienting responses, tracking, placing, and jumping were investigated for each eye separately, while the other eye was covered with a black contact lens. In addition, visual fixation was tested both by looking at the kittens' eyes, while targets were moved at a distance of 30 to 50 cm across the visual field and by taking photographs while the kittens were confronted with a target light bulb 3 m away. From these photographs the approximate angle of squint could also be determined according to the method described by Sherman (1972). The squint angle was determined once more after paralysis at the beginning of the electrophysiological experiment by direct measurements of eye position with a fundus camera. Within a range of $10°$ these angles were comparable to those measured in the alert kittens.

In four of the five kittens with symmetrical squint, visually guided behaviour was indistinguishable from that of normal cats. These kittens were obviously capable of alternating fixation since they could align the visual axis of either eye with the visual targets. Squint angle between the visual axes of both eyes was estimated as not exceeding $30°$. The remaining kitten with divergent squint showed, however, highly abnormal visual behaviour. Orienting responses to moving targets and visually guided reaching and placing were inaccurate. Moreover, this cat was definitely unable to fixate targets moving in front of its nose. As in animals with lateral eyes the visual axes were always pointing laterally, although large amplitude eye movements could be executed in any direction. When the kitten was alert the angle between the axes of both eyes varied between 60° and 80°. After paralysis the squint angle was 80°. In relation with other experiments this cat was also tested for visual acuity in a jumping stand according to the method designed by Mitchell et al. (1977). Unlike other cats with squint, this animal had difficulties learning to distinguish between a grating and a flux-equated homogeneous surface and therefore had to be excluded from further behavioural analysis. The behavioural difference between this and the other four kittens was clearly reflected by the electrophysiological data. In the four kittens with normal visual behaviour the ocular dominance distribution of cortical units was as skewed as described by Hubel and Wiesel (1965) for kittens with asymmetric strabismus (Fig. 3(A-D)). Only 25% of the 184 responsive cells had remained binocular. The percentage of units excitable with the usual spectrum of light stimuli was in the normal range (84% out of 226 cells) and their RFs showed no abnormalities. Both eyes were equally effective in driving cortical cells (Fig. 1(A,C)), but there was a rather pronounced dominance of the respective contralateral eye, especially in the kittens with convergent squint (Fig. 3(B)).

In the kitten with abnormal visual behaviour, by contrast, only 67% of 62 analyzed cells could be driven with light stimuli and a large fraction (68%) of these responsive cells had remained binocular (Fig. 3(E,F)). As in the other kittens both eyes were about equally effective in driving cortical cells but again, the respective contralateral eye was more potent. Many of the cells still reacting to light gave only weak and sluggish responses to optimally aligned stimuli. Rated in the subjective scale for response quality the average index for responses to the respective dominant eye was 2.2 ± 0.9. In the kittens with normal behaviour this index was 3.2 ± 0.8. In spite of the difficulty in obtaining responses to visual stimuli in this last kitten, 87% of the responsive cells showed a clear preference for stimulus orientation. This excludes the possibilities that the large squint angle or some other peripheral factors have prevented retinal responses to contours. Binocularly deprived cats have a similarly low score of responsive units but most of the cells also lose orientation preference (Wiesel and Hubel, 1965; Singer and Tretter, 1976; Watkins et al., 1978). Thus, although retinal signals from a normal environment were available to the striate cortex, they failed to develop or maintain normal responsiveness of the cortical units. The close correlation between the impairment of visual behaviour and cortical physiology suggests that the signals from the two deviating eyes could no longer be accommodated in the context of visuomotor integration and have therefore been prevented from developing or maintaining normal response properties of cortical neurons. This interpretation is in agreement with the results obtained from the kittens with rotated eyes. When taken together, these results suggest that experience-dependent changes in cortical circuitry are not solely dependent on local interactions between the afferents from the two eyes and the common cortical target cells. These processes seem to be gated in addition by control systems that enable or disable local changes

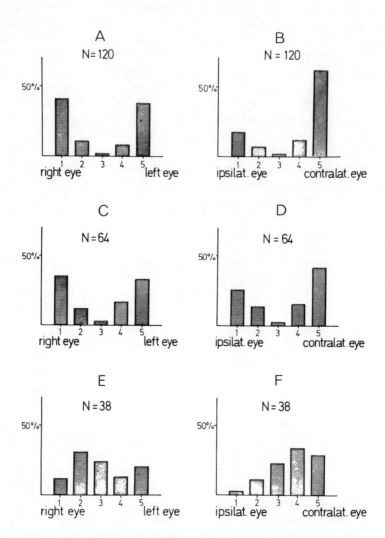

FIGURE 3 Ocular dominance (OD) distributions from kittens raised with symmetrical convergent (A,B) and divergent (C-F) squint. Histograms A/B and C/D are from the four kittens whose visually guided behavior was normal, histograms E/F are from the fifth kitten with abnormal visual behaviour. In the kittens with normal visual behaviour, cortical binocularity is markedly reduced irrespective of whether the strabismus is convergent (A/B) or divergent (C/D). In the kitten with abnormal vision and divergent squint, 68% of the responsive units have remained binocular (E,F). Since both hemispheres were investigated, the OD distributions are plotted not only for the right and left eyes (A,C,E), but also with respect to the laterality of the eyes (B,D,F). This shows that both eyes (right/left) have remained about equally effective in driving cortical cells, although the respective contralateral eye is dominant, particularly in the two cats with convergent squint and the cat with abnormal visual behaviour. (From Singer et al., 1979c.)

in circuitry as a function of the behavioral adequacy of the respective retinal signals. When these signals can no longer be coped with—as indicated by impaired visuomotor behavior—they are not used for the induction of specific changes in cortical circuitry.

In the present case of squint as well as in the experiments with eye rotation, maintenance of binocularity was always associated with a marked impairment of excitatory transmission. This suggests a causal relationship between the consolidation of synaptic transmission and competitive interactions: the failure to increase the safety factor of synaptic transmission has apparently also reduced competitive interactions between the converging afferents from the two eyes.

The present results show that the persistence of binocularity is consistently associated only with disturbed visuomotor behaviour and not with a particular type of surgical intervention. This suggests that it is the behavioural adequacy of retinal signals rather than the symmetry of proprioceptive signals from extraocular muscles that determines whether incongruent binocular signals lead to competitive disruption of cortical binocularity. For the present it is unknown where and through which mechanisms the adequacy of retinal signals is evaluated and how the signals are generated that apparently gate neuronal plasticity in striate cortex. However, recently we have obtained some evidence that the action of this postulated gating system is not confined to the critical period of early postnatal development. The results presented in the next paragraph provide evidence that this gating system might play an important role in determining the functional state of the mature brain as well.

Five cats that had normal visual experience from birth were operated on in the same way as the kittens but only after they were at least 2 years old. One eye was sutured closed and the open eye was rotated by 180° ($n = 2$) or 90° ($n = 3$). Survival time between eye rotation and recording was at least six months.

The behavioural and electrophysiological investigation of these cats revealed that inverted vision impairs the functional state of visual cortex also beyond the classical critical period. After an initial period of head nystagmus and false orienting responses, the adult cats, like the kittens described previously, no longer relied on their open rotated eye for visual orienting. As in the kittens the responsiveness of cortical units to light stimulation was markedly reduced. The percentage of light reactive units and the score of response quality were even lower than in the kittens operated on at the beginning of the critical period (Fig. 1(C,E)). On the average the decrease in responsiveness was equally severe for both eyes (Fig. 1(C), hatched columns), although in individual animals either the closed ($n = 2$) or the rotated eye ($n = 3$) could be slighly dominant. These experiments provide further support for the hypothesis that the decrease in responsiveness cannot be attributed to damage of the rotated nerve: if this were the case, then the responsiveness to stimulation of the closed eye should have remained normal since the lids were merely closed well beyond the critical period. Comparison with corresponding data from binocularly deprived animals (Fig. 1(D)) shows that the reduction of cortical responsiveness is more severe after eye rotation than after complete deprivation of contour vision.

A number of observations suggest further that this decrease in responsiveness is not general but is selective for cortical cells with particular properties. In the cats operated on as adults there was a significant reduction of binocular units; the ocular dominance distributions were flat and clearly different from those in normal cats. As described in detail elsewhere (Singer et al., 1979b), the most surprising observation was that in animals with 180° eye rotation (both for kittens and adults) cells with vertically oriented receptive fields were remarkably underrepresented. Figure 4 indicates that this bias in the orientation distribution is seen only in cells receiving binocular input and is absent in the population of cells that are excitable only from either the rotated or the closed eye. In addition to this distortion in the orientation domain, direction selectivity was also influenced in the sense that there was a reduction in the number of units that encode movements towards the vertical meridian. These rather selective effects of inverted vision provide further evidence that the observed changes are not due to damage of the rotated nerve.

We hypothesize that the same gating signals that have prevented the shift in ocular dominance towards the open rotated eye in the monocularly deprived kittens have also led to severe alteration in the functional state of previously normal, mature cortex.

Both in the developing as well as in the mature brain these signals seem to interfere with the safety factor of excitatory transmission. In the kitten they prevent the consolidation of still labile connections and thereby disable competitive interactions that otherwise would have led to a disruption of binocularity. In the adult they reduce the efficiency of previously normal synaptic connections. This functional disconnection shows a certain degree of selectivity, whereby binocular cells, especially those encoding horizontal movements, were more severely affected than monocular cells. This suggests that the normal response properties of these cells become particularly disturbed when the eye is rotated—a hypothesis that receives some support from the following arguments. With respect to visuomotor coordination, horizontal image displacements are distinguished from all others: horizontal eye and head movements are the most frequent and special reflex loops are reserved for horizontal optokinetic and vestibulo-ocular nystagmus. Also, vergence movements are exclusively performed in the horizontal plane. Although inverted vision leads to mismatch along all meridians, it is conceivable that the most pronounced effects are observed in those subsystems which serve as substrate for the most frequent and most important visuomotor performances. The discord induced in the large number of systems involved in the guidance of horizontal eye movements could explain the preferential disconnection of cells which convey signals about horizontal displacements of retinal images. The finding that disconnection had affected mainly the cells with symmetrical binocular input might be related to the fact that these cells rather than cells with monocular receptive fields are mediating cortical output activity.

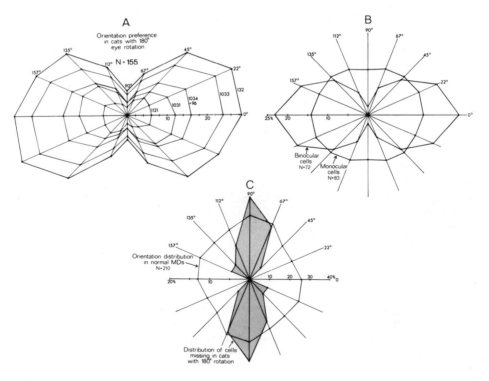

FIGURE 4 (A) Polar plot of orientation preferences (0° = horizontal, 90° = vertical). The length of the respective vectors corresponding to the number of cells with corresponding orientation preference (scale on 0° vector). The five annuli were obtained by cumulative plotting of individual data, results from cats 1034 and 96 were pooled because of the small number of responsive cells in these two animals. The trend for an underrepresentation of orientations close to the vertical is apparent in every animal. (B) Polar plots of orientation preferences in monocular and binocular cells. The two populations of monocular cells that are driven from the rotated and the deprived eyes, respectively, do not differ with respect to their orientation preference and are therefore added together. Data are pooled from all six animals and expressed as percentages. (C) Polar plot of orientation preferences in cats with conventional monocular deprivation (MD) and eye immobilization. The vectors within the stippled area indicate the percentage of cells in the respective orientation classes that are missing in cats with inverted vision. These percentages were calculated from the differences between the orientation distribution of cats with monocular inverted vision and conventional MDs, the 0° vector being taken as the reference. All plots are smoothed for clearer presentation according to the formula

$$\frac{N_{(\alpha - 22.5 deg)} + 2N_\alpha + N_{(\alpha + 22.5 deg)}}{4} = N$$

where N is the number of cells in a particular orientation class and α the angle of orientation. (From Singer et al., 1979b.)

In summary the present results lead to the conclusion that experience-dependent changes of neuronal connections during early development are not solely dependent on local interactions between afferents from the two eyes and the common cortical target cells. This conclusion receives further support from previous experiments (Singer, 1979) which attempted to induce changes in ocular dominance in anesthetized and paralyzed kittens. Although monocular stimulation with textured stimuli was maintained over as many as four days, no clear changes in ocular dominance towards the open eye were detectable. Exceptions were seen only when the mesencephalic reticular formation was stimulated in contingency with the presentation of the light stimulus. Thus, activity-dependent consolidation and disruption of excitatory connections in striate cortex seems to be controlled by internally generated gating signals that enable or disable changes in local circuitry with respect to the behavioural relevance or adequacy of the retinal signals. To date, nothing is known about the neuronal substrate of this internal control system. But its crucial role in developmental plasticity is evident. It reduces the hazard for the animal that inadequate sensory signals might lead to long-lasting, sometimes irreversible but behaviourally inappropriate modifications of neuronal circuits. The developing brain can select among the continuous stream of afferent activity and enable only those signals which have been identified as adequate in the context of highly integrated sensory motor processes to induce specific changes in neuronal connectivity.

Acknowledgements The experiments described in this review were performed with M. von Gruenau, J. Rauschecker, F. Tretter, and U. Yinon.

References

Cynader, M., and D. E. Mitchell (1977). Monocular astigmatism effects on kitten visual cortex development. Nature 270:177-178.

Guillery, R. W. (1972). Binocular competition in the control of geniculate cell growth. J. Comp. Neurol. 144:117-127.

Holländer, H., and H. Vanegas (1977). The projection from the lateral geniculate nucleus onto the visual cortex in the cat. A quantitative study with horseradish peroxidase. J. Comp. Neurol. 173:519-536.

Hubel, D. H., and T. N. Wiesel (1962). Receptive fields, binocular interaction, and functional architecture in the cat's visual cortex. J. Physiol. (Lond.) 160:106-154.

Hubel, D. H., and T. N. Wiesel (1965). Binocular interaction in striate cortex of kittens reared with artificial squint. 28:1041-1059.

Maffei, L. (1978). Binocular interaction in strabismic kittens and adult cats deprived of vision. Arch. Ital. 116:390-392.

Maffei, L., and S. Bisti (1976). Binocular interaction in strabismic kittens deprived of vision. Science 191:579-580.

Maffei, L., and A. Fiorentini (1976). Asymmetry of motility of the eyes and change of binocular properties of cortical cells in adult cats. Brain Res. 105:73-78.

Mitchell, D. E., F. Giffin, and B. Timney (1977). A behavioural technique for the rapid assessment of the visual capabilities of kittens. Perception 6:181-193.

Rauschecker, J., and W. Singer (1978). Changes in the circuitry of the kitten visual cortex are gated by postsynaptic activity. (Submitted to Nature).

Sherman, S. M. (1972). Development of interocular alignment in cats. Brain Res. 37:187-203.

Singer, W. (1976). Modification of orientation and direction selectivity of cortical cells in kittens with monocular vision. Brain Res. 118:460-468.

Singer, W. (1977). Effects of monocular deprivation on excitatory and inhibitory pathways in cat striate cortex. Exp. Brain Res. 30:25-41.

Singer, W. (1979). Central-core control of visual cortex functions. In: The Neurosciences; 4. Study Program (in press).

Singer, W., and F. Tretter (1976). Receptive field properties and neuronal connectivity in striate and parastriate cortex of contour-deprived cats. J. Neurophys. 39:613-630.

Singer, W., J. Rauschecker, and R. Werth (1977). The effect of monocular exposure to temporal contrasts on ocular dominance in kittens. Brain Res. 134:568-572.

Singer, W., F. Tretter, and U. Yinon (1979a). Inverted monocular vision prevents ocular dominance shift in kittens and impairs the functional state of visual cortex in adult cats. Brain Res. 164:294-299.

Singer, W., F. Tretter, and U. Yinon (1979b). Inverted vision causes selective loss of striate cortex neurons with binocular, vertically oriented receptive fields (Brain Res., in press).

Singer, W., M. v. Gruenau, and J. Rauschecker (1979c). Requirements for the disruption of binocularity in the visual cortex of strabismic kittens (in preparation).

van Hof-van Duin, J. (1976). Early and permanent effects of monocular deprivation on pattern discrimination and visuomotor behavior in cats. Brain Res. 111:261-276.

Watkins, D. W., J. R. Wilson, and S. M. Sherman (1978). Receptive-field properties of neurons in binocular and monocular segments of striate cortex in cats raised with binocular lid suture. J. Neurophysiol. 41:322-337.

Wiesel, T. N., and D. H. Hubel (1963). Single-cell responses in striate cortex of kittens deprived of vision in one eye. J. Neurophysiol. 26:1003-1017.

Wiesel, T. N., and D. H. Hubel (1965). Comparison of the effects of unilateral and bilateral eye closure on cortical unit responses in kittens. J. Neurophysiol. 28:1029-1040.

Behavioural Recovery from Visual Deprivation: Comments on the Critical Period

BRIAN TIMNEY

Department of Psychology
University of Western Ontario
London, Ontario, Canada

DONALD E. MITCHELL

National Vision Research
Institute of Australia
Carlton, Victoria, Australia

Monocular deprivation imposed early in the life of kittens can have profound effects on the functional properties of neurons in the visual cortex (Wiesel and Hubel, 1963). These physiological consequences of deprivation have been described by other contributors to this volume, so we will not detail them here. Instead, we wish to concentrate on the behavioural consequences of monocular deprivation, and, in particular, upon the upper age limits of susceptibility to such deprivation. The experiments we shall describe form part of a series of studies of the visual critical period carried out in conjunction with Max Cynader at Dalhousie University.

The concept of a critical period in visual development is well established. Wiesel and Hubel (1963) reported that the physiological effects of unilateral eyelid suture were less severe when the operation was performed after kittens had been allowed several weeks of normal visual experience than if it had been done at the time of natural eye opening. By suturing the eyelids of kittens for varying periods of time, beginning at different ages, they (Hubel and Wiesel, 1970) were able to map out the approximate time course of susceptibility to monocular deprivation. They concluded that susceptibility "begins suddenly near the end of the fourth week, remains high until sometime between the sixth and eighth weeks, and then declines, disappearing around the end of the third month."

These conclusions were supported by Blakemore and Van Sluyters (1974) who showed that if the sutured eyelids of one eye were opened and, at the same time, those of the other eye were closed, then it was possible to reverse the monocular deprivation effect, either partially or completely. The extent to which the previously deprived eye could recapture control of cortical neurons was dependent upon the age at which the reversal was instituted. If it was done at 4 or 5 weeks of age, the initially deprived eye eventually would come to dominate the overwhelming majority of cortical cells. However, if reverse suturing was delayed until 14 weeks, no recovery was evident at all and the initially experienced eye maintained its control of cortical neurons.

Behaviourally, when a monocularly deprived cat is tested through the deprived eye, it appears initially to be completely blind (e.g., Ganz and Fitch, 1968; Dews and Wiesel, 1970; Chow and Stewart, 1972). However, there is usually some recovery, the degree of which seems to depend upon both the length of the deprivation period and the age at which it is imposed. Mitchell and his co-workers (Mitchell, Cynader, and Movshon, 1977; Giffin and Mitchell, 1978) have made extensive measurements of the recovery of vision in cats following various periods of monocular deprivation imposed at different ages. They observed that, in all cases in which kittens were deprived from birth, there was a period of apparent blindness which increased progressively with longer deprivation. However, substantial recovery occurred invariably, even in a cat which had been deprived until the age of 4 months; that is, throughout the generally accepted critical period. In the present paper we shall describe additional observations we have made which suggest that the critical period may not be completely over by the end of the third month.

Although different investigators have used a variety of different measures of visual function in their assessment of the abilities of deprived cats, including estimates of visuomotor behaviour (Movshon, 1976), pattern discrimination (Rizzolatti and Tradardi, 1971) and interocular transfer of learned pattern discriminations (Ganz and Fitch, 1968), we have concentrated mainly upon obtaining measures of visual acuity. Our assumption is that visual acuity is more likely to be correlated with the response characteristics of single neurons than the other measures.

The technique we use to measure visual acuity has been described in detail elsewhere (Mitchell, Giffin, and Timney, 1977), so we will describe it only briefly here. The apparatus is shown in Fig. 1. It consists of a box containing two trapdoors separated by a narrow divider. These trapdoors are latched and may be released by the experimenter through a switch-operated solenoid. On one is placed a high-contrast, square-wave grating and on the other a homogeneous gray field of the same space-averaged luminance. The animals are placed in a box, located on a platform some distance above the trapdoors, and they are trained to jump to the striped pattern. Correct jumps are rewarded with fondling and small amounts of beef baby food. Errors are punished by withholding these rewards and occasionally by releasing the trapdoor so that the cat falls about 30 cm to the floor.

Typically, an animal is run in blocks of five or ten trials. If it achieves a criterion of at least 70% correctness, then the spatial frequency of the grating is

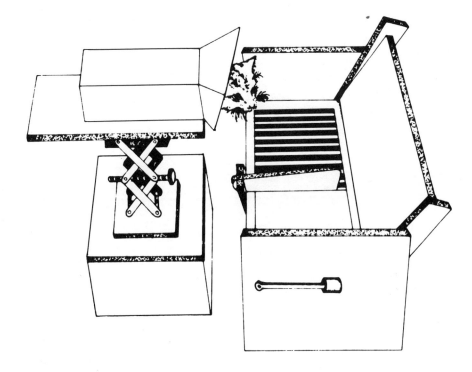

FIGURE 1 An illustration of the jumping stand used in the measurements of visual acuity. Details of the procedure in the text. (After Mitchell, Giffin and Timney, 1977).

increased and another block of trials run. This sequence is repeated until the animal fails to meet criterion. The task is then made very much easier by increasing the width of the stripes and the whole procedure is repeated at least once more. Threshold is taken as the highest spatial frequency at which the cat is correct consistently on at least 70% of the trials.

Recovery from Long-term Monocular Deprivation

As mentioned above, Giffin and Mitchell (1978) reported substantial recovery of vision in the deprived eye of a cat which was monocularly sutured for four months. We have obtained data from several additional cats raised under similar conditions and these are shown in Fig. 2. The cats shown in this figure were all deprived for periods of four months. At that time, the cats illustrated in the upper two panels were reverse sutured, while the one represented in the lower panel simply had its deprived eye opened. In all cases, the pattern of recovery was similar. There was a period of about one month when the animals appeared to be completely blind. Thereafter, there was a fairly rapid increase in visual acuity, which soon reached an asymptotic level. The final acuity level for the two reverse-sutured cats was about 4 cyc/deg and for the

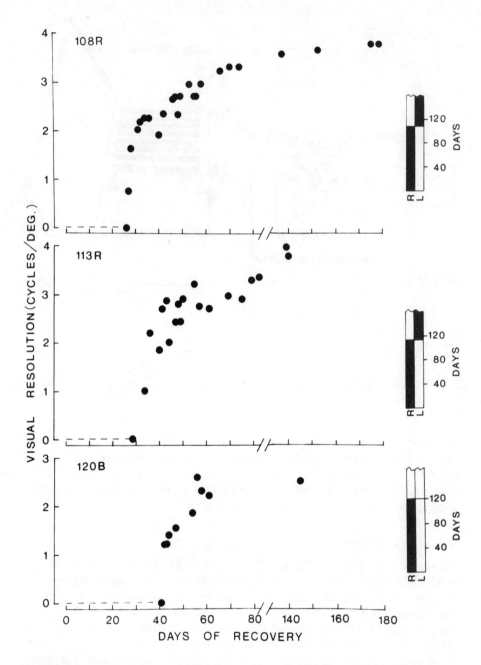

FIGURE 2 Recovery of visual acuity in the deprived eye of three cats which were monocularly deprived for four months. Rearing history is depicted schematically on the right, with the black segment indicating the deprivation period. The dashed line represents a period during which the cats were tested but there were no signs of vision.

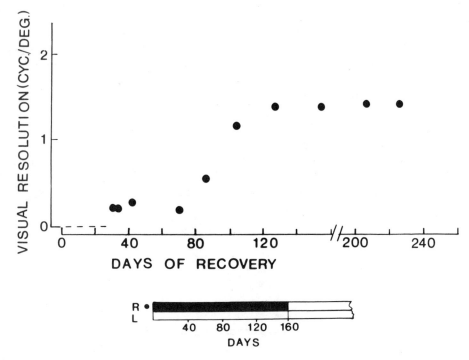

FIGURE 3 Recovery of visual acuity in the deprived eye of one cat which was monocularly sutured for five months, then had the deprived eye opened.

binocularly opened cat about 3 cyc/deg. Figure 3 shows data from another cat which was monocularly deprived until 5 months of age and thereafter had both eyes open. This cat showed a longer period of apparent blindness, about 40 days, but eventually it did recover use of the deprived eye, achieving a maximum acuity of 1.4 cyc/deg. Although there are other interpretations, these results suggest that the kitten visual system retains some plasticity beyond the age of 4 months. This conclusion is reinforced further by studies of the recovery from prolonged binocular visual deprivation.

Recovery from Prolonged Dark-rearing

Physiologically, one of the striking differences between monocularly deprived and binocularly deprived cats is that the binocularly deprived cats suffer less of a deficit than their monocularly deprived counterparts (Wiesel and Hubel, 1965). Although neurons from binocularly deprived cats do not show normal response properties, a large proportion may be driven by visual stimulation. Furthermore, after exposure to a normally illuminated environment, considerable recovery may be observed (Cynader et al., 1976). This is in contrast to the monocularly deprived cat where few, if any, neurons can be influenced at all by stimulation of the deprived eye, and recovery appears to be minimal.

Given these results, we wondered whether binocularly deprived cats would show a greater degree of recovery than those monocularly deprived. Some of these data have already been reported (Timney et al., 1978). Cats were reared in total darkness from before their eyes were open until various ages. After they were removed from the dark we examined the time course of recovery of their visual abilities.

Figure 4 shows data from cats which were dark-reared for four, six, eight, and ten months, respectively. For comparison, we show also the development of acuity in normally reared kittens. Several points emerge from this figure.

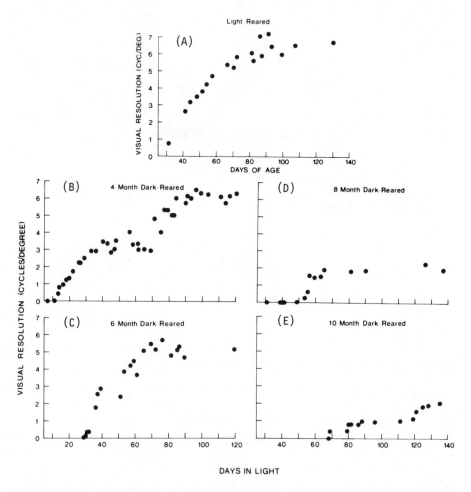

FIGURE 4 The development of visual acuity in cats raised under different conditions of visual deprivation. (A) Normal kittens raised in a lighted environment. The data represent combined results from a number of kittens and are plotted as a function of chronological age. (B)-(E) Cats dark-reared for the periods indicated in each panel. Visual acuity is plotted as a function of number of days since removal from the dark.

At first, all the dark-reared cats appeared to be completely blind. However, after a period of time which depended upon the duration of the deprivation, they began to show some signs of vision. Thereafter, they showed a systematic recovery of visual function. It is worth noting that, at least for cats dark-reared up to the age of 6 months, the time course of recovery, after the initial period of blindness, is very like that of the normal developmental function. The final acuities achieved by our 4-month dark-reared cats were well within the range of normals, between 6 and 7 cyc/deg, and the final acuity of the 6-month dark-reared (5.7 cyc/deg) is probably not significantly less than that of normals.

The results for the cats deprived for eight months and ten months are somewhat different. The initial periods of apparent blindness were substantially longer (39 days for the 8-month, 67 days for the 10-month animal), although there were indications of weak visual tracking before they were able to perform a discrimination on the jumping stand. The extent of recovery in these animals was rather less than that of the cats deprived for shorter periods. Neither cat achieved acuities greater than about 2 cyc/deg, even after more than four months in a normally illuminated environment. Although these values are low compared to normal animals, they do represent a substantial recovery of visual function. In fact, to casual inspection these animal were not distinguishable from normals when allowed to run freely among other cats in the colony.

The fact that dark-reared cats show such a large amount of recovery from their deprivation is another piece of evidence suggesting that plasticity in the cat visual system is not restricted to the first three months of life. Before discussing this point in detail, we wish to turn to a different aspect of plasticity in older cats.

Consequences of Monocular Deprivation Following Dark-rearing

The results described above indicate that the visual system of the dark-reared cat retains plasticity in the sense that it is able to show substantial recovery following deprivation. These findings are in agreement with physiological data of Cynader et al. (1976) who showed that in dark-reared cats allowed subsequent normal visual experience, there were many neurons which possessed nearly normal response properties.

Given that such extensive recovery is possible in dark-reared cats, it is natural to ask whether this also implies a greater degree of susceptibility to monocular deprivation than would be observed in normal animals of the same age. The following experiment was directed towards this question. Dark-reared cats* were subject to monocular eyelid suture at various times after being removed from the dark, then visual acuity was measured in the usual way following eye-opening. Figure 5 shows results from three cats which were dark-reared until 4 months of age. The top portion of each panel shows the deprivation history of the cat following dark-rearing. The cat represented in the upper

---

*These cats were some of the same animals used in the previous experiment. In that study all the data presented were from the non-deprived eye.

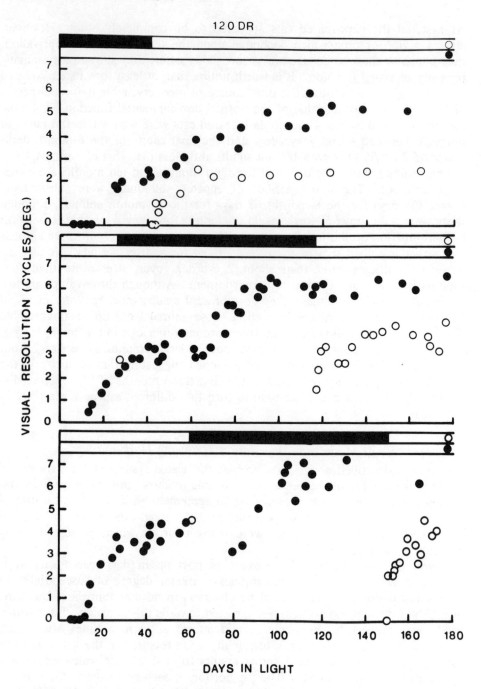

FIGURE 5 Recovery of vision in the deprived eyes (open circles) and the normal eyes (filled circles) of three cats which were dark-reared for the first four months of life and then were monocularly sutured at different times following removal from the dark. Rearing history following removal from the dark is depicted schematically above each panel.

panel had its right eyelid sutured for a period of six weeks immediately following removal from the dark. It can be seen that the short period of monocular
deprivation had an effect which was both large and long-lasting. After two days
of apparent blindness there was some recovery over the next three weeks, but
the final acuity obtained was only about 2 cyc/deg. The other two cats were
both monocularly deprived for three months after having been allowed, respectively, one (middle panel) and two (lower panel) months of normal binocular
visual experience. Here again, substantial deficits were observed in the
deprived eyes of these animals, although they were not as severe as that of the
cat sutured immediately upon removal from the dark. For the cat allowed only
one month of binocular experience, it appeared that the deficit was permanent.
For the cat allowed two months of normal vision there were indications that
thresholds were beginning to approach normal levels, although testing had to be
terminated before this could be established with any degree of confidence.

Figure 6 shows data from a cat dark-reared to 6 months of age, then monocularly deprived for a further three months. This cat exhibited no signs of
vision through the deprived eye when it was first opened.

However, within a week, it was able to discriminate about 0.9 cyc/deg. Some
slight improvement occurred thereafter, but the maximum acuity achieved in
the deprived eye was only 2.2 cyc/deg, well below that of the non-deprived eye.

It remains to be demonstrated whether there is an upper age limit beyond
which monocular deprivation does not have an effect on animals dark-reared

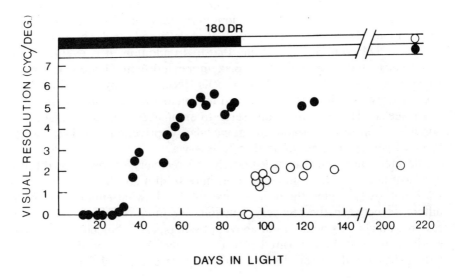

FIGURE 6 Recovery of vision in the deprived eye (open circles) and the normal eye (filled
circles) of a cat which was dark-reared until the age of 6 months, then immediately had one
eye sutured closed for a further three months.

for extensive periods of a year or more. Cynader (1977, unpublished observations) has recorded from the cortex of cats which had been dark-reared for up to ten months before being monocularly deprived and these animals still showed a substantial monocular deprivation effect.

Discussion

We have presented three lines of evidence which suggest that the visual system of the cat, tested behaviourally, retains plasticity for longer than was previously thought. These are (1) cats monocularly deprived until at least 4 months, i.e., beyond the limits of the typically defined critical period, show substantial recovery of visual function; (2) cats dark-reared to 6 months of age recover to essentially normal levels of visual acuity, although less recovery occurs in cats deprived for longer periods; (3) dark-reared cats are highly susceptible to the effects of monocular deprivation, even at the age of 6 months.*

The concept of a sensitive or critical period for monocular deprivation was developed initially to deal with *susceptibility* to the effects of such deprivation (Hubel and Wiesel, 1970). However, this notion was extended by Blakemore and Van Sluyters (1974) on the basis of their reverse suturing experiments. They demonstrated that cortical neurons were not only susceptible to deprivation but were also capable of regaining their connections with the deprived eye. These authors therefore characterized the critical period as "a time when the afferent connections of cortical cells are utterly plastic." Our results from cats which were monocularly deprived for four or five months suggest that this sensitive period is not entirely over by the age of 4 months.

The finding is not altogether unexpected. Several workers have observed some degree of recovery following long-term monocular deprivation (Ganz and Fitch, 1968; Dews and Wiesel, 1970; Rizzolatti and Tradardi, 1971; Ganz and Haffner, 1974; van Hof-van Duin, 1976). In addition, it is even evident from the reversal index calculated by Blakemore and Van Sluyters (1974, Fig. 3) which is based on the proportion of cells "recaptured" by the initially deprived eye, that reversibility has not declined completely to zero when reversal is done at 14 weeks. The present results serve to emphasize that substantial recovery is possible even after prolonged monocular deprivation, a point which has received perhaps less attention than it deserves.

The mechanisms which may underly the recovery of vision in the deprived eye have been discussed in detail elsewhere (Giffin and Mitchell, 1978) and we shall not dwell upon them here, except to draw attention to the marked difference between the recovery functions for monocularly deprived and dark-reared animals. As pointed out above, physiologically, binocularly deprived cats are in some ways less impaired than those which have been deprived monocularly (Wiesel and Hubel, 1965). This may be accounted for in terms of the

---

*We have also measured the effect of three months of monocular deprivation in a normally reared 4-month-old cat. Immediately following eye opening, there was a reduction of acuity in the deprived eye of about one octave. Following this there was a systematic recovery, with normal levels being approached after about three weeks.

binocular competition hypothesis. There is much evidence to suggest that there is an active competition between afferent fibres from each eye for synaptic space on cortical neurons (e.g., Hubel and Wiesel, 1965; Sherman et al., 1974; Blakemore et al., 1976). Thus, when a kitten has one eye sutured closed, that eye is placed at a competitive disadvantage and loses its cortical connectivity. In a binocularly deprived cat, neither eye is at an advantage. Here, each eye is capable of driving cortical neurons, but the response properties tend to be abnormal and resemble those of immature kittens. These differences in the initial state of the cortex immediately following the deprivation period suggest that the mechanisms of recovery might well be different in the two cases. In monocularly deprived cats it is probable that functional connections in the cortex first have to be reestablished before refinement of their receptive field properties can occur, whereas in binocularly deprived cats it appears that the connections are already present and only require experience for their further development.

From the results of the third experiment it is evident that one effect of dark-rearing, at least up to 6 months of age, is to maintain the visual system in a highly modifiable state. Further, it appears that there is a systematic decline in plasticity with increasing time in the light, as demonstrated by the finding that the monocular deprivation effect is somewhat less in dark-reared cats which have been allowed a period of binocular visual experience before unilateral eyelid suture.

The possibility that the visual system is simply "frozen" at an immature level by extended periods of dark-rearing is still a matter of conjecture. There are several lines of evidence to indicate that such a hypothesis is too simplistic. Physiologically, there is a degradation of response quality in dark-reared kitten cortex (Frégnac and Imbert, 1978). Also, our own results, which indicate an increasing time to the first signs of vision with longer deprivation periods, and the failure of cats deprived of vision for more than six months to achieve complete recovery, suggest that there are changes occurring in the absence of any visual experience. Nevertheless, the results of all the present experiments, when taken together, do suggest that the concept of a chronologically bound sensitive period terminating around the end of the third month does not do full justice to the plastic properties of the cat visual system.

Acknowledgements This research was supported by Grant No. A7660 from the National Research Council of Canada to D. E. Mitchell, and the work was carried out at the Department of Psychology, Dalhousie University.

References

Blakemore, C., and R. C. Van Sluyters (1974). Reversal of the physiological effects of monocular deprivation in kittens: further evidence for a sensitive period. J. Physiol. (Lond.) 237:195-216.

Blakemore, C., R. C. Van Sluyters, and J. A. Movshon (1976). Synaptic competition in the kitten's visual cortex. Cold Spring Harbor Symposia on Quantitative Biology 40:601-609.

Chow, K. L., and D. L. Stewart (1972). Reversal of structural and functional effects of long term visual deprivation in cats. Exp. Neurol. 34:409-433.

Cynader, M. (1977). Extension of the critical period in cat visual cortex. Presented at ARVO, Sarasota, Florida.

Cynader, M., N. Berman, and A. Hein (1976). Recovery of function in cat visual cortex following prolonged visual deprivation. Exp. Brain Res. 25:139-156.

Dews, P. B., and T. N. Wiesel (1970). Consequences of monocular deprivation on visual behaviour in kittens. J. Physiol. (Lond.) 206:437-455.

Frégnac, Y., and M. Imbert (1978). Early development of visual cortical cells in normal and dark-reared kittens: Relationship between orientation selectivity and ocular dominance. J. Physiol. (Lond.) 278:27-44.

Ganz, L., and M. Fitch (1968). The effect of visual deprivation on perceptual behavior. Exp. Neurol. 22:639-660.

Ganz, L., and M. E. Hafner (1974). Permanent perceptual and neurophysiological effects of visual deprivation in the cat. Brain Res. 20:67-87.

Giffin, F., and D. E. Mitchell (1978). The rate of recovery of vision after early monocular deprivation in kittens. J. Physiol. (Lond.) 274:511-537.

Hubel, D. H., and T. N. Wiesel (1970). The period of susceptibility to the physiological effects of unilateral eye closure in kittens. J. Physiol. (Lond.) 206:419-436.

Mitchell, D. E., M. Cynader, and J. A. Movshon (1977). Recovery from the effects of monocular deprivation in kittens. J. Comp. Neurol. 176:53-64.

Mitchell, D. E., F. Giffin, and B. Timney (1977). A behavioural technique for the rapid assessment of the visual capabilities of kittens. Perception 6:181-193.

Movshon, J. A. (1976). Reversal of the behavioural effects of monocular deprivation in the kitten. J. Physiol. (Lond.) 261:175-187.

Rizzolatti, G., and V. Tradardi (1971). Pattern discrimination in monocularly reared cats. Exp. Neurol. 33:181-194.

Sherman, S. M., R. W. Guillery, J. H. Kaas, and K. J. Sanderson (1974). Behavioural, electrophysiological and morphological studies of binocular competition in the development of the geniculo-cortical pathways of cats. J. Comp. Neurol. 158:1-18.

Timney, B., D. E. Mitchell, and F. Giffin (1978). The development of vision in cats after extended periods of dark-rearing. Exp. Brain Res. 31:547-560.

van Hof-van Duin, J. (1976). Early and permanent effects of monocular deprivation on pattern discrimination and visuomotor behavior in cats. Brain Res. 111:261- 276.

Wiesel, T. N., and D. H. Hubel (1963). Single-unit response in striate cortex of kittens deprived of vision in one eye. J. Neurophysiol. 26:1003-1017.

Wiesel, T. N., and D. H. Hubel (1965). Comparison of the effects of unilateral and bilateral eye closure on cortical unit responses in kittens. J. Neurophysiol. 28:1029-1040.

STUDIES OF THE CAT'S VISUAL SYSTEM

Lability of Directional Tuning
and Ocular Dominance of Complex Cells
in the Cat's Visual Cortex

P. HAMMOND

Department of Communication & Neuroscience
University of Keele
Keele, Staffordshire, England

Abstract Directional specificity and ocular dominance for motion of bar stimuli against stationary textured backgrounds, and for motion of the same random texture alone, were assessed in 62 complex cells from the infragranular layers of the striate cortex in normal adult cats, lightly anesthetized with N_2O/O_2 and pentobarbitone.

Directional bias for preferred versus opposite directions of motion was enhanced with texture; two-thirds of cells directionally *biased* for bars were directionally *selective* for texture.

A majority of cells (52) showed substantial differences in preferred directions for bar and texture motion. Tuning for texture was typically broader than for bars; 22 cells showed bimodal tuning for texture, with depressed sensitivity in directions preferred for bars. Bar tuning was frequently broader on the flank of the tuning curve nearest the preferred direction for texture. Many cells, especially those with large receptive fields, were more responsive to texture than to bar motion.

Eleven cells showed interocular differences in sharpness and bias of directional tuning for texture; bar/texture tuning relationships were otherwise replicated in each eye.

Ocular dominance for bars and texture was compared in 31 cells; 14 showed stimulus-dependent shifts of up to three ocular dominance groups, with reversal of eye preference in three cases. There were no trends favoring ipsilateral or contralateral inputs, or increased binocularity for texture motion.

The results are interpreted as evidence that directional and orientational sensitivity are mediated by separate mechanisms, not necessarily in register for the two eyes.

163

Introduction

The extreme plasticity of the developing nervous system in early postnatal life is well established. The susceptibility of the visual system, in particular, to deprivation and to selective visual experience within the "critical period" is especially well documented. In cat the critical period spans the first three months of life, reaching its height some four weeks after birth, beyond which it is generally assumed that the organization of the adult visual system, whether arising from normal visual development or as the result of environmental deprivation, is immutable by subsequent visual experience. Thus, properties such as the directional and orientational preferences of cells in the adult visual cortex are held to be invariant (Andrews et al., 1975; Hammond et al., 1975). The same is true for ocular dominance: in the visual cortex the pattern of alternating ocular dominance bands established in early life, whether resulting from normal or abnormal visual experience, is not further modifiable in the adult (e.g., Wiesel and Hubel, 1963, 1965; Hubel and Wiesel, 1970; Wiesel et al., 1974; Hubel et al., 1975, 1976; LeVay et al., 1975, 1978).

Such conclusions are, however, based on a restrictive set of stimuli (bars and edges of variable orientation). Moreover, the "orientation" tuning of cortical cells is usually assessed with *moving* bar stimuli and, as pointed out by Henry et al. (1974), direction and orientation cannot be dissociated since, for long stimuli, the effective direction of motion is necessarily orthogonal to orientation. We have previously established that complex cells (especially those in the infragranular layers, subsequently referred to as "deep-layer" complex cells), but not simple cells, are sensitive to texture motion (Hammond and MacKay, 1975a,b, 1976, 1977). Motion of a field of random texture (static visual noise: for example, see Fig. 1 of Hammond and MacKay, 1977) is used here to investigate directional tuning in deep-layer complex cells, since it lacks inherent orientation and overcomes some of the summative limitations encountered with moving single spots.

Monocular and interocular comparisons of the tuning and ocular dominance of binocularly driven complex cells for bar and for texture motion indicate that the directional and orientational sensitivity of a high proportion of such cells are mediated by separate inputs. Moreover, in normally reared adult cats, beyond the critical period, the directional preferences of single cells, interocular differences in directional preferences, and even ocular dominance may be stimulus-dependent, being profoundly influenced by the configuration of the visual input (Groos, Hammond, and MacKay, 1976; Hammond and MacKay, 1977; Hammond, 1978c,d,e,f).

Methods

Full details of methods are given elsewhere (Hammond and MacKay, 1977; Hammond, 1978e,f). Complex cells were recorded from the infragranular layers in the striate cortex of adult cats, lightly anesthetized with 72.5%:27.5% $N_2O:O_2$ supplemented with intravenous pentobarbitone ($1mg \cdot kg^{-1} \cdot hr^{-1}$), with monitoring of EEG, blood pressure or ECG and pulse, end-tidal CO_2, temperature and unitary discharge rate (Hammond, 1978a,b). Acute (Hammond and MacKay, 1977) or semi-chronic (Hammond, 1978a,b) preparations with recovery between recording sessions were used. Eye preparation was conventional (immobilization during recording (i.v. gallamine triethiodide, $7-10mg \cdot kg^{-1} \cdot hr^{-1}$), mydriasis, 5-mm diameter artificial pupils, neutral contact lenses 5-mm diameter artificial pupils, neutral contact lenses with appropriate focal correction, average luminance 0.9 log $cd \cdot m^{-2}$). Recording, spike processing, and monitoring were also conventional, during normal penetrations through A17 (verified histologically) with 4M-NaCl micropipettes. Receptive fields of cells were within 12° below the area centralis, close to the vertical meridian.

Stimuli were presented on a CRT display at 50 cm, and were either a 10° × 10° frame of static visual noise moved back and forth in different directions, or a long dark bar of optimal width, moving against the same stationary textured background. Frame orientation changed with direction but frame outline remained stationary. Velocity and excursion of bar or texture motion were always matched (see also MacKay and Yates, 1975; Hammond and MacKay, 1977), and specified by direction: motion upwards (0°), to the right (90°), downwards (180°), to the left (270°). Monocular tuning curves were obtained separately for bar and texture motion for the dominant eye and, where possible, also for the other eye. Every block of 10 or 20 trials provided data for two directions of motion, 180° apart; direction was changed systematically in 30° or 40° steps; intervening points were taken in reverse order, to take account of response variability and hysteresis, and additional points were established close to the preferred direction.

Polar diagrams represent response vectorially, after subtracting equivalent resting discharge for a comparable period with only the stationary textured background present. In conventional plots, firing is indicated above and below the resting discharge. For comparison, tuning curves for bar and texture motion are normalized, equivalent firing in the preferred direction being indicated in $imp \cdot sec^{-1}$.

Results

Directional tuning for bar and texture motion Reliable tuning curves for both bar and texture motion were obtained for the dominant eye receptive field of 62/84 deep-layer complex cells.

Consistent with the sharpening of orientation tuning associated with increase in line length in the majority of cells (Henry et al., 1974a,b; Gilbert, 1977; Hammond and Andrews, 1978a,b), tuning for texture motion was almost always appreciably broader than for bar motion (Fig. 1(A)). In cells asymmetrically tuned for bar motion (Hammond and Andrews, 1978a), bar tuning was broader on the flank of the tuning curve nearest the preferred direction for texture motion (Fig. 1(B)).

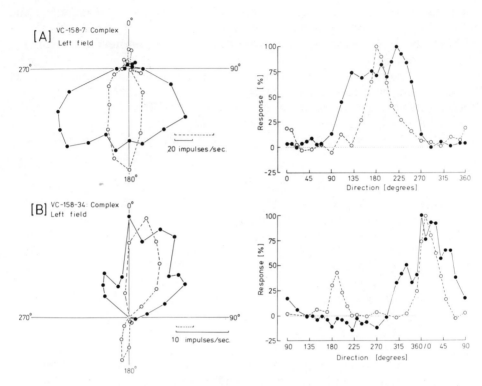

FIGURE 1 Differences in directional tuning of complex cells for texture motion (filled circles) and for bar motion against the same stationary textured background (open circles). Note the broader tuning (A) and enhanced directional bias (B) for texture motion; also, the asymmetrical bar tuning in (B) with the broader flank of the tuning curve towards the preferred direction for noise. Average of ten trials; resting discharge indicated by dotted line. Polar diagrams and graphs normalized for comparison: optimal response for bars and noise, respectively, 50.9 and 113.3 imp·sec^{-1} in (A), 58.7 and 18.9 imp·sec^{-1} in (B).

Of 72 cells, two-thirds (49) were *directionally biased* for bar motion, responding differentially to preferred and opposite ("null") directions; one-third (23) were *directionally selective* with no response or with firing suppressed in the null direction. Texture motion invariably enhanced the directional bias, evoking directionally selective responses from two-thirds (33) of the cells directionally biased for bar motion (Fig. 1(B)).

Only 10/62 cells showed similar preferred directions and broadness of tuning for bar and texture motion. The majority showed substantial differences in preferred directions for bar and for texture motion, typically 40-50° (Fig. 2(A)) and exceptionally 90°. Cells were either *unimodally* tuned for texture motion, with a single lobe in the tuning curve corresponding to each lobe of the bar tuning curve (40/62 cells, e.g., Fig. 2(A)), or *bimodally* tuned for texture motion, with a pair of lobes distributed more or less symmetrically about each

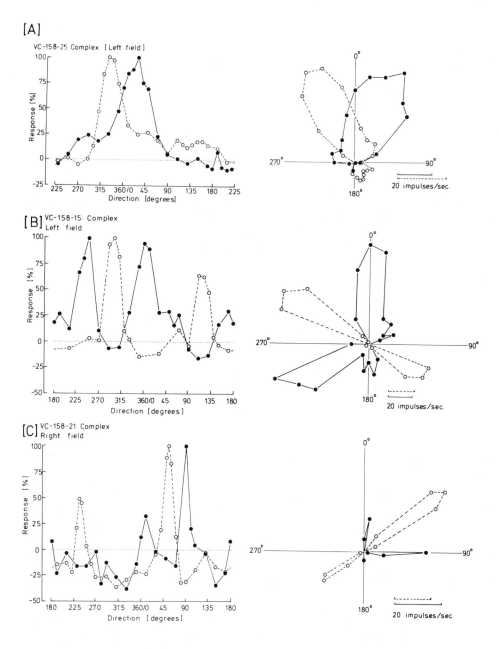

FIGURE 2 Dissimilar directional tuning for texture (filled circles) and bar (open circles) motion: complex cells unimodally (A) and bimodally (B, C) tuned for texture motion. In bimodally tuned cells, note the depression of sensitivity to texture motion in directions preferred for bar. Conventions as in Fig. 1. Optimal response for bar and texture motion, respectively: 59.8 and 40.5 imp·sec^{-1} in (A); 72.8 and 81.3 imp·sec^{-1} in (B); 54.1 and 41.5 imp·sec^{-1} in (C).

lobe of the bar tuning curve (22/62 cells; Fig. 2(B), (C)). In the latter group, sensitivity to texture motion was depressed in directions of maximal sensitivity for bar motion.

The distribution of differences in preferred directions for bar and texture motion are shown in Fig. 3 for the total sample of cells, and for the unimodally and bimodally tuned groups. With the exception that, amongst bimodally tuned cells there are no representatives in the small-difference or 80-90° difference bins, the distribution of differences is similar to that for unimodally tuned cells.

Since the orientation of the texture frame changed systematically with direction (although the frame outline remained stationary), the following controls established that neither the frame outline nor spurious orientational components of the sample of texture within the frame contributed to the observed differences in directional tuning: (a) regularly changing the texture sample within the frame; additional trials in which (b) frame orientation was held constant whilst varying direction, (c) direction was held constant whilst varying frame orientation. Moreover, bar tuning was not influenced by the presence of the stationary textured background.

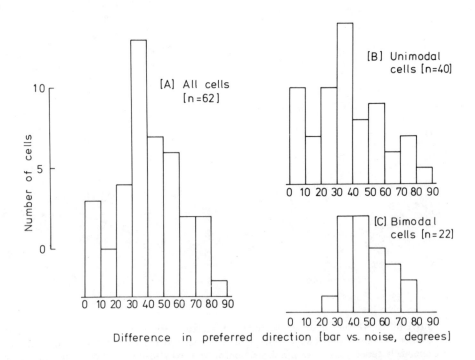

FIGURE 3 Distribution of differences in preferred directions for texture and bar motion (A) all cells, (B) cells unimodally, and (C) bimodally tuned for texture motion. No distinction is made between clockwise and counterclockwise differences.

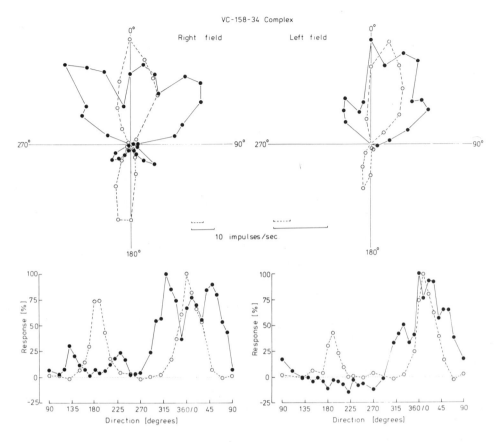

FIGURE 4 Tuning for texture (filled circles) and bar (open circles) motion for each monocular receptive field of a complex cell, showing interocular differences in directional bias and in distribution and broadness of tuning for texture. Details as in Fig. 1. Optimal response for bar and texture motion, respectively: right eye, 83.6 and 46.1 imp·sec⁻¹; left eye, 58.7 and 18.9 imp·sec⁻¹.

Interocular comparisons of directional tuning Interocular comparisons of tuning for bar and texture motion were established in 31 cells. With two exceptions (one cell poorly responsive to bar motion through one eye, the other with opposite preferred directions for the two eyes), bar tuning was similar through either eye. Tuning for bar and texture motion was generally similar for each eye, or any differences in tuning in one eye were replicated in the other. However, 11 cells showed interocular differences in sharpness of tuning and/or directional bias for texture motion. Figure 4 gives one such example, in which texture motion induced sharper tuning through the more weakly driving (left) eye, with enhanced directional bias.

Ocular dominance for bar and texture motion was compared for the same group of 31 cells. Figure 5(A) shows the ocular dominance distribution for the

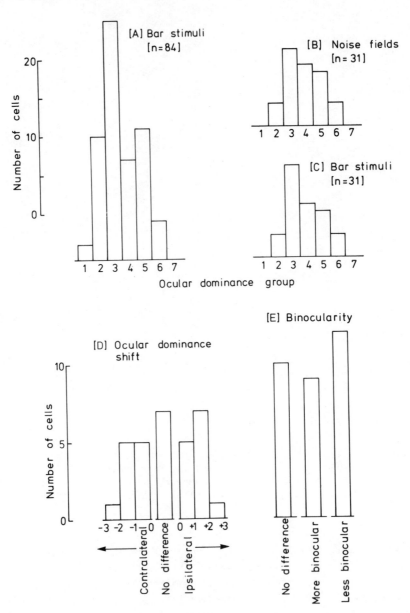

FIGURE 5 Ocular dominance distribution and stimulus-dependent changes in ocular dominance of deep-layer complex cells. Ocular dominance distribution of 84 cells for bar motion (A), in a subset of 31 cells for texture (B), and bar motion (C). Ocular dominance classification: groups 1 and 7, contralateral and ipsilateral monocular drive, respectively; groups 2-6, binocular (groups 2-3, contralateral dominance; groups 5-6, ipsilateral dominance; group 4, equal drive from either eye). For the same subset of cells, (D) gives the distribution of ocular dominance *shifts* of individual cells for texture motion compared with bar motion; (E) indicates the resultant changes in binocularity for texture.

total sample of 84 cells for bar motion, with the anticipated preponderance of cells with strong binocular drive (groups 3-5). The superficially similar distributions for the subset of 31 cells are shown in Fig. 5(B) and (C) and, as can be seen from Fig. 5(D), they conceal substantial shifts in dominance of up to three ocular dominance groups in almost half the cells (14/31). In three of the cells with good binocular drive, the shift led to a *reversal* of eye preference although, overall, there were no obvious trends in favour of the contralateral or ipsilateral eye.

In view of the disparate receptive fields of many cortical cells, it was of interest to see whether the shifts in ocular dominance might reflect increased binocularity for a positionally non-specific stimulus such as texture, compared with a positionally specific bar stimulus. Regrouping the data of Fig. 5(D), according to whether shifts in ocular dominance resulted in cells being more or less binocular for texture motion compared with bar motion (Fig. 5(E)), shows that this expectation was not fulfilled; there was actually an insignificant trend in the sense of decreased binocularity for texture.

Discussion

Monocular differences in tuning, and in directional bias for preferred and null directions, for bar and texture motion are strong evidence for direction- and orientation-sensitivity being mediated by different mechanisms. The narrower tuning for bars compared with texture motion is not unexpected, in view of the known sharpening of orientation tuning which is frequently associated with increase in line length. This and the depressed sensitivity of bimodally tuned cells to texture motion in the vicinity of the preferred directions for bar stimuli indicate inhibitory interaction between the directional and orientational mechanisms.

To some extent, texture motion must influence the orientational mechanism, according to the extent of correlation between neighbouring texture-elements in any random sample. Likewise, bar motion must also influence the directional mechanism. These factors are consistent with the broadening of bar tuning towards the preferred direction for texture motion cells dissimilarly tuned for bar and texture motion, and might also account for the asymmetric tuning frequently observed for bar motion (Hammond and Andrews, 1978a,b).

The presence of *systematic* differences in preferred directions for bar and texture motion within an orientation column remains an open question. There was some indication of common differences in preferred directions from cell to cell, but abrupt transitions and profound changes in texture sensitivity between closely neighbouring cells were also seen.

The interocular differences in tuning of some cells for texture motion and the marked shifts in ocular dominance of many cells for bars and texture are also consistent with dissociation of orientation- and direction-sensitive inputs to complex cells and indicate that there is frequently a stimulus-dependent mismatch between each monocular input of such binocularly driven cells.

Whether the same holds for the more weakly noise-sensitive complex cells of the supragranular layers and for noise-insensitive simple cells remains to be determined.

The application of bicuculline, which blocks GABA-mediated intracortical inhibition, differentially influences the directional and orientational sensitivity of complex cells found primarily in the infragranular layers of the adult cat (Sillito, 1975, 1977). Beyond the critical period, the same procedure changes the ocular dominance of almost half the cells in the striate cortex (Duffy et al., 1976; Patel and Sillito, 1978); likewise, in monocularly deprived cats, there is some recovery of cortical influence from the deprived eye in the adult, following enucleation of the experienced eye (Kratz et al., 1976). These results uniformly demonstrate the importance of intracortical inhibition for ocular dominance and for directional and orientational sensitivity, and suggest that the stimulus-dependence of such parameters in the adult related to the extent to which different classes of stimuli (e.g., bars and texture) invoke intracortical inhibitory mechanisms.

Acknowledgements The support and encouragement of D. M. MacKay who, together with G. A. Groos, Laboratorium voor Fysiologie en Fysiol Fysica, Leiden, and H.-Chr. Nothdruft (supported by E.T.P.B.B.R. Traineeship), Max- Planck-Institut für Biophysikalische Chemie, Göttingen, collaborated in early experiments, and the technical assistance of D. J. Scott, are greatly appreciated. A Beckman LB-2 gas analyzer and Kowa RC-2 fundus camera were provided by generous grants G975/665/N and G977/774/N from the Medical Research Council to P. H. Approval for the reproduction of illustrations from the Physiological Society (Figs. 1-3) and the publishers and editors of Experimental Brain Research (Figs. 4-5) is gratefully acknowledged.

References

Andrews, D. P., P. Hammond, and C. R. James (1975). Absence of spontaneous variability of orientational and directional tuning in cat visual cortical cells. J. Physiol. 251:49-50P.

Duffy, F. H., S. R. Snodgrass, J. L. Burchfiel, and J. L. Conway (1976). Bicuculline reversal of deprivation amblyopia in the cat. Nature 260:256-257.

Gilbert, C. D. (1977). Laminar differences in receptive field properties of cells in cat primary visual cortex. J. Physiol. 268:391-421.

Groos, G. A., P. Hammond, and D. M. MacKay (1976). Polar responsiveness of complex cells in cat striate cortex to motion of bars and of textured patterns. J. Physiol. 260: 47-48P.

Hammond, P. (1978a). On the use of nitrous oxide/oxygen mixtures for anaesthesia in cats. J. Physiol. 275:64P.

Hammond, P. (1978b). Inadequacy of nitrous oxide/oxygen mixtures for maintaining anaesthesia in cats: satisfactory alternatives. Pain 5:143-151.

Hammond, P. (1978c). Lability of ocular dominance of complex cells in cat striate cortex. Exp. Brain Res. 32:R18.

Hammond, P. (1978d). Directional tuning of complex cells in cat striate cortex. Neurosci. Letters, Suppl. 1:S373.

Hammond, P. (l978e). Directional tuning of complex cells in area 17 of the feline visual cortex. J. Physiol. 285:479-491.

Hammond, P. (1979). Stimulus-dependence of ocular dominance and directional tuning of complex cells in area 17 of the feline visual cortex. Exp. Brain Res. 35 (in press).

Hammond, P., and D. P. Andrews (1978a). Orientation tuning of cells in areas 17 and 18 of the cat's visual cortex. Exp. Brain Res. 31:341-351.

Hammond, P., and D. P. Andrews (1978b). Collinearity tolerance of cells in areas 17 and 18 of the cat's visual cortex: relative sensitivity to straight lines and chevrons. Exp. Brain. Res. 31:329-339.

Hammond, P., D. P. Andrews, and C. R. James (1975). Invariance of orientational and directional tuning in visual cortical cells of the adult cat. Brain Res. 96: 56-59.

Hammond, P., and D. M. MacKay (1975a). Differential responses of cat visual cortical cells to textured stimuli. Exp. Brain Res. 23:427-430.

Hammond, P., and D. M. MacKay (1975b). Responses of cat visual cortical cells to kinetic contours and static noise. J. Physiol. 252:43-44P.

Hammond, P., and D. M. MacKay (1976). Interrelations between cat visual cortical cells revealed by use of textured stimuli. Exp. Brain Res., Suppl. 1:397-402.

Hammond, P., and D. M. MacKay (1977). Differential responsiveness of simple and complex cells in cat striate cortex to visual texture. Exp. Brain Res. 30:275-296.

Henry, G. H., P. O. Bishop, and B. Dreher (1974a). Orientation, axis, and direction as stimulus parameters for striate cells. Vision Res. 14:767-778.

Henry, G. H., B. Dreher, and P. O. Bishop (1974b). Orientation specificity of cells in cat striate cortex. J. Neurophysiol. 37:1394-1409.

Hubel, D. H., and T. N. Wiesel (1970). The period of susceptibility to the physiological effects of unilateral eye closure in kittens. J. Physiol. 206:419-436.

Hubel, D. H., T. N. Wiesel, and S. LeVay (1975). Functional architecture of area 17 in normal and monocularly deprived macaque monkeys. Cold Spring Harb. Symp. Quant. Biol. 40: 581-590.

Hubel, D. H., T. N. Wiesel, and S. LeVay (1976). Columnar organization of area 17 in normal and monocularly deprived macaque monkeys. Exp. Brain Res., Suppl. 1:356-361.

Kratz, K. E., P. D. Spear, and D. C. Smith (1976). Post-critical period reversal of effects of monocular deprivation on striate cortex cells in the cat. J. Neurophysiol. 39: 501-511.

LeVay, S., D. H. Hubel, and T. N. Wiesel (1975). The pattern of ocular dominance columns in macaque visual cortex revealed by reduced silver stain. J. Comp. Neurol. 159: 559-576.

LeVay, S., M. P. Stryker, and C. J. Shatz (1978). Ocular dominance columns and their development in layer IV of the cat's visual cortex: a quantitative study. J. Comp. Neurol. 179:223-244.

MacKay, D. M., and S. R. Yates (1975). Textured kinetic stimuli for use in visual neurophysiology: an inexpensive and versatile electronic display. J. Physiol. 252:10-11p.

Patel, H. H., and A. M. Sillito (1978). Inhibition and the normal ocular dominance distribution in cat visual cortex. J. Physiol. 280:48-49P.

Sillito, A. M. (1975). The contribution of inhibitory mechanisms to the receptive field properties of neurones in the striate cortex of the cat. J. Physiol. 250:305-329.

Sillito, A. M. (1977). Inhibitory processes underlying the directional specificity of simple, complex, and hypercomplex cells in the cat's visual cortex. J. Physiol. 271:699-720.

Wiesel, T. N., and D. H. Hubel (1963). Single-cell responses in striate cortex of kittens deprived of vision in one eye. J. Neurophysiol. 26:1003-1017.

Wiesel, T. N., and D. H. Hubel (1965). Comparison of the effects of unilateral and bilateral eye closure on cortical unit responses in kittens. J. Neurophysiol. 28:1029-1040.

Wiesel, T. N., D. H. Hubel, and D. M. K. Lam (1974). Autoradiographic demonstration of ocular-dominance columns in the monkey striate cortex by means of transneuronal transport. Brain Res. 79:273-279.

Intrinsic Connectivity in Area 18 of the Cat

GÜNTHER NEUMANN

Max-Planck-Institut für Psychiatrie
Munich, West Germany

Abstract In order to determine the organization of afferent and intrinsic connectivity in area 18 of the cat, latencies and scatter of single-unit responses to electrical stimulation of the primary afferents were analyzed. The units' laminar positions were assessed by evaluating current source densities along the same recording track on the way back to the cortical surface.

Mono-, di-, and trisynaptic responses are discernible in the latency histogram after stimulation of the lateral geniculate nucleus. The monosynaptic responses can be subdivided into two classes according to their latency scatter. The low scatter of the disynaptic responses suggests that they arise almost exclusively from the low scatter monosynaptic activity.

The most prominent feature of layers I-III is the preponderance of disynaptic responses which are followed by trisynaptic activity in about 30% of the cells. The most striking properties of the infragranular layers are the high percentage of multispike responses and the large scatter of latencies between cells.

The wiring diagram that accounts for the temporal and spatial properties of the analyzed responses shows strong and local connections from the input stage cells in layer IV to cells in the supragranular layers and in layer V. In addition, the supragranular and, probably also, the infragranular layers project onto themselves forming positive feedback loops.

Intrinsic connectivity in cat visual cortex was recently explored by Mitzdorf and Singer (1978) using the current source density method which is based upon the evaluation of evoked potentials. The present study was intended to complement these results with single-unit data. It was confined to area 18 because of the homogeneous properties of the primary afferents of this area.

Single-unit responses following electrical stimulation of the lateral geniculate nucleus (LGN) and the optic chiasm (OX) were analyzed. Unit activity was recorded with glass micropipettes filled with potassium citrate whose impedances at 300 Hz ranged from 10 to 30 megohms. Histograms for each

175

cell were compiled from the responses to 50 double shocks. The two shocks of each pair were separated by 20 msec. The latency and scatter of individual unit discharges were subsequently determined from the stored histograms. In cells responding with several spikes to a single stimulus, every action potential was counted as an independent response. A number of tests were performed to distinguish between activity from nerve cells and axons, the latter being discarded from the analysis. The main criterion for discrimination between fibers and cells was the reaction to current injected through the recording electrode. Profiles which exhibited a marked change in spontaneous activity following current injection were classified as cells. Profiles which did not react to injected current but showed the typical injury discharge after impalement were also regarded as cells. All profiles which did not fulfill the criteria above were classified as fibers and discarded.

The depth distribution of the units with respect to the cortical layers was assessed by measuring current source densities (Mitzdorf and Singer, 1978) along the same recording track on the way back to the cortical surface. The analysis presented here is based upon data obtained only from those recording tracks which were sufficiently linear to allow superposition of responses obtained at various depths from different penetrations. In these seven tracks, which were selected from four (out of thirteen) experiments, a total of 172 cells were analyzed. To correct for variations in cortical thickness and penetration angle between experiments, the tracks were expanded or compressed until the respective layers identified with the current source density method came into register with one another. The variation in mean latencies, produced by changes in the position of the stimulating electrodes between experiments was also corrected (maximal correction 0.1 msec for LGN-stimulation, 0.4 msec for OX-stimulation).

Figure 1(A) shows the histogram of response latency increase that occurs when the site of stimulation is changed from LGN to OX. The monomodal distribution and the short absolute values of this latency increase indicate that the activity relayed to area 18 is conveyed by a homogeneous group of fast-conducting retinal axons. This finding is in accordance with earlier results of Stone and Dreher (1973) and Tretter et al. (1975) who have shown that the main input to area 18 is provided by afferents of the Y-type. Also, the geniculo-cortical fibers projecting to area 18 form a homogeneous group with respect to conduction velocity (Cleland et al., 1971; Hoffmann et al., 1972; Singer and Bedworth, 1973; Stone and Dreher, 1973). The latency histogram of shortest latencies after LGN stimulation shown in Fig. 1(B) exhibits two prominent peaks; a third peak is slightly indicated. Since the afferent fibers have uniform conduction velocities, these different peaks most likely represent responses recorded at different synaptic distances from the stimulation site. In accordance with other authors (Tretter et. al., 1975) they are interpreted as reflecting mono-, di-, and trisynaptic responses, respectively.

On the left side of Fig. 2 the latency scatter of the individual responses (expressed by the standard deviations) is plotted separately for monosynaptic,

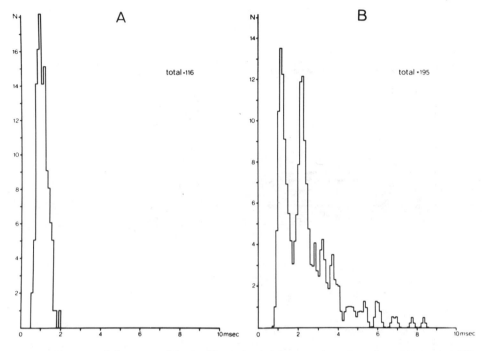

FIGURE 1 (A) Histogram of differences in shortest latencies after OX- and LGN-stimulation. (B) Histogram of shortest latencies after LGN-stimulation. The three peaks are interpreted as the latency ranges of mono-, di-, and trisynaptic responses, respectively. The latency histogram is smoothed by three-point filtering.

disynaptic, and later responses. While the standard deviations of the large majority of disynaptic responses form a rather homogeneous block between 0.05 and 0.16 msec, the standard deviations of the monosynaptic responses form a bimodal distribution indicating two classes of monosynaptic responses. The distributions of low-scatter and high-scatter monosynaptic responses within supragranular, granular, and infragranular layers are different, as is shown on the right side of Fig. 2: the low-scatter monosynaptic responses are almost exclusively from cells situated in layer IV, whereas the cells with high-scatter monosynaptic responses are found in all of the three subdivisions with a slight preference for the supragranular layers. It is unlikely, therefore, that the bimodal distribution of monosynaptic standard deviations is merely due to a sampling artifact.

The structural basis for the differentiation of monosynaptic responses into two classes is not clear yet. There is a marked tendency for multispike responses in cells showing a high-scatter monosynaptic action potential, whereas cells exhibiting a low-scatter monosynaptic action potential more often have a single response. This might indicate a higher degree of excitatory convergence, both mono- and polysynaptic, onto cells with high-scatter monosynaptic responses. Since a high degree of excitatory convergence is thought to be characteristic of pyramidal cells (Van Essen and Kelly, 1973; Singer et al.,

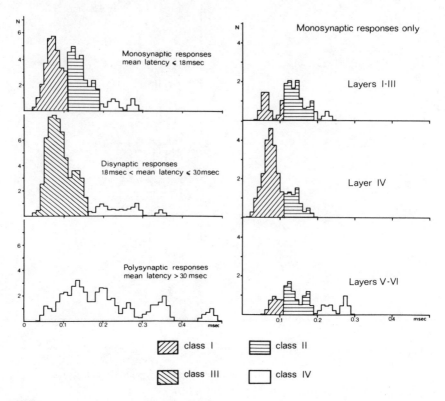

FIGURE 2 Left side: Frequency distribution of standard deviations of monosynaptic, disynaptic, and later responses. Right side: Frequency distribution of standard deviations of monosynaptic responses found in supragranular layers, layer IV, and infragranular layers. Abscissa: Standard deviations of the responses. Ordinate: Number of responses showing the respective standard deviation. All distributions are smoothed by three-point filtering. Classes I-III are marked by different hatchings.

1975), it might be speculated that the low-scatter monosynaptic responses are from stellate cells, while the high-scatter monosynaptic responses are from pyramidal cells. This is further substantiated by the finding that the low-scatter monosynaptic responses occur almost exclusively in layer IV where most stellate cells are located.

On the basis of their latencies and scatter, the responses were divided into four classes. Class I contains the low-scatter monosynaptic responses with standard deviations up to 0.1 msec. The second class is formed by the high-scatter monosynaptic activity with standard deviations up to 0.18 msec. Class III contains the bulk of disynaptic responses with low scatter, i.e., with standard deviations up to 0.16 msec. All responses that do not fulfill the criteria of classes I-III are grouped together in class IV.

The scatter of class III disynaptic responses is only slightly higher than that of low-scatter monosynaptic activity. Since it is hard to imagine how convergence of responses showing high scatter could produce low-scatter responses, it is very likely that the class III disynaptic responses arise almost exclusively from the low-scatter monosynaptic activity. In addition, the monosynaptic cells whose axons converge onto class III disynaptic cells should be located rather close to each other, because the low-latency scatter of disynaptic responses does not allow for convergence of fibers running over significantly different conduction distances.

The safety factor of transmission is significantly higher for low-scatter activity (mean value for 26 cells is 44.2 ± 9.3 spikes per 50 stimuli) than for high-scatter activity (29.5 ± 15.0, .IN $= 31$), suggesting a stronger coupling of synapses mediating class I responses. The respective values for class III disynaptic activity are 38.3 ± 14.6, .IN $= 62$. This indicates a high safety factor of transmission also from class I monosynaptic cells to their disynaptic targets.

Figure 3 shows how the responses of the different classes are distributed over cortical depth. Layer IV, as determined from current source density measurements, is indicated on the left margin. No attempt has been made to determine the exact laminar borders within the supra- and infragranular subdivisions. For simplification, the upper and lower halves of the supra- and infragranular segments will be referred to as layers II and III, and layers V and VI, respectively. The different symbols indicate earliest responses of the different classes.

The bulk of class I low-scatter monosynaptic responses is located in layer IV, whereas the class II high-scatter monosynaptic responses concentrate in layers III, IV, and VI. This distribution of monosynaptic responses correlates well with LeVay and Gilbert's (1976) distribution of afferent terminals originating from geniculate layers A and A_1. The bulk of disynaptically responding cells lies in layers I-III. These cells are dispersed throughout the supragranular layers, but are densest in layer III. Some cells exhibiting disynaptic responses are also found in layers IV and V.

In the supragranular layers, virtually all cells exhibit class III disynaptic responses, which, in about 30% of the cells, are followed by responses in the trisynaptic latency range. The current source density data (Mitzdorf and Singer, 1978) suggest that these trisynaptic responses are mediated by the supragranular disynaptic activity. Together with the present finding, that di- and trisynaptic responses occur in the same cells, this suggests that the group of supragranular cells projects onto itself forming a positive feedback loop.

The most prominent features of the infragranular layers are the high percentage of cells responding with several spikes and the large variation of latencies in the late responses.

The right half of Fig. 3 shows the responses to the second stimulus delivered 20 msec after the first. Onset latencies and rise times of electrically evoked IPSP's (e.g., Singer et al., 1975; Tretter et al., 1975) suggest that at that time the inhibition induced by the first stimulus is nearly maximal. In the upper

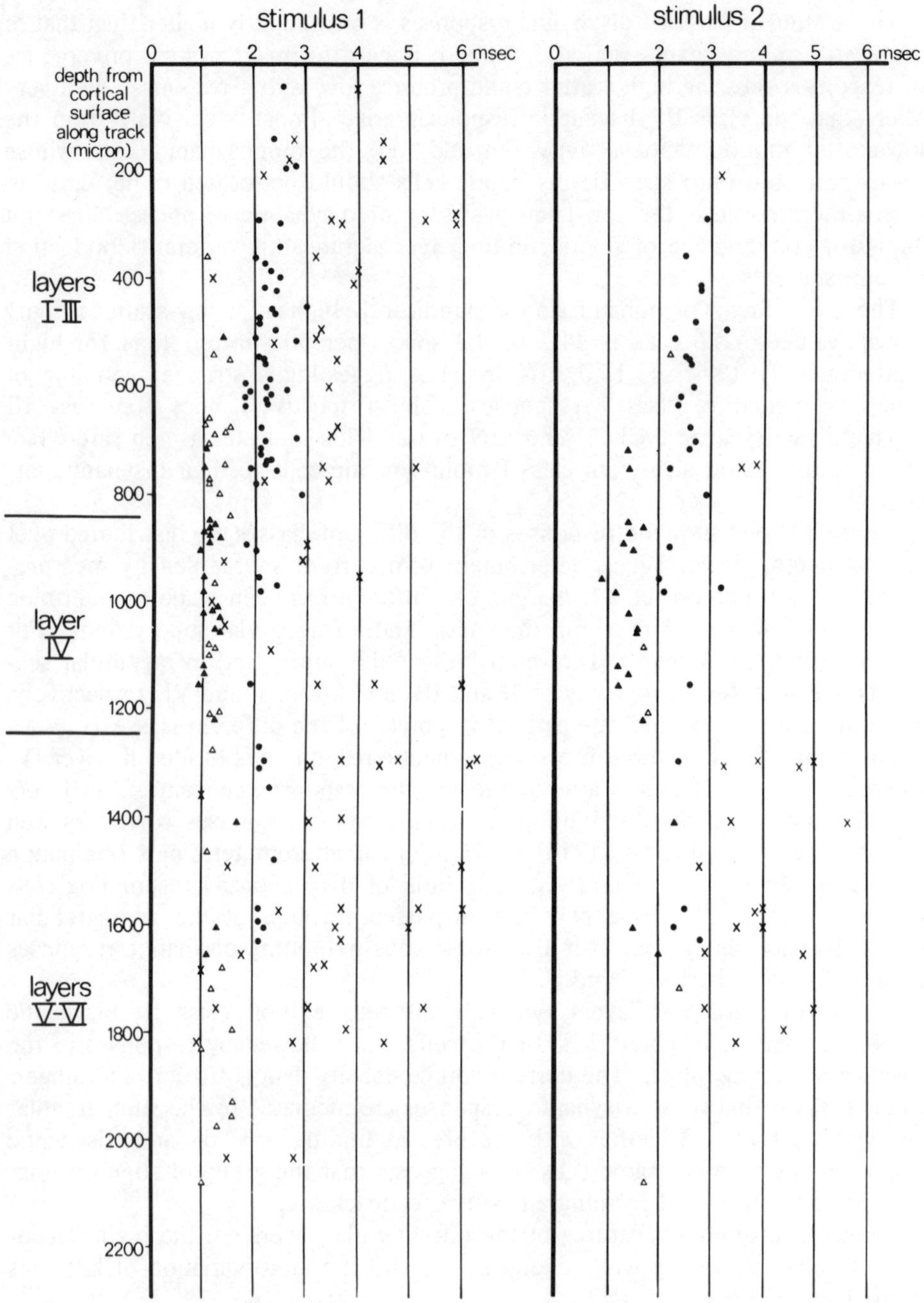

FIGURE 3 Distribution of earliest responses of classes I-IV over depth from the cortical surface. Filled triangles: Class I low-scatter monosynaptic responses. Open triangles: Class II high-scatter monosynaptic responses. Filled circles: Class III disynaptic responses. Crosses: Class IV, all other responses. Abscissa: Shortest latencies after LGN-stimulation. Ordinate: Depth from the cortical surface along recording track. Left side: Responses to first stimulus. Right side: Responses to second stimulus. The responses to the second stimulus were classified according to the respective responses.

layers, responses to the second stimulus are obtained essentially from cells that respond with low-scatter mono- or disynaptic activity to the first stimulus; virtually no responses are elicitable from cells showing only high-scatter monosynaptic or trisynaptic responses to the first stimulus. By contrast, in the infragranular layers a higher percentage of reactive cells responds with more than one spike to the second (54%) than to the first stimulus (45%). Similarly, the mean number of responses per reactive cell is not reduced either; it is 1.8 both in response to the first and to the second stimulus. This suggests that in infragranular layers the excitatory drive is counterbalanced less by inhibition.

There are late responses in the infragranular layers, particularly after the second stimulus, which can hardly be accounted for by activity arriving from outside the infragranular layers: only very few responses in the supragranular and granular layers precede that late activity by a time interval compatible with monosynaptic transmission. These responses are thus likely to arise from other infragranular responses. This finding, taken together with the fact that a large percentage of infragranular cells respond with multiple spikes to a single stimulus, suggests that the majority of trisynaptic and later responses in the infragranular layers is mediated via positive feedback loops inside these layers. It cannot be ruled out, however, that at least part of the late infragranular responses are induced by supragranular disynaptic activity. A projection from supragranular layers to layer V, which could mediate such an influence, has been suggested by several authors (e.g., Nauta, Butler, and Jane, 1973; Lund and Boothe, 1975). In order to account for the considerable and differing time lags of infragranular responses in relation to supragranular disynaptic responses, one has to assume that the descending fibers disperse in horizontal directions making synaptic contacts at various distances. If 1 msec is allowed for one synaptic delay, a maximum of about 1.5 msec has to be attributed to conduction time for the very late responses to the second stimulus. The mean value of standard deviations of late infragranular responses with latencies between 4 and 8 msec (first stimulus) is 0.25 ± 0.14 msec, .IN = 10. This scatter is smaller than one would expect, if fibers converged onto single infragranular cells irrespective of their conduction distances. This finding suggests that only fibers of rather similar lengths contribute to a single response. Moreover, since infragranular cells often exhibit several responses to a single stimulus, selective convergence of several groups of fibers running over different distances would have to be assumed in order to attribute late infragranular responses to the influence of supragranular activity. In view of these considerations the simplest assumption is that the infragranular response pattern results from positive feedback loops inside the infragranular layers. Further work is needed to determine the exact extent of supragranular influences onto infragranular layers.

Figure 4(A) represents in a summary diagram the afferent and intrinsic connections of area 18 that are required to account for the temporal and spatial distribution of the responses identified in this study. The afferent fibers are coupled with high safety factor to the input stage cells in layer IV, which in turn are strongly and locally connected to disynaptic cells in the supragranular layers. These cells, which are dispersed throughout layers I-III, with greatest density in

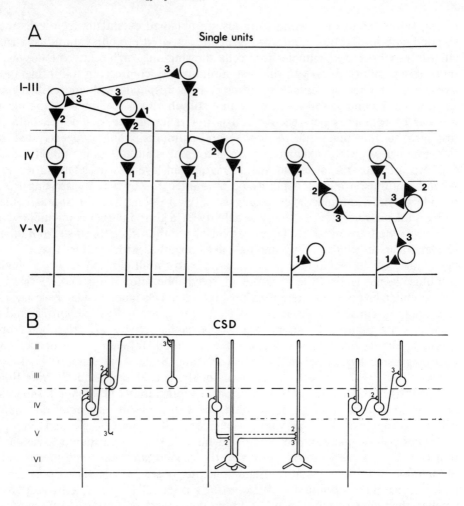

FIGURE 4 (A) Intrinsic wiring scheme suggested by single-unit analysis. Strongly coupled synapses are indicated by large synapse symbols. Left half depicts intrinsic connectivity in supragranular layers, right half in infragranular layers. (B) Intrinsic wiring scheme suggested by current source density analysis (from Mitzdorf and Singer, 1978). In both diagrams the numbers at the synapse symbols indicate the synaptic distance from the stimulation site.

III, connect to cells of their own type forming a positive feedback loop inside the supragranular layers. The disynaptic cells lying just above layer IV often receive additional monosynaptic input of the high-scatter type.

The input stage neurons make additional, strongly coupled, disynaptic contacts with neurons lying in layers IV and V. The infragranular layers receive a monosynaptic input almost exclusively of the high-scatter type, which extends throughout these layers but is the most dense in layer VI. Also, the infragranular cells probably project back onto cells of their own type.

For comparison, the summary diagram of a current source density study of area 18 (Mitzdorf and Singer, 1978) is reproduced in Fig. 4(B). The main flow of supragranular activity found by the current source density method proceeds from the input stage cells in layer IV to cells situated in layer III and then, via long-distance connections, to the distal parts of apical dendrites of pyramidal cells also lying in layer III. The synapses mediating mono- and disynaptic activity, both being located near the cells' somata, show strong coupling and little scatter. The single-unit results harmonize well with these findings and demonstrate that it is the same cell population that receives disynaptic and trisynaptic synapses. The cell somata, however, are not quite as concentrated in layer III as the current source density method would suggest. This disagreement might be the result of a masking of layer II disynaptic sinks by the sources of layer III sinks which are much stronger because of the higher density of disynaptic cells in layer III.

In the infragranular layers the axons of the input stage neurons synapse on the apical dendrites of pyramidal cells lying mainly in layer VI according to current source density analysis. These cells in turn project to cells of their own type partly via long-distance connections. The monosynaptic input to layer VI is visible but very weak in area 18 and is thus not shown. The single-unit analysis again ascertains that disynaptic and polysynaptic responses occur in the same cells. However, almost all of the cells receiving disynaptic input were found in the upper half of the infragranular layers in this study. It is difficult to assess whether this result is partly due to the small sample of cells of this category or whether there is a real contradiction between the two methods.

There is a third, less prominent pathway identified with the current source density method, which relays activity from the input stage cells, via disynaptic cells also situated in layer IV, to neurons lying in layer III. Single units exhibiting disynaptic responses are found in layer IV. It was, however, impossible to discriminate two classes of trisynaptic activity in the supragranular layers. Since the trisynaptic synapses of the first pathway shown are situated remotely from the cell somata, their influence on these cells is hard to assess. The current source density method suggests that the trisynaptic EPSP's mediated by this pathway remain mainly subthreshold (Mitzdorf and Singer, 1978). Therefore, the possibility cannot be completely excluded that the trisynaptic responses seen in supragranular cells are mediated by the third pathway identified with the current source density method.

In summary, the two methods of analyzing area 18 intrinsic connectivity by electrical stimulation have yielded very similar results. They partly complement each other because the current source density method assesses the location of excitatory synapses, while the single-unit analysis with the cell/fiber discrimination criteria used indicates the approximate location of the cell somata.

References

Cleland, B. G., M. W. Dubin, and W. R. Levick (1971). Sustained and transient neurones in the cat's retina and lateral geniculate nucleus. J. Physiol., Lond. 217:473-496.

Hoffmann, K.-P., J. Stone, and S. M. Sherman (1972). Relay of receptive-field properties in dorsal lateral geniculate nucleus of the cat. J. Neurophysiol. 35:518-531.

LeVay, S., and C. D. Gilbert (1976). Laminar patterns of geniculo-cortical projection in the cat. Brain Res. 113: 1-19.

Lund, J. S., and R. G. Boothe (1975). Interlaminar connections and pyramidal neuron organisation in the visual cortex, area 17, of the macaque monkey. J. Comp. Neurol. 159:305-334.

Mitzdorf, U., and W. Singer (1978). Prominent excitatory pathways in the cat visual cortex (A17 and A18): a current source density analysis of electrically evoked potentials. Exp. Brain Res. 33:371-394.

Nauta, H. J. W., A. B. Butler, and J. A. Jane (1973). Some observations on axonal degeneration resulting from superficial lesions of the cerebral cortex. J. Comp. Neurol. 150:349-360.

Singer, W., and N. Bedworth (1973). Inhibitory interaction between X and Y units in the cat lateral geniculate nucleus. Brain Res. 49:291-307.

Singer, W., F. Tretter, and M. Cynader (1975). Organization of cat striate cortex: a correlation of receptive-field properties with afferent and efferent connections. J. Neurophysiol. 38:1080-1098.

Stone, J., and B. Dreher (1973). Projection of X- and Y-cells of the cat's lateral geniculate nucleus to areas 17 and 18 of visual cortex. J. Neurophysiol. 36:551-567.

Tretter, F., M. Cynader, and W. Singer (1975). Cat parastriate cortex: a primary or secondary visual area? J. Neurophysiol. 38:1099-1113.

Van Essen, D., and J. Kelly (1973). Correlation of cell shape and function in the visual cortex of the cat. Nature 241:403-405.

Visual Cell X/Y Classifications:
Characteristics and Correlations

SHAUL HOCHSTEIN

The Hebrew University of Jerusalem
Jerusalem, Israel

Abstract Investigators have used a battery of tests for a variety of characteristics to classify retinal ganglion and lateral geniculate nucleus (LGN) cells as X or Y. The characteristics that have been tested include: linearity of spatial summation, response dynamics (temporal frequency tuning), action potential conduction velocity (axon diameter), receptive field center size (spatial frequency tuning), dendritic field size, cell body size, and receptive field eccentricity in the visual field. The methods and results of these measurements are reviewed here, with special attention given to evidence on two issues. Are the results clearly bimodal separating distinct groups, or is the grouping a result of an arbitrary division of a continuum? What are the correlations among the classifications according to different characteristics—that is, are there really two distinct groups (X/Y) or are many cells found with mixed characteristics? Evidence is presented suggesting that in general the distributions are not strictly bimodal. Furthermore, the results of new experiments where many of the characteristics above were tested for each LGN unit from which recordings were made, indicate that the correlations among the different characteristics are weak, though positive. Thus, although a tendency for correlation does appear, a cell that is identified as X by a single characteristic test may be classified as Y when tested for another characteristic.

Introduction

A dozen years ago, Enroth-Cugell and Robson (1966) gave the names X and Y to retinal ganglion cells which performed, respectively, linear and nonlinear spatial summation of light falling on their receptive fields. Since then, there have appeared many reports categorizing cells as X or Y by related, or by unrelated but seemingly correlated, characteristics. (A separate group, termed W-cells, will be largely ignored here since the studies on the preferential effects of deprivation (see Sherman, 1979) also relate only to X and Y.) The various

185

techniques used for characterizing cells as X or Y will be reviewed here and the results of previous work and my own new experiments (Hochstein, 1979) analyzed with regard to (1) the separation into groups of visual units on the basis of a clearly bimodal distribution on a chosen parameter and (2) the correlation between the results when different parameters are measured on the same cells.

These questions are demonstrated in Fig. 1. The three graphs on the left are theoretical histograms relating number of cells found versus values of a test parameter. It is justified to claim that cells belong to separate classes that serve different functions—that is, belong to distinct "modes"—only if they are distinguishable by a classifying test—that is, if their distribution over the test parameter is of the form of Fig. 1(A), bimodal. If the distribution is like 1(C), it is

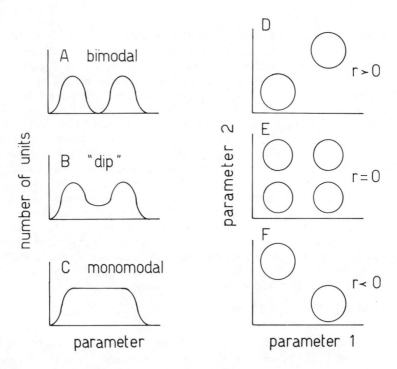

FIGURE 1 (Left) Theoretical frequency distribution histograms showing (A) bimodal, (B) "dip," and (C) monomodal distributions. A separation into two classes is justified only for the bimodal distribution. The "dip" distribution may represent two types of units with an overlap region. (Right) Correlations between two parameters having bimodal distributions. The correlation may be (D) positive, (E) zero, or (F) negative. Bimodality over two parameters is clearly insufficient to prove correlation.

monomodal and any separation into classes is arbitrary. Figure 1(B) is an inter-mediate form since it has a dip but no zero between the two peaks. It may be interpreted as a monomodal distribution or as a bimodal distribution with some degree of overlap in the central region where both modes are represented. Alternatively, a line may be drawn (e.g., at the lowest point of the dip) and the modes defined as separated by this line. The three graphs on the right relate values found for two parameters. Each point on such a graph would represent a single visual unit. For demonstration purposes no points are drawn, instead cir-cles define the areas where points would be found. For all three graphs, the histograms of the number of cells versus each of the two parameters would be bimodal, but the parameters are positively correlated in Fig. 1(D), negatively correlated in 1(F), and have zero correlation in 1(E).

In general, results of X/Y characteristic studies have had histograms similar to Fig. 1(B) and correlations somewhat between (D) and (E), i.e., positive but weak correlations. A number of different tests are now discussed individually.

Conduction Time (related to axon diameter)

This parameter was measured long before Enroth-Cugell and Robson (1966) coined the terms X and Y. In 1940 and 1942, George Bishop and O'Leary found four groups of fibers with different conduction times measured using electrical stimulation of the optic tract following eye removal. In 1951 and 1952, Chang found three fiber groups (and related them to the trichromatic theory of vision). In 1953, Peter Bishop reported that both anatomical and electrophysiological measurements gave only two distinct groups, and he assumed a positive correlation between size and conduction velocity. In 1955, George Bishop, now with Clare, repeated his finding of four groups, and in 1966, Fukada et al. reported only two. Finally, in 1967, Donovan repeated the anatomical study and found a monomodal distribution of fiber diameters in the cat optic tract, and in 1971, Fukada measured conduction velocity and also found a monomodal distribution, shown here in Fig. 2, but nevertheless reported a tendency for a relationship between conduction velocity and the results of other tests by which he characterized Type I (phasic—later related to Y) and Type II (tonic—later related to X) cells.

When conduction times of different units are compared, a number of factors must be considered: stimulation technique (visual or electrical), conduction distance, conduction differences in myelinated and unmyelinated segments, conduction dependence on direction (ortho- or antidromic), and stimulation and/or synaptic latencies. When using visual stimulation, two recording sites are necessary to subtract out the extremely long pre-ganglion cell latency (Cle-land et al., 1971). When recording or stimulating at the retina, the intraretinal nonmyelinated segment conduction time must be taken into account. This time is considerable, although the distance is short, since the conduction is very slow. The best separation into classes is indeed on the basis of intraretinal con-duction velocity. The slope of the plot of conduction distance versus

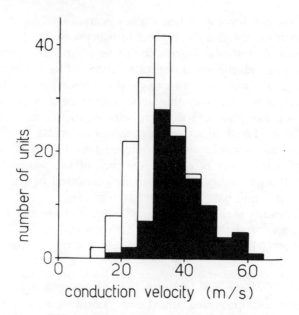

FIGURE 2 Frequency distribution of conduction velocity. Filled bars represent ON-Type I (phasic) units and open bars ON-Type II (tonic) units. (Only one OFF-Type II was found in this study compared with over a hundred OFF-Type I, so only the ON unit results are displayed.) Stimulation at optic tract near LGN; recording at optic chiasm. Velocity is distance divided by measured latency of onset of antidromic spike. Distribution is monomodal but Type I axons tend to have higher velocities than Type II axons. (Adapted from Fukada (1971).)

conduction time is 2.9 mm/ms for X cells and 4.9 mm/ms for Y cells (Stone and Fukuda, 1974; Rowe and Stone, 1976). The points on these graphs lie along a straight, though jittery, line indicating a constant *intra*-retinal conduction *velocity* independent of retinal eccentricity. The jitter in the data points making up these graphs is sufficient, however, to make questionable the significance of the difference in *extra*-retinal conduction time, which would average less than 1 ms. In fact, the only Type A (non-overlapping bimodal) distributions found have been for conduction measurements including this intraretinal segment (Cleland et al., 1971; Cleland and Levick, 1974; a Type B dipped histogram was found by Rowe and Stone, 1976). Rowe and Stone (1976) point out that for antidromic electrical stimulation there is a considerable slowing of the conduction near the soma, so that axon and cell body recordings must not be compared. This problem was avoided by Cleland et al. (1971), who used orthodromic conduction from visual stimulation and recorded simultaneously at the retina and the LGN.

  Measurement of optic nerve (ON) or optic chiasm (OX) to lateral geniculate nucleus (LGN) conduction time gives a monomodal (Fig. 1(C)) distribution, although a tendency for fast Y and slow X cells (Fukada, 1971; Stone and Hoffmann, 1971; Fukada and Saito, 1972; Hoffmann et al., 1972; and Wilson et al., 1976). A dip in the histogram is introduced and the X/Y separation improved by using the difference between the conduction times for two different stimulation sites (e.g., ON to LGN less OX to LGN, Wilson et al., 1976). Although the conduction distance is reduced, it is now the same for all cells. This same procedure subtracts out stimulation and synaptic latencies which may add to the improvement in the bimodality.

In summary, conduction time is a tricky characteristic which must be measured very carefully; otherwise, the dichotomy may be blurred. My own measurements, using the simple OX to LGN conduction times, confirm that the distribution is monomodal and the correlation with results of other X-Y tests is positive but not complete.

Linearity of Spatial Summation

The characteristic originally tested by Enroth-Cugell and Robson (1966) was the linearity of spatial summation. They used two characterizing tests: introduction and removal of a grating pattern and response to a drifting grating. For X cells a "Null" position could be found for the grating pattern where no response to its contrast modulation was seen. These cells also always had a modulated response to the drifting grating. For Y cells, no such "Null" position could be found, and the response to drifting high-spatial frequency gratings was an unmodulated increase in maintained firing rate. Hochstein and Shapley (1976a,b) extended these tests, demonstrated the need to use "high" spatial frequency gratings also for the "Null" test, proposed a model to describe the mechanism responsible for the complex responses of Y cells, and defined a "nonlinearity index" for spatial summation of a unit. The frequency distribution of cells in this index is reproduced in Fig. 3. I have continued these tests of

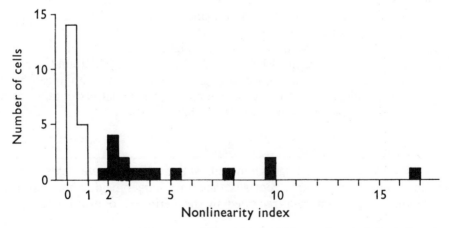

FIGURE 3 Frequency distribution of nonlinearity index. The nonlinearity index is the maximum value of the second harmonic (nonlinear component/first harmonic (linear component)) ratio. All the X cells (those with linear spatial summation —open bars) are in the peak below 1; all the Y cells (those which also have nonlinear spatial summation—filled bars) had values not less than 2. (From Hochstein and Shapley (1976a).)

LGN cells and find a similar bimodal distribution, but the correlation with other characteristics is weak. However, Cleland et al. (1971 and 1973), using the results of modifications of the drifting grating test, claim good correlation with other tests.

Response Dynamics (temporal frequency tuning)

Cleland et al. and Fukada, in 1971, independently introduced classifications on the basis of response dynamics. Units with a large phasic component and no tonic component were called Transient or Type I cells and associated with Y cells, while units with a substantial tonic component were called Sustained or Type II cells and associated with X cells. Cleland et al. (1973) showed that the actual level of the phasic and the tonic components depended on the background illumination but that a bimodality is found for any fixed illumination. (However, no one has ever published the distribution histograms for the ratio of phasic/tonic response amplitudes.*) As to the correlation between this and other measures, Cleland et al. (1971) say that "whereas a cell at the retinal or geniculate level can generally be classified as ON center or OFF center by a few seconds' work with hand-held wands, the same is not true for the sustained/transient classification. We found it necessary to apply several tests and even then units with intermediate properties were encountered." Hochstein and Shapley (1976a) also report peripheral X (linearly summating) cells with transient responses, and I have now found that at the geniculate the correlation is even weaker.

Receptive Field Center Size (related to dendritic field size?)

Enroth-Cugell and Robson (1966) reported that the receptive field centers of Y cells tended to be larger than those of X cells. One might be led to say that conduction velocity is related to axon diameter which is related to cell body size which is related to dendritic field size which is related to receptive field size. However, any or all of these relations might be weak or absent. In fact, Enroth-Cugell and Robson (1966) and Cleland et al. (1973) found overlapping correlations of X/Y or sustained/transient classifications with receptive field size, and Fukada (1971; see Fig. 4) found a monomodal distribution of sizes and an overlapping correlation with other characteristics.

Eccentricity (on retina or in visual field)

Wiesel (1960) states that "ganglion cells with small receptive fields were most often recorded in the area centralis" and that "larger field centers were most common for ganglion cells recorded in the periphery of the retina." Enroth-

*Recently, Bullier and Norton (1979) measured the ratio of the tonic/phasic portions of the response (their PTI) as a quantitative measure of response dynamics. Their data show a monomodal distribution over this parameter and monomodal or dip distributions over other new or old parameters. They therefore use a multivariate analysis to separate LGN cells into two modes. Obviously, this analysis would be superfluous if one parameter would suffice to form a bimodal distribution.

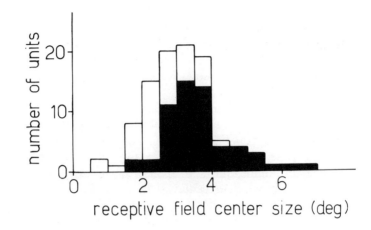

FIGURE 4 Frequency distribution of size of receptive field center. Filled bars represent ON-Type I and open bars ON-Type II. Size of receptive field center estimated by change of response to small flashing light spot from ON to OFF with distance from center point. Distribution is monomodal but Type I fields tend to be larger than Type II fields. (Adapted from Fukada (1971).)

Cugell and Robson found that the same was true for their sample of 20 cells. However, this weak correlation of size with eccentricity and the above-mentioned weak correlation of size with X/Y classification does not necessarily imply a correlation of eccentricity with X/Y classification. As mentioned above, Stone and Fukuda (1974), Cleland and Levick (1974), and Rowe and Stone (1976) show an increase in conduction time with retinal eccentricity, but a constant conduction velocity for each cell type. Fukada (1971) found that "neither conduction velocity nor the size of receptive field center increased with distance from *area centralis*," but Hoffmann et al. (1972) and Ikeda and Wright (1972) found an increase in the proportion of Y cells with eccentricity, and Enroth-Cugell and Robson (1966) report that the "receptive field centers of Y cells...had less tendency to be located centrally in the visual field" than those of X cells. Hochstein and Shapley (1976a), on the other hand, show similar retinal distributions for X and Y retinal ganglion cells, and I have found the same result for X and Y LGN cells. All would agree, in this case, that knowing the eccentricity of the receptive field of a cell would not be sufficient information to determine its X/Y characteristics.

Conclusion

The X/Y dichotomy has been based on numerous characterizing tests, including conduction time, linearity of spatial summation, response dynamics, receptive field center size, and eccentricity. Of these, the only ones giving bimodal distributions (Fig. 1(A)) are intraretinal conduction velocity (or intraretinal conduction time for cells at a fixed retinal locus) and linearity of spatial summation. Extraretinal conduction velocity (evaluated by a two-stimulating-sites technique) gives a dip distribution (Fig. 1(B)) and probably so does response dynamics for a fixed background illumination, although no quantitative measure of this parameter has been reported.* Receptive field center size and eccentricity give monomodal distributions.

As to the correlations between the results of tests, much has been claimed but little actually shown. Linearity of spatial summation has not been correlated with *intra*-retinal conduction time and has a weak correlation with *extra*-retinal conduction time, improved by a two-stimulation-site technique (So and Shapley, personal communication). Response dynamics for a fixed background illumination and receptive field center size for a fixed retinal location may correlate well with other parameters. A number of studies used a "battery of tests" and claimed a high degree of agreement between the results, but failed to show that agreement quantitatively. As quoted above, Cleland et al. (1971) say some mixed types were found and that it was hard work to classify a cell by all tests. In addition, "In general, it was more difficult to obtain clear-cut results at the geniculate level" than at the retina. A disturbing feature of numerous studies (e.g., Hoffmann et al., 1972; Sherman et al., 1972; Wilson et al., 1976) is the display of a distribution histogram of one test parameter, indicating X and Y cells by light and dark bars, respectively, showing a good correlation of the test parameter with the X/Y dichotomy but neglecting to indicate which test was used to determine the X/Y classification for the purposes of the histogram. One assumes that it was not the test parameter of the histogram.

I have now studied over 100 geniculate units from 16 cats and classified each one by most of these X/Y classification tests (Hochstein, 1979). Over 85 cells were tested for any one characteristic and over 65 for each pair of tests. I find the correlations between the results of the different tests to be positive but weak (generally less than 0.5). Thus, if a cell is judged X by one test, there is less than 75% chance of its being judged X also by a second test.

Perhaps the best summary of the question of correlations must use the term employed by Enroth-Cugell and Robson: "tendency." They originated the X/Y terminology on the basis of linearity of spatial summation only. They then concluded that "Receptive field centers of Y cells *tended* to be larger than those of X cells and had less *tendency* to be located centrally in the visual field... This makes it likely that Y cell axons generally are larger than X cell axons." We need add only a *tendency* for X cell responses to be more tonic than Y cell responses. However, these are only tendencies and one would be wise to state in every case which classification is being used, and to be cautious with drawing conclusions about one classification following study using another.

*See footnote on p. 190.

Acknowledgements I thank Professors Peter Hillman and Robert Shapley for the stimulating discussions and helpful comments on this study. This work was supported by grants from the U.S.-Israel Binational Science Foundation (BSF), Jerusalem, Israel, and the Israel Academy of Sciences.

References

Bishop, G. H., and J. L. O'Leary (1940). Electrical activity of the lateral geniculate nucleus of the cat following optic nerve stimuli. J. Neurophysiol. 3:308-322.

Bishop, G. H., and J. L. O'Leary (1942). Factors determining the form of the potential record with vicinity of the synapses of the dorsal nucleus of the lateral geniculate body. J. Cell. Comp. Physiol. 19:315-331.

Bishop, G. H., and M. H. Clare (1955). Organization and distribution of fibers in the optic tract of the cat. J. Comp. Neurol. 103:269-304.

Bishop, P. O. (1953). Synaptic transmission. An analysis of the electrical activity of the lateral geniculate nucleus in the cat after optic nerve stimulation. Proc. Roy. Soc. B 141:362-392.

Bullier, J., and T. T. Norton (1979). X and Y relay cells in cat lateral geniculate nucleus: quantitative analysis of receptive field properties and classification. J. Neurophysiol. 42:244-273.

Chang, H. T. (1951). Triple conducting pathway in the visual system and trichromatic vision. Annee psycol. 50:135-144.

Chang, H. T. (1952). Functional organization of central visual pathways. Res. Publ. Ass. Nerv. Ment. Dis. 30:430-453.

Cleland, B. G., M. W. Dubin, and W. R. Levick (1971). Sustained and transient neurones in the cat's retina and lateral geniculate nucleus. J. Physiol. (Lond.) 217:473-496.

Cleland, B. G., and W. R. Levick (1974). Brisk and sluggish concentrically organized ganglion cells in the cat's retina. J. Physiol. (Lond.) 240:421-456.

Cleland, B. G., W. R. Levick, and K. J. Sanderson (1973). Properties of sustained and transient ganglion cells in the cat retina. J. Physiol. (Lond.) 228:649-680.

Donovan, A. (1967). The nerve fiber composition of the cat's optic nerve. J. Anat. 101:1-11.

Enroth-Cugell, C., and J. G. Robson (1966). The contrast sensitivity of retinal ganglion cells of the cat. J. Physiol. (Lond.) 187:517-552.

Fukada, Y. (1971). Receptive field organization of cat optic nerve fibers with special reference to conduction velocity. Vis. Res. 11:209-226.

Fukada, Y., A. C. Norton, K. Motokawa, and K. Tasaki (1966). Functional significance of conduction velocity in the transfer of flicker information in the optic nerve of the cat. J. Neurophysiol. 29:698-714.

Fukada, Y., and H. Saito (1972). Phasic and tonic cells in the cat's lateral geniculate nucleus. Tohoku J. Exptl. Med. 106:209-210.

Hochstein, S., and R. Shapley (1976a). Quantitative-analysis of retinal ganglion cell classifications. J. Physiol. (Lond.) 262:237-264.

Hochstein, S., and R. Shapley (1976b). Linear and nonlinear spatial subunits in Y cat retinal ganglion cells. J. Physiol. (Lond.) 262:265-284.

Hochstein, S. (1979). Cat LGN X and Y units: a multifarious distribution. In preparation.

Hoffmann, K.-P., J. Stone, and S. M. Sherman (1972). Relay of receptive field properties in dorsal lateral geniculate nucleus of the cat. J. Neurophysiol. 35:518-531.

Ikeda, H., and M. J. Wright (1972). Receptive field organization of "sustained" and "transient" retinal ganglion cells which subserve different functional roles. J. Physiol. 227:769-800.

Rowe, M. H., and J. Stone (1976). Conduction velocity groupings among axons of cat retinal ganglion cells and their relationship to retinal topography. Exp. Br. Res. 25:339-357.

Sherman, S. M. (1979). Differential susceptibility to deprivation of X and Y systems. This volume.

Sherman, S. M., K.-P. Hoffmann, and J. Stone (1972). Loss of a specific cell type from dorsal lateral geniculate nucleus in visually deprived cats. J. Neurophysiol. 35:532-541.

Stone, J., and Y. Fukuda (1974). Properties of cat retinal ganglion cells: a comparison of W-cells with X- and Y-cells. J. Neurophysiol. 37:722-748.

Stone, J., and K.-P. Hoffmann (1971). Conduction velocity as a parameter in the organization of the afferent relay in the cat's lateral geniculate nucleus. Br. Res. 32:454-459.

Wiesel, T. (1960). Receptive fields of ganglion cells in cat's retina. J. Physiol. 153:583-594.

Wilson, P. D., M. H. Rowe, and J. Stone (1976). Properties of relay cells in cat's lateral geniculate nucleus—comparison of W-cells with X- and Y-cells. J. Neurophysiol. 39:1193-1209.

Functional Plasticity in the Mature Visual System: Changes of the Retino-geniculate Topography After Chronic Visual Deafferentation

ULF Th. EYSEL
ULRICH MAYER
Institut für Physiologie
Universitätsklinikum Essen
Essen, Germany

Abstract Photocoagulation of that area of the retina nasal to the optic disc resulted in chronic visual deafferentation of the lateral part of layer A in the contralateral lateral geniculate nucleus (LGN) of adult cats. Pre- and post-lesion fundus photography revealed well-preserved retinal geometry with negligible changes in the size and position of the lesion. The histology of the coagulated retina displayed a steep decay of all retinal layers at the border and total destruction of the retina within the lesion.

During the first ten days of visual deafferentation, single-cell recordings, using tungsten microelectrodes, from layers A and A_1 of the LGN contralateral to the photocoagulated eye yielded a border of light-excitability of layer A cells corresponding to the normal projection of the lesion onto the LGN. Beginning about 20 days after deafferentation, an increased lateral spread of excitation could be demonstrated upon careful investigation of the retino-geniculate topography near the border of deafferentation within the LGN. Light-excitable cells were detected in layer A which had receptive fields horizontally displaced by more than 1° of visual angle with respect to the normal retinotopy of the receptive fields of layer A_1 cells. After 30 to 40 days the maximal receptive field displacements reached values of up to 5°.

Considering the magnification factor in the LGN for the horizontal eccentricity of the retinal lesions (20°), the results suggest a lateral spread of excitation within the LGN developing with time after deafferentation and exceeding the normal lateral spread of excitation by up to 250 μm.

195

Monocular visual deafferentation by enucleation or by application of total or restricted retinal lesions might be a powerful tool for revealing the extent of plasticity in the subcortical visual system. Enucleation of one eye elicited translaminar collateral sprouting of optic tract axons in the lateral geniculate nucleus (LGN) of the kitten (Guillery, 1972; Kalil, 1972; Hickey, 1975) and electrophysiological experiments with retinal lesions suggested translaminar reinnervation of deafferented cells in the developing cat (Eysel and Mayer, 1978). With the same preparation in the adult cat neither electrophysiological (Eysel, 1979) nor anatomical (Guillery, 1972; Hickey, 1975) evidence of visual reinnervation in the LGN could be provided. Such findings have led to the widely accepted view that the mature projections of the retinal ganglion cells in the cat are no longer capable of functional or morphological plasticity.

Taking into account the complete separation of the projections of the two eyes in different layers of the LGN of the cat, monocular deafferentation produces very unfavorable conditions for possible reinnervation: the deafferented cells and the remaining potentially reinnervating visual inputs are far apart and the specific input from the same eye, which might have the highest priority for reinnervation (Goodman et al., 1973), is completely destroyed.

By contrast, restricted retinal lesions cause deafferentation of only part of a given layer in the LGN. Within this layer the conditions for detecting residual plasticity are much more promising since the remaining specific input fibers from the partially destroyed retina are directly adjacent to the deafferented LGN cells with extensive overlap of their axon arborizations and dendritic trees.

Accordingly, in the present investigations monocular nasal retinal lesions were applied to study possible reorganization or reinnervation of deafferented cells in the LGN of adult cats.

Materials and Methods

Thirty-two cats (1.9 to 3.5 kg) were used. Two animals served as normal controls. In 30 cats the nasal part of the left retina was photocoagulated with a xenon-photocoagulator (LOG-2, Clinitex Inc., Beverly, Mass.) under barbiturate anesthesia (32 mg/kg). Fundus photographs were taken before and after coagulation, and later at intervals of ten days under short-acting anesthesia (10 mg/kg ketamine hydrochloride and 3 mg/kg xylazine hydrochloride). The electrophysiological experiments were performed between 6 hours and 373 days after photocoagulation. The animals were anesthetized with barbiturate and were atraumatically fixed in a stereotactic head holder. The average length of experimental sessions was 36 hours. During this time a continuous infusion of a mixture of d-tubocurarine (0.3mg/kg·h) and gallamine triethiodide (4.0mg/kg·h) in a glucose and Ringer solution was administered to paralyze the eye muscles and to prevent dehydration. The animals were artificially respired with air and the end-expiratory pCO_2 was kept constant at between 3.6% and 3.8%. Arterial blood pressure, epidural electro-corticogram, and body temperature were continuously monitored. Body temperature was automatically maintained at 38.5°C.

The corneae were protected by clear contact lenses; eye refraction was measured by refractoscopy and corrected with additional lenses in front of both eyes. Mydriasis and retraction of the nictitating membranes were achieved by local application of atropine sulphate and phenylephrine hydrochloride.

Using tungsten microelectrodes with tip diameters of less than 1 μm, recordings were made from layers A and A_1 of the LGN contralateral to the photocoagulated eye at Horsley-Clarke coordinates A 5.0 to 6.0 and L 8.5 to 10.5.

Receptive field (RF) centers were located by means of monocular stimulation with small spots of light (0.3° to 1.0° in diameter) and light bars (0.2° × 0.5°, 0.2° × 1.0°) with variable intensity, as well as with black dots (0.5° and 1.0° in diameter) which were moved by hand on a white tangent screen placed 50 cm in front of the cat's eyes. The optic discs and vessels of both eyes were back-projected with an ophthalmoscope onto the tangent screen and mapped in order to control for small eye movements. The RF center positions and the border of the lesion were drawn on the same map. After the experiments the locations of the areae centrales were calculated according to the method described by Sanderson and Sherman (1971) and the RF maps were redrawn with the two eyes aligned to their common fixation point. At the same time, the RF positions were recomputed in terms of azimuth and elevation (Bishop, Kozak, and Vakkur, 1962) within the left visual hemi-field (Fig. 1). The RFs of more than 1000 single units from layers A and A_1 were mapped in the 32 experiments.

At the end of most of the experiments the animals were perfused with a mixture of glutaraldehyde and paraformaldehyde (Karnowsky, 1965) through the abdominal aorta (Gonzalez-Aguilar and De Robertis, 1963), and the eyes and brains were preserved to be used as anatomical controls.

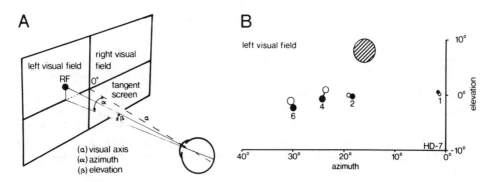

FIGURE 1 (A) Localization of the receptive fields (RFs) on the tangent screen. RF positions on the tangent screen were converted into degrees of visual angles of azimuth and elevation for the reconstruction of the RF maps. (B) Normal RF center map of layer A and A_1 cells from the binocular segment of the left visual hemifield. Open circles represent RF centers from layer A; filled circles represent RF centers from layer A_1. The left optic disc is marked by the hatched area. The different penetrations are labeled by numbers.

Results

Electrophysiological control experiments yielded the well-known retinotopy of RFs that were obtained for single LGN cells during vertical penetrations at different lateral positions (Bishop, Kozak, Levick, and Vakkur, 1962; Sanderson, 1971a, 1971b; Sanderson and Sherman, 1971; Kaas et al., 1972). This is demonstrated in Fig. 1(B) for a series of vertical penetrations at an anterior-posterior position corresponding to the projection of the areae centrales onto the LGN. The correspondence between azimuth with an accuracy of better than 1° of visual angle for the RFs from both eyes was always obtained when all penetrations were made with the same electrode and when only the last cells of layer A and the first ones of layer A_1 (not more than 200 μm apart) were considered. Since this correspondence was used as a reference point in the present investigations, these conditions were strictly followed for the construction of all RF maps. Single RF center positions in these maps were, as a rule, representative of overlapping RFs from several cells recorded at the border between layers A and A_1.

For the investigation of changes in the LGN after partial visual deafferentation, photocoagulator lesions were applied to the left nasal retina causing deafferentation of the lateral portion of layer A in the right LGN (Fig. 2(A)). In order to draw valid conclusions from electrophysiological investigations of the retino-geniculate topography within the LGN, it was at first necessary to control for possible changes in the size of the lesion and in the retinal geometry of an eye with a chronic photocoagulator lesion. Typically, lesions were restricted to the area nasal to the left optic disc, with a temporal border 18° to 22° nasal to the area centralis (Fig. 2(B)). The lesions which were 10° to 15° wide and 20° to 30° high usually contracted in size between 5 and 15 days after coagulation by not more than 0.5° to 1° of visual angle and showed no longer-term changes. The retinal geometry between the area centralis and the border of the lesion displayed minimal concomitant displacements consisting of small tractions of the retina towards the lesion which primarily affected retinal tissue immediately adjacent to the lesion (Fig. 2(B)).

The histology of the photocoagulated retinae displayed a steep decline from normal structure to complete loss of all retinal layers within 100-200 μm at the border of the lesions and no potentially excitable tissue was left within the regions which appear black in the fundus photographs (see also Fig. 3(C,D)).

Six and 24 hours after photocoagulation of the left nasal retina in electrophysiological experiments, the border of light-activation in layer A of the right LGN followed the normal topography of the projection of the border of the lesion with an accuracy of 1° of visual angle. This was also found after five days of deafferentation (Fig. 3(A)). During all penetrations, for RFs between the area centralis and the border of the lesion, a good topographical correspondence was found between layer A and A_1. When one recorded more laterally in the LGN at projections of further temporal regions of the visual field, only RFs from the normal eye could be detected.

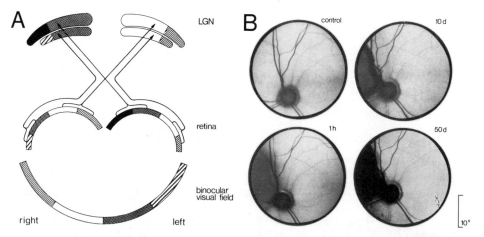

FIGURE 2 (A) Schematic diagram of the experimental paradigm. Photocoagulation of the left nasal retina (black) leads to deafferentation of part of layer A in the right LGN (black). The lateral part of the left visual hemifield is still represented in layer A_1 from the corresponding part of the right retina (coarse hatching). (B) Fundus photographs of a left eye with photocoagulator lesion. Normal fundus photograph (control) and photographs of the same fundus 1 hour, 10 and 50 days after photocoagulation (cat HD-38). The retinal geometry changes only minimally between 1 hour and 10 days, remaining constant thereafter.

Significantly different results were found 30 days or more after photocoagulation, as demonstrated in Fig. 3(B,D). A normal correspondence of the retinotopy of layers A and A_1, for RFs between the area centralis and the border of the lesion, was still present 30 and 98 days after deafferentation. Recordings that were made further laterally in the LGN (penetrations 13 and 16 in Fig. 3(B) and penetrations 3 and 6 in Fig. 3(D)) revealed RFs in layer A_1 to be more temporally located in the left visual hemifield, as expected, while during the same vertical electrode penetrations, the cells in layer A were light-excitable from the retinal area immediately adjacent to the border of the lesion in the left eye. These layer A cells thus had RFs horizontally displaced with respect to those in layer A_1 and were termed "displaced field (DF)" neurons. RFs at the border of the lesion were displaced by up to 5° of visual angle.

RF displacements of more than 1° were first observed after 20 days and the effect appeared to be fully developed 30 to 40 days after photocoagulation. DF neurons at the border of the lesion could still be found in a cat 373 days after deafferentation. A more detailed description of the results is in preparation (Eysel et al., 1979).

The DF neurons had lower spontaneous discharge rates than did normal neurons in the LGN and displayed the irregular burst activity that has been observed in visually deafferented LGN cells in other studies (Eysel et al., 1974; Eysel and Grüsser, 1975; Eysel, 1977; Eysel and Grüsser, 1978). The DF neurons responded irregularly to electrical stimulation of the optic tract near the

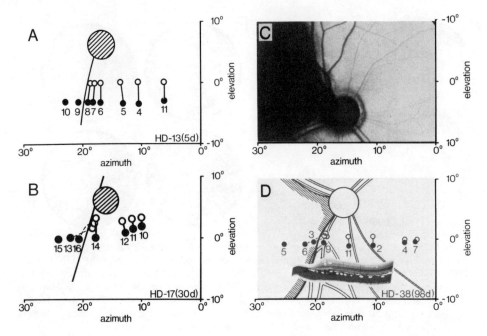

FIGURE 3 (A) RF center map of a cat five days after photocoagulation. Map of the left visual field with RF centers, left optic disc, and border of the lesion after five days. RFs found during the same penetration are linked by solid lines. The different penetrations are labeled by numbers. The symbols are the same as in Fig. 1(B). (B) RF center map of a cat 30 days after photocoagulation. Broken lines between layer A and A_1 RFs indicate recordings from the same vertical penetration but with horizontally displaced RFs. (C) Left fundus photograph of cat HD-38 taken 98 days after photocoagulation. (D) RF center map from the same cat (HD-38) after 98 days. Conventions same as in (A) and (B). The inset shows a retinal section corresponding approximately to the area in which the RFs of the left eye were found.

optic chiasm with latencies of 1.3 to 2.2 msec. The light-excitability of these cells was, as a rule, lower than in normal LGN neurons that were studied under comparable conditions.

Discussion

Retinal histology revealed no potentially excitable tissue remaining within the retinal lesions which could account for the observed lateral spread of excitation. The contraction of the retinal scar actually displaced the retina in the opposite direction to the RF shifts observed in DF neurons. Since the ortho- and antidromic response latencies of those DF cells that were studied by means of electrical stimulation of the optic tract and visual cortex, respectively, indicated a weak but direct input from the optic tract, the underlying mechanisms must be sought in the LGN itself.

RF displacements of 4° to 5° at 20° azimuth correspond to 200-250 μm within the LGN according to the retino-geniculate magnification factor (Sanderson, 1971b).

For an explanation of the described effect of deafferentation in the LGN, we may consider several factors which have already been discussed by Wall and Egger (1971) and Wall (1975, 1976) with respect to plasticity in the somato-sensory system.

One functionally nonsignificant mechanism could be the migration of cells in the LGN from the normally innervated portion of layer A into the part which was deafferented and which suffered subsequent transneuronal degeneration and cell shrinkage, as described by Cook et al. (1951). Against this possibility is the abrupt discontinuity of the lateral spread of excitation at the border of the lesion, rather than a gradual shift in the topography of layer A with respect to layer A_1, and the special properties of the cells with displaced RFs which resemble those of deafferented cells in other studies (Eysel et al., 1974; Eysel and Grüsser, 1975; Basbaum and Wall, 1976; Millar et al., 1976; Eysel, 1977; Eysel and Grüsser, 1978; Eysel, 1979).

More plausible hypotheses, which explain the electrophysiological evidence, are the following: (a) intralaminar collateral sprouting of the optic tract axon terminals—axonal sprouting has been observed in the kitten LGN (Guillery, 1972; Kalil, 1972; Hickey, 1975), as well as in other subcortical structures of adult animals (McCouch et al., 1958; Raisman, 1969; Raisman and Field, 1973; Goldberger and Murray, 1974; Tsukahara et al., 1975); (b) expansion of the dendritic trees of LGN cells; (c) activation of already existing but normally ineffective excitatory synaptic connections, as discussed by Wall (1976) and Basbaum and Wall (1976), for the spinal cord. The latter could be due to translocation of synapses from the periphery of the dendrites towards the cell soma and the spike trigger zone, or postsynaptic denervation hyperexcitability of the deafferented cells (Cannon and Haimovici, 1939; Loeser and Ward, 1967; Anderson et al., 1971; Bird and Aghajanian, 1975; reviewed by Sharpless, 1975).

From indirect electrophysiological evidence, Rowe (1976) discussed the possibility of intralaminar reinnervation in the kitten LGN after roughly homonymous binocular retinal lesions. Stelzner and Keating (1977), however, failed to observe intralaminar sprouting of retinal axons in the adult monkey LGN after combined retinal and cortical lesions, but in their autoradiographic study the possibility of short axonal sprouts (100-200 µm) could not be ruled out completely.

Our present results suggest a small but nevertheless definite degree of intralaminar plasticity in the LGN of the adult cat in response to partial lesions of the retina. Thus, it seems premature to exclude the LGN of the cat when the subcortical potential for plastic changes is discussed in the mature afferent visual system.

Acknowledgements. The authors are indebted to Dr. P. Hammond and Dr. K.-P. Hoffmann for their critical reading of the manuscript. We want to thank Dr. F. Gonzalez-Aguilar for the histological controls and Miss G. Kirchner for technical assistance. This work was supported by the Deutsche Forschungsgemeinschaft (Ey 8/7).

References

Anderson, L. S., R. G. Black, J. Abraham, and A. A. Ward (1971). Neuronal hypersensitivity in experimental trigeminal deafferentation. J. Neurosurg. 35:444-452.

Basbaum, A. I., and P. D. Wall (1976). Chronic changes in the response of cells in adult cat dorsal horn following partial deafferentation: the appearance of responding cells in a previously non-responsive region. Brain Res. 116:181-204.

Bird, S. J., and G. J. Aghajanian (1975). Denervation supersensitivity in the cholinergic septohippocampal pathway: a microiontophoretic study. Brain Res. 100:355-370.

Bishop, P. O., W. Kozak, W. R. Levick, and G. J. Vakkur (1962). The determination of the projection of the visual field onto the lateral geniculate nucleus in the cat. J. Physiol. 163:503-539.

Bishop, P. O., W. Kozak, and G. J. Vakkur (1962). Some quantitative aspects of the cat's eye: axis and plane of reference, visual field coordinates and optics. J. Physiol. 163:466-502.

Cannon, W. B., and H. Haimovici (1939). The sensitization of motoneurones by partial "denervation." Amer. J. Physiol. 126:731-740.

Cook, W. H., J. H. Walker, and M. L. Barr (1951). A cytological study of transneuronal atrophy in the cat and rabbit. J. Comp. Neur. 94:267-291.

Eysel, U. Th. (1977). Functional changes of cat lateral geniculate cells after chronic monocular deafferentation. Brain Res. 127:363-364.

Eysel, U. Th. (1979). Maintained activity, excitation, and inhibition of lateral geniculate neurons after monocular deafferentation in the adult cat. Brain Res. 166:259-271.

Eysel, U. Th., F. Gonzalez-Aguilar, and U. Mayer (1979). A functional sign of reorganization in the visual system of adult cats: lateral geniculate neurons with displaced receptive fields after lesions of the nasal retina. In preparation.

Eysel, U. Th., and O.-J. Grüsser (1975). Intracellular postsynaptic potentials of cat lateral geniculate cells and the effects of degeneration of the optic tract terminals. Brain Res. 98:441-455.

Eysel, U. Th., and O.-J. Grüsser (1978). Increased transneuronal excitation of the cat lateral geniculate nucleus after acute deafferentation. Brain Res. 158:107-128.

Eysel, U. Th., O.-J. Grüsser, and J. Pecci Saavedra (1974). Signal transmission through degenerating synapses in the lateral geniculate body of the cat. Brain Res. 76:49-70.

Eysel, U. Th., and U. Mayer (1978). Lateral geniculate nucleus recordings after partial visual deafferentation during the "sensitive period" in the cat. Exp. Brain Res. 32:R13-R14.

Goldberger, M. E., and M. Murray (1974). Restitution of function and collateral sprouting in the cat spinal cord: the deafferented animal. J. Comp. Neur. 158:37-54.

Gonzalez-Aguilar, F., and E. De Robertis (1963). A formalin-perfusion fixation method for histophysiological study of the central nervous system with the electron microscope. Neurology 13:758-777.

Goodman, D. C., R. S. Bogdasarian, and J. A. Horel (1973). Axonal sprouting of ipsilateral optic tract following opposite eye removal. Brain Behav. Evol. 8:27-50.

Guillery, R. W. (1972). Experiments to determine whether retinogeniculate axons can form translaminar collateral sprouts in the dorsal lateral geniculate nucleus of the cat. J. Comp. Neur. 146:407-420.

Hickey, T. L. (1975). Translaminar growth of axons in the kitten dorsal lateral geniculate nucleus after removal of one eye. J. Comp. Neur. 161:359-382.

Kaas, J. H., R. W. Guillery, and J. M. Allman (1972). Some principles of organization in the dorsal lateral geniculate nucleus. Brain Behav. Evol. 6:253-299.

Kalil, R. E. (1972). Formation of new retinogeniculate connections in kittens after removal of one eye. Anat. Rec. 172:339-340.

Karnovsky, M. J. (1965). A formaldehyde-glutaraldehyde fixative of high osmolality for use in electron microscopy. J. Cell. Biol. 27:137A.

Loeser, J. D., and A. A. Ward (1967). Some effects of deafferentation on neurons of the cat spinal cord. Arch. Neur. 17:629-636.

McCouch, G. P., G. M. Austin, C. N. Liu, and C. Y. Liu (1958). Sprouting as a cause of spasticity. J. Neurophysiol. 21:205-216.

Millar, J., A. I. Basbaum, and P. D. Wall (1976). Restructuring of the somatotopic map and appearance of abnormal neuronal activity in the gracile nucleus after partial deafferentation. Exp. Neur. 50:658-672.

Raisman, G. (1969). Neuronal plasticity in the septal nuclei of the adult rat. Brain Res. 14:25-48.

Raisman, G., and P. M. Field (1973). A quantitative investigation of the development of collateral reinnervation after partial deafferentation of the septal nuclei. Brain Res. 50:241-264.

Rowe, M. H. (1976). Effects of early retinal lesions on conduction velocity relationships in the dorsal lateral geniculate nucleus of the cat. Brain Res. 118:27-44.

Sanderson, K. J. (1971a). The projection of the visual field to the lateral geniculate and medial interlaminar nuclei in the cat. J. Comp. Neur. 143:101-118.

Sanderson, K. J. (1971b). Visual field projection columns and magnification factors in the lateral geniculate nucleus of the cat. Exp. Brain Res. 13:159-177.

Sanderson, K. J., and S. M. Sherman (1971). Nasotemporal overlap in visual field projected to lateral geniculate nucleus in the cat. J. Neurophysiol. 34:453-466.

Sharpless, S. K. (1975). Disuse supersensitivity. In: The developmental neuropsychology of sensory deprivation. A. H. Riesen (ed.). Academic Press, New York, pp. 125-152.

Stelzner, D. J., and E. G. Keating (1977). Lack of intralaminar sprouting of retinal axons in monkey LGN. Brain Res. 126: 201-210.

Tsukahara, N., A. Hultborn, F. Murakami, and Y. Fujita (1975). Electrophysiological study of formation of new synapses and collateral sprouting in red nucleus neurons after partial denervation. J. Neurophysiol. 38:1359-1372.

Wall, P. D. (1975). Signs of plasticity and reconnection in spinal cord damage. In: Outcome of severe damage to the central nervous system. Ciba Foundation Symposium 34 (new series) Elsevier, Amsterdam, pp. 35-54.

Wall, P. D. (1976). Plasticity in the adult mammalian central nervous system. Progr. in Brain Res. 45:359-379.

Wall, P. D., and M. D. Egger (1971). Formation of new connections in adult rat brains after partial deafferentation. Nature, 232:542-545.

Integration of Visual and Nonvisual Information in Nucleus Reticularis Thalami of the Cat

FRITZ SCHMIELAU

Institute for Medical Psychology
Ludwig-Maximilians University of Munich
Munich, Germany

Abstract In cats with grey fur extracellular recordings from 83 neurons of the nucleus reticularis thalami were obtained. Spontaneous activity and responses to electrical and light stimuli were investigated. Electrical stimuli were applied to each optic nerve, optic chiasm, visual cortex, superior colliculus, and to the mesencephalic reticular formation. Stationary light spots of various diameters were presented monocularly and binocularly. The corticofugal influence was assessed by reversibly inactivating the exposed parts of visual areas 17 and 18. Anatomical material (Szentágothai et al., 1972) and recent neurophysiological data (Dubin and Cleland, 1977) suggest that neurons of the perigeniculate nucleus are involved in inhibition of geniculate relay cells; therefore, special attention has been given to the possible role of nucleus reticularis thalami neurons as *extrinsic* interneurons of the dorsal lateral geniculate nucleus (LGN).

The perigeniculate nucleus apparently is a functional visual substructure of the nucleus reticularis thalami, receiving predominantly monosynaptic input from Y-cells of both retinae, whereas the rest of the nucleus reticularis thalami above the LGN is mainly innervated polysynaptically. Neurons of the nucleus reticularis thalami are identified as interneurons by their transsynaptic activation due to visual cortical stimulation and their typical response scatter to optic nerve, optic chiasm, visual cortex, and superior colliculus stimulation. Each of these stimuli usually elicits a burst of spikes (primary excitation) which, in most cases, is followed by a period of inhibition. Stimulation of the mesencephalic reticular formation causes inhibition in 63% (n = 52) and activation of 11% (n = 9) of nucleus reticularis thalami cells. Inactivation of the visual cortex prolongs the primary excitation caused by optical chiasm, visual cortex, and superior colliculus stimulation and reduces the strength of postexcitatory inhibition reversibly.

The model that is proposed explains binocular inhibition and disinhibitory effects of mesencephalic reticular formation stimulation in LGN relay cells via a pathway including the nucleus reticularis thalami. In addition, corticofugal effects that were described recently (Schmielau and Singer, 1977a; Schmielau and Singer, 1977b) are in accordance with the hypothesis.

Recent neurophysiological investigations of neurons in the perigeniculate nucleus in cat (Dubin and Cleland, 1977) stressed the importance of these cells as possible extrinsic interneurons of the dorsal lateral geniculate nucleus. Dubin and Cleland (1977) identified these cells as interneurons on the basis of their transsynaptic activation caused by stimulation of the visual cortex; and they showed that nucleus perigeniculate cell responses scatter significantly more than LGN relay cells.

Interneurons

Geniculate interneurons as an inhibitory link between adjacent relay cells have been discussed for approximately ten years. Suzuki and Kato (1966) were able to demonstrate their existence in cat; I-cells in the rat LGN were first reported by Burke and Sefton (1966). In both reports the responses to stimulation of the optic nerves and visual cortex were found to consist of repetitive discharges (bursts). Short axon cells in the LGN of the cat, as described by O'Leary (1940), Szentágothai et al. (1972), Tömböl (1969), have been a candidate for the anatomical substrate for the electrophysiologically identified interneurons. Recent neurophysiological investigations, however, also point towards an additional *extrageniculate* location for inhibitory interneurons. The data from Dubin and Cleland (1977) in cat demonstrate two populations of cells fulfilling the functional criteria of interneurons: one inside the LGN (intrinsic interneuron) and the other close to the dorsal border of geniculate lamina A (extrinsic interneuron) in the perigeniculate nucleus.

Nucleus Reticularis Thalami ⟷ Nucleus Perigeniculatus

Thuma (1928) and O'Leary (1940) have demonstrated by Golgi methods that cells lying close to the LGN in the "nucleus perigeniculatus anterior" receive visual input. However, precise terminology, location, magnitude, input-output relations, and also the function of this cell group are not known or generally accepted.

Sanderson (1971) recorded from binocular, ipsilaterally and contralaterally excitable neurons in the "nucleus perigeniculatus," reaching up to 2 mm from the dorsal surface of lamina A and extending over the whole rostro-caudal length of LGN. "Perigeniculate interneurons," described by Dubin and Cleland (1977), were found in a region up to 500 μm above the main dorsal geniculate nucleus. They were reported to be generally binocular. Negishi et al. (1961) recorded from neurons of the "nucleus reticularis of the thalamus"; although they do not specifically mention it, it seems they recorded from the caudal part of the reticularis thalami. They refer to the anatomical data of Rose (1952) and Ranson and Clarke (1953), who were able to demonstrate that the "nucleus reticularis" extends from the anterior to the posterior part of the thalamus. According to Rose (1952) and Ranson and Clarke (1953), the nucleus is sectionally organized, each section being in close relation to an underlying thalamic nucleus. Negishi et al. (1961) showed statistically significant differences of spontaneous and evoked discharge patterns of neurons in "nucleus reticularis thalami" and in the LGN. The anatomical extension of the nucleus reticularis thalami, according to Rose (1952) and Ranson and Clarke (1953), has been confirmed by Scheibel and Scheibel (1966), Jones (1975), and Holländer (1978). Both Szentágothai (1972) and Holländer (1978) stress the anatomical separation between the nucleus perigeniculatus and the nucleus reticularis thalami.

Contrary to these opinions, Garey (1978) and Singer (1978) place the nucleus reticularis thalami rather anterior with respect to LGN. In the stereotactic atlas of the cat brain by Snider and Niemer (1970) the terminus "nucleus reticularis thalami" does not exist. They limit the "nucleus reticularis" to Horseley-Clarke locations ranging from anterior (A) 13 to 8. Furthermore, eight caudal nuclei are mentioned between A1 and posterior (P) 2.5 (nucleus pontis oralis reticularis) and P8-P11 (nucleus reticularis lateralis). Thus, according to the atlas, there exists a gap between A7.5 and A1.5, corresponding approximately to the rostro-caudal extension of the LGN. Snider and Niemer (1970) mention, however, that not all structures are listed in the index.

Function of Interneurons

The possible function of geniculate interneurons in general and of perigeniculate and/or thalamic reticular neurons in particular have been discussed extensively in the context of feedback and feedforward inhibition of LGN relay cells, both in cat and rat (c.f. reference Szentágothai (1972) and Dubin and Cleland (1977)). Vastola (1960) puts forward the hypothesis of recurrent inhibition in order to explain his data on binocular inhibition in cat relay cells. Dubin and Cleland (1977) strongly favour the hypothesis that intrinsic interneurons are engaged in feedforward inhibition by receiving direct excitatory retinal input and acting upon neighbouring relay cells, whereas perigeniculate cells are supposed to function as extrinsic interneurons. According to Dubin and Cleland (1977) perigeniculate cells should hypothetically receive axon collaterals from relay cells.

Axon collaterals of LGN relay cells have been described by O'Leary (1940), Szentágothai et al. (1966), Tömböl (1967), and Guillery (1971). However, these collaterals seem to be restricted to the laminae from which they arise. Tömböl (1978), however, points out that collaterals outside of the main laminae might be shown by using the rapid Golgi method. Very recently, it was demonstrated that relay cells send axon collaterals into the nucleus perigeniculatus (Ahlsen and Lindström, 1978a,b). This was done by injecting horseradish peroxidase (HRP) into a small area of layer A_1 of LGN (Ahlsen and

Lindström, 1978) utilizing the retrograde transport of HRP, and by injecting HRP intracellularly into physiologically identified geniculate relay cells and neurons of the perigeniculate nucleus (Lindström et al., 1978). The latter method stains the whole neuron, including its axonal and dendritic branching. It was also shown that nucleus perigeniculatus cells project to both lamina A and A_1, where the axonal branches show a great extent of divergence. The existence of recurrent collaterals of perigeniculate nucleus efferents has already been mentioned before by O'Leary (1940) and Szentágothai (1972). Both authors, however, described axons of nucleus perigeniculatus cells joining the optic radiation. This fact is in contradiction to the electrophysiological data of Dubin and Cleland (1977).

The "status quo" of the nucleus perigeniculatus versus the nucleus reticularis thalami discussion is best characterized by Szentágothai (1972): "Certain similarities between the perigeniculate nucleus and the thalamic reticular nucleus may be of interest. The relation of the perigeniculate nucleus to the main laminae of the lateral geniculate body is virtually the same as shown by Scheibel and Scheibel (1966) to be the case between reticular nucleus and the main thalamic nuclei. One might, therefore, speculate whether the perigeniculate nucleus could not be considered as a part of the reticularis related specially to the lateral geniculate body."

Cortical Influence on Nucleus Reticularis Thalami

In the perigeniculate of the thalamic reticular nucleus of rat, Sumimoto et al. (1976) were able to identify two types of I-cells. They called them "perigeniculate reticular neurons." One type is innervated monosynaptically by retinal fibers, whereas the second type receives a disynaptic retinal input via axon collaterals of geniculate relay cells. In the same animal, after visual cortex ablation, only the monosynaptic I-cells were left (Sumimoto et al., 1977). This effect was probably due to the loss of relay cell (P-cell) input from type 2 I-cells, caused by retrograde degeneration of P-cells. In addition, the residual type 1 I-cells showed markedly less inhibition which was interpreted by the authors as due to the missing inhibitory input from type 2 I-cells, rather than as a consequence of a missing corticofugal input.

In the work of Dubin and Cleland (1977), in cat, too, a corticofugal influence onto the nucleus perigeniculatus cells has nearly been ruled out, although anatomical data both in cat and rat (and in several other species) demonstrate a massive projection of thick corticofugal fibres penetrating the neuropil field of the thalamic reticular nucleus (Scheibel and Scheibel, 1966; Tömböl, 1978). Recently, in rabbit, additional support has been given by Montero et al. (1977), who used a double-tracer technique. They clearly demonstrate a strictly retinotopic organized corticofugal projection onto the thalamic reticular nucleus which "in accuracy and in orientation of the major field axis...correspond(s) closely to the map established in the dorsal geniculate nucleus."

Thus, the baseline of the work presented here seemed to be good evidence of a reciprocal relationship between peripheral visual input to the nucleus reticularis thalami and a corticofugal, topographically organized projection onto the same structure, as well as a close relationship of the nucleus reticularis thalami to the LGN.

The Influence of Mesencephalic Reticular Formation on Nucleus Reticularis Thalami

Singer and Dräger (1972) showed that cells in the main geniculate layers are *disinhibited*, when the mesencephalic reticular formation is stimulated. The effect of the mesencephalic reticular formation stimuli upon complex neuronal field activity had already been shown by Moruzzi and Magoun (1949). Their results indicate a desynchronization of the EEG elicited by the mesencephalic reticular formation stimulus. Since that time many investigations have been performed showing influences of the mesencephalic reticular formation stimulation upon various parts of the brain, both at the cortical and subcortical levels (c.f. reference Yingling and Skinner, 1975; Sakai et al., 1976; Sasaki et al., 1976; Singer, 1977). These effects have been discussed with respect to problems of recruiting responses, arousal, attention, and motivation.

Within the primary visual system of cat, Suzuki and Taira (1961) and Singer (1973) demonstrated an influence of the mesencephalic reticular formation stimulation upon the synaptic transmission rate between retina and the lateral geniculate nucleus, and upon single-cell activity in the visual cortex (Singer et al., 1976). The convergence of retinal afferents and those of the nonspecific reticular system onto geniculate cells had already been shown by Suzuki and Taira (1961). Akimoto and Creutzfeldt (1958) and Creutzfeldt and Akimoto (1958) suggested an influence of the mesencephalic reticular formation onto the visual cortex.

Since an influence of visual cortex on the LGN has been demonstrated, the question arises whether corticofugal activity might mediate the mesencephalic reticular influence onto LGN. It was demonstrated by Satinsky (1966) that cortical ablation does not abolish the effect of mesencephalic reticular formation stimulation upon geniculate relay cell activity. The effect of mesencephalic reticular formation stimulation upon geniculate relay cells with a reversibly inactivated visual cortex was demonstrated by Schmielau and Singer (1977b). They suggested that the efficiency of a mesencephalic reticular formation stimulus under both the "closed-loop condition" (active visual cortex) and the "open-loop condition" (inactive visual cortex) mainly depends on the strength of intrageniculate inhibition. Thus, either different cortical structures or a subcortical structure (or both) have to be involved in mediating the influence of the mesencephalic reticular formation stimulation to LGN. A *direct* influence from the mesencephalic reticular formation upon LGN is still open; only in one work (Bowsher, 1970) has a monosynaptic input been suggested (degeneration study).

On the basis of the special considerations outlined above, the nucleus reticularis thalami is assumed to be the structure involved in transmitting the mesencephalic reticular formation output to LGN.

Methods

Extracellular recordings from 83 neurons of the feline nucleus reticularis thalami were obtained. Eight female cats of similar weight (2.5 kg) and with grey striped fur were used since these parameters are regarded to be critical for the reproducibility of the results with respect to (a) the location of brain structures and (b) topographical organisation in different phenotypes of cats.

Animals As was known from earlier recordings of lateral geniculate neurons (Schmielau, 1975) a special phenotype of cat is reflected in a special structural organisation of visual pathway neurons. Besides general physiological differences demonstrating the utility of a special phenotype of cats (i.e., the higher mortality of cats with black fur) as experimental animals, it was observed that cats with red striped fur, the so-called "Ginger Tom" cats, are, at least at the geniculate level of neuronal organisation, different from "normal" cats. They resemble Siamese cats insofar as within a projection column, usually containing only cells with similar receptive field (RF) locations throughout all LGN laminae, a sudden jump of the receptive field location into the "wrong" visual hemifield was noticed, when the electrode passed over from one lamina to another. Recent confirmation of genetically coupled differences in cats was given by Todd (1977). He showed that the sex-linked orange mutant is nearly exclusively male. In contrast, cats with grey fur are, i.e., either a nonmutant type (wild type) or the mutant "blotched tabby." These two types were used in the experiments presented here.

Anesthesia and general methods For surgery animals were initially anesthetized by intramuscular application of 25 mg of Ketanest. During the course of the preparation, 0.5 to 1 ml of a Ketanest:Ringer solution (15:85) was intravenously applied every 5 to 10 minutes, depending on the depth of anesthesia, as judged from the conventional criteria (e.g., electrocardiogram (EKG), reflexes, size of the pupil). In addition, local anesthesia of pressure points was secured by application of Xylocain. Further details of surgery and stereotactic holding of the animal might be derived from an earlier publication (Schmielau and Singer, 1977a). During the course of the experiment, animals were artificially respirated with 25 to 30 cm^3 of a nitrous oxide-oxygen mixture (7:3) at 25 strokes per minute. Fluid balance and paralysis were secured by intravenous application of a Ringer-Laevulose-Flaxedil-D-Tubocurarin solution (31.5:31.5:10:27) throughout the experiment. Experiments with animals in good physiological state (judged by the EEG, EKG, state of the eyes, retinal blood supply, etc.) were usually carried out for two or three days. Eyes were protected by black contact lenses with an artificial pupil that was 2 mm in diameter. At the beginning of the experiment, Atropine (5%) and Neosynephrine were applied to their eyes in order to dilate the pupil and withdraw the nictitating membranes. The contact lenses were removed hourly and the corneae were cleaned and moistened if it seemed necessary. The eyes were refracted and the locations of areae centralis and blind spots were plotted on the tangent screen. These positions were checked at regular intervals throughout the course of the experiment. The local electroencephalogram (EEG) (at the border of visual areas 17 and 18) and body temperature were monitored continuously. Body temperature was kept at 37.5°C by means of an electronically controlled heating blanket.

Cell recording Recordings of single cells were obtained by means of glass micro-pipettes (R = 5-10 MΩ), filled with K-citrate (1.5 mol), from the left hem-isphere at Horsley-Clarke coordinates A6-7, L9-10. Cells were discriminated from axons according to the common criteria. Absolute depth of cells (with respect to the cortical surface) and relative distance to the most dorsal relay cell in lamina A of the LGN were recorded. Cells from 10 mm below the cortical surface down to the ventral part of LGN lamina A_1 were investigated. Recep-tive fields of LGN relay cells were located in the central visual field. Only those cells which were identified as interneurons according to the criteria of Dubin and Cleland (1977) are presented. They were transsynaptically activated by visual cortical stimulation and showed a typical latency scatter of these responses (see stimulation).

Stimulation Each neuron was tested with light and various electrical stimuli. In general, the methods that were applied were similar, if not identical, to those of Dubin and Cleland (1977) in order to make the data comparable.

Stationary light spots of various diameters, ranging from 0.5-10° were presented monocularly and binocularly onto a tangent screen at a distance of 171 cm in front of the cat's eyes.

Single electrical stimuli of 50 μsec width were applied to either of the optical nerves, the optical chiasm (A14; L1, L-1; H5), the superior colliculus (A2; L1.5, L3; H13) and the mesencephalic reticular formation (A2; L3; H8). In addition, double shocks were applied to each structure in order to test inhibi-tory processes. Combinations of stimuli at various interstimulus time intervals which were delivered to both of the optic nerves were used to investigate bino-cular inhibition. For each neuron, repetitive stimulation of the mesencephalic reticular formation with five pulses of 50 μsec width and at intervals of 15 msec were used. The optic nerves were stimulated with bifurcating electrodes. For the other stimulation sites bipolar concentric electrodes were used (tip separa-tion, 1 mm). In cases of excitatory stimuli, thresholds were determined and recordings were taken at values of two times the threshold. The mesencephalic reticular formation was stimulated with an intensity sufficient to desynchronize EEG. For visual cortex stimulation, stainless steel needles were inserted into the white matter (4-5 mm) at Horsley-Clarke coordinates L2, A2, P3, P8. Stimulation was performed between two adjacent electrodes. In order to inves-tigate the effect of mesencephalic reticular formation stimulation on post-excitatory inhibition elicited by optical nerve, optical chiasm, superior colli-culus, visual cortex, or light stimulus, this test stimulus was preceded by a con-ditioning mesencephalic reticular formation stimulus at various interstimulus time intervals. Usuallly, interstimulus time intervals of 60-100 msec were used; they have been shown to evoke disinhibition in LGN cells (Singer, 1973; Schmielau and Singer, 1977b).

Cortical cooling The influence of visual cortex onto nucleus reticularis thalami cells was investigated by cooling the exposed parts of visual areas 17 and 18. The technique and control measurements were the same as those described by Schmielau and Singer (1977a). In each cat the number of cooling cycles was restricted to three in order to avoid cortical damage. Properties of a given neuron of the nucleus reticularis thalami were investigated before (pre-cool), during (cool), and after (post-cool) cooling in order to assure the reversibility of inactivation effects.

Recording of data and evaluation Evoked potentials, EEG, and EKG were recorded and Polaroid photographs were taken at regular intervals throughout the experiment. Cell responses were registered in the same way. Poststimulus time histograms of cell responses evoked by light and electrical stimuli were registered with the help of a PDP 11/34 computer.

Histology Recording and stimulation sites were identified by a series of 40 μm thick coronary sections of both hemispheres. Nissl stains were used to demonstrate segregations of cell laminae within the nucleus reticularis thalami and the LGN. A more detailed description of the methods used will be given elsewhere (Schmielau, 1979).

Results

Cells of the nucleus reticularis thalami were identified as interneurons in the classical sense (see above) by two properties: (a) they were transsynaptically activated by visual cortex stimulation and (b) their response patterns to electrical stimulation showed a significant higher scatter than those of geniculate relay cells. At stimulus strength of twice the threshold the characteristic response of nucleus reticularis thalami neurons to optical nerve, optical chiasm, visual cortex, and superior colliculus stimuli was a grouped discharge consisting of up to seven spikes. Threshold values for eliciting nucleus reticularis thalami responses were in general much lower than those of layer A, A_1 geniculate relay cells. As has been pointed out recently (Dubin and Cleland, 1977), LGN neurons are antidromically activated with latencies to visual cortical stimulation of about 0.5 msec showing hardly any scatter.

Primary excitation As can be seen from Fig. 1, 76 (92%) of a sample of 83 neurons were driven by cortical stimulation. No differences with respect to excitability were observed when the stimulus was applied between A2 and P3 or P3 and P8. Fifty-three (64%) cells responded to visual cortical stimulation with more than one spike (two to seven). The latency of the first spike ranged between 0.8 and 4.0 msec. In 19 (23%) cells the response to a visual cortical stimulus consisted only of a single spike (latency range: 0.8-4.9 msec).

After stimulation of the optical chiasm, 71 (86%) cells were activated; 46 of these cells responded with two to six spikes in contrast to 25 in which only *one*

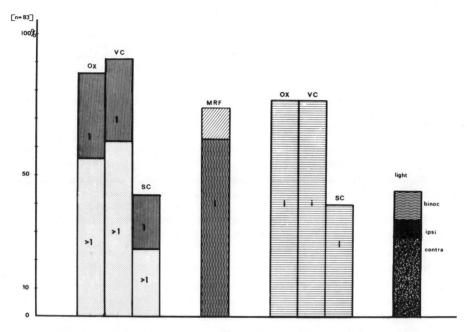

FIGURE 1 Response properties of nucleus reticularis thalami neurons. Data derived from 83 neurons. See text for details.

spike was elicited. The total latency range in both groups extended from 0.6 to 6.0 msec. For comparison the latencies of Y-cells in the geniculate nucleus range between 0.9 and 1.4 msec and those of X-type cells range between 1.5 and 2.5 msec (Singer and Bedworth, 1973; Schmielau, 1976).

Thirty-six (43%) of the nucleus reticularis thalami cells were excited by superior colliculus stimulation with several (n = 20) or with a single (n = 14) spike(s) ranging from 1.0 to 4.5 msec. This is probably primarily an antidromical activation of bifurcating Y-type ganglion cell axons projecting both to the superior colliculus and LGN.

Postexcitatory inhibition From the third block of data in Fig. 1, it can be seen that nucleus reticularis thalami neurons are inhibited as well by optical chiasm, visual cortical, and superior colliculus stimulation (single stimulus at stimulus strength twice the threshold). This inhibition of spontaneous activity which follows a primary excitation is present in most cells, however, at different percentage values with respect to the stimulation site. Whereas optical chiasm and visual cortical stimuli inhibit 64 (77%) cells, a stimulus applied to the superior colliculus elicits inhibition only in 33 (40%) of the neurons of the nucleus reticularis thalami. This disinhibitory effect is seen very clearly even without applying averaging methods such as poststimulus time histograms, since the spontaneous activity in nucleus reticularis thalami cells is usually very high.

The average value for all neurons recorded is 30 per sec (range: 5-100 per sec). From occasional quasi-intracellular recordings it was possible to show that the postexcitatory inhibition elicited by optical chiasm, visual cortical, and superior colliculus stimulation is postsynaptic, since inhibitory postsynaptic potentials were elicited. The same conclusions were derived from double-shock experiments at various interstimulus time intervals. The postexcitatory inhibition of nucleus reticularis thalami cells due to optical chiasm, visual cortical, and superior colliculus stimulation is considerably longer than in LGN cells of cats at the same level of anesthesia (Schmielau, 1977). On average, optical chiasm elicited inhibition in nucleus reticularis thalami cells lasted for 161 msec (range: 30-500 msec), whereas inhibition after visual cortical stimulation had a duration of 185 msec (range: 40-330 msec). A superior colliculus stimulus caused inhibition of 121 msec duration (range: 30-300 msec).

Mesencephalic reticular formation stimulation Repetitive mesencephalic reticular formation stimulation, which has been shown to be *disinhibitory* in LGN relay cells (Singer and Dräger, 1972; Schmielau, 1977) and mainly *excitatory* in neurons of the visual cortex (Singer et al., 1976), is apparently mainly *inhibitory* in nucleus reticularis thalami cells. In Fig. 1 it can be seen that 61 (74%) of the nucleus reticularis thalami cells were influenced by the mesencephalic reticular formation stimulus. Fifty-two (63%) cells were inhibited, nine (11%) were activated. In 50 neurons the average duration of inhibition was 223 msec (range: 50-600 msec). Even single mesencephalic reticular formation stimuli were able to elicit a pronounced inhibition. However, in these cases the duration of inhibition was generally shorter and showed a somewhat different pattern. As in the case of optical chiasm, visual cortical, and superior colliculus elicited inhibition, mesencephalic reticular formation induced inhibition is also *postsynaptic*. Occasionally, when a neuron could be held long enough to apply the entire test program mentioned above, a single mesencephalic reticular formation stimulus was found to evoke an orthodromic short latency response in nucleus reticularis thalami cells consisting of several spikes with a latency range of the first spike of 0.5-3.4 msec (see below). The activation found in nine nucleus reticularis thalami cells by repetitive stimulation of mesencephalic reticular formation lasted from 100 msec to 1 sec.

Light stimulation As can be seen from the right-most column in Fig. 1, 37 (45%) neurons of the nucleus reticularis thalami, *including* the nucleus perigeniculatus, were found to be excitable by light stimulation. Most of them (n = 19) were activated only by a light stimulus applied to the contralateral eye, five neurons were excited by the ipsilateral eye, and 13 neurons were driven binocularly. As will be shown elsewhere (Schmielau, 1979) receptive fields of nucleus reticularis thalami neurons, however, do not possess uniform RFs. The light-evoked response varies with the size of the spot and with the location of the spot in the RF. In addition, even binocular nucleus reticularis thalami cells do show preference for one eye. Thus, several binocular cells might have been

identified as monocular ones, because much time is needed to show binocularity with small spots projected into different areas of the RF of a nucleus reticularis thalami cell, and this procedure could only be carried out in a few neurons.

*Properties of nucleus reticularis thalami neurons as a function of the location within
the nucleus* Since the size of the perigeniculate and the thalamic reticular
nucleus that is reported in neurophysiological investigations varies from one
author to another, special effort was taken to deal with this problem. The location of each nucleus reticularis thalami cell was recorded with respect to its distance from (a) the cortical surface and (b) the most dorsal relay cell in lamina
A (see Methods). In order to get reproducible results without a large intra-
individual scatter among brains, cats with special features were used. This
method proved to be successful since the lateral geniculate nucleus was always
found at approximately the same location within the brain. This was confirmed
both by neurophysiological and histological methods. The dorsal border of the
LGN at electrode penetration sites (Horsely-Clarke) A6-7, L9-10 was found in
a depth of about 14 ($\pm$1) mm.

At this electrode position the nucleus perigeniculatus was situated immediately dorsal to lamina A. This nucleus is regarded as part (a specialized subgroup) of the nucleus reticularis thalami. This will be shown below. The rest
of the nucleus reticularis thalami is located dorsal to the nucleus perigeniculatus
and separated from it by a cell-free layer of approximately 50 μm. These
findings are confirmed both by neurophysiological and histological methods.

In the upper part of Fig. 2 the response characteristics of nucleus reticularis
thalami neurons to light stimulation and electrical stimulation of the optic
chiasm, visual cortex, and superior colliculus are plotted as a function of the
distance to the very first relay cell encountered in lamina A. In contrast to cells
of the nucleus reticularis thalami, relay cells in lamina A were only excited by
stimulation of the contralateral eye. They were identified as relay cells since
they responded to electrical stimulation of the visual cortex with a single spike
of about 0.5 msec with hardly any scatter ($<$ 0.1 msec in a given cell). This
spike is elicited by antidromic invasion of the relay cell soma. Cells of the
nucleus reticularis thalami, however, were transsynaptically activated by visual
cortical stimulation; most of them responded with several spikes. The
transsynaptic character of these responses was demonstrated by collision tests.
The same method as that of Dubin and Cleland (1977) was used.

As might be seen from Fig. 2, nucleus reticularis thalami neurons were found
from locations more than 4 mm above LGN down to the close vicinity of LGN
lamina A (minimal distance 10 μm). In Fig. 2(A) the ocularity of nucleus reticularis thalami cells is plotted as a function of the height h above LGN> cells
is plotted as a function of the height h above LGN. Binocular cells (filled
boxes) are found as far as 1500 μm from lamina A, however, concentrating in
the nucleus perigeniculatus of the nucleus reticularis thalami. Besides that,
monocular cells are found as well, both contralateral (filled circles) and ipsilateral (open circles) ones. The most dorsal cell excited by stimulation of the

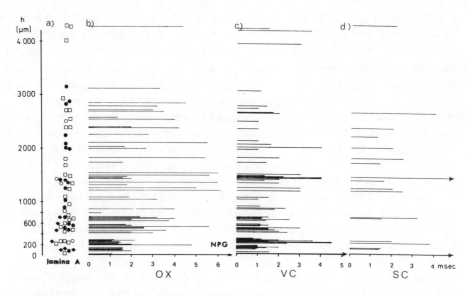

FIGURE 2 Ocular dominance and latency distribution of nucleus reticularis thalami neurons as a function of the height above LGN lamina A. Data derived from 61 neurons. See text for details.

contralateral eye was found at h = 3000 μm. The most distant ipsilateral one was located at about 2500 μm from lamina A. For the monocular cells there is a tendency to be distributed across the whole magnitude of the nucleus reticularis thalami. Besides that, many cells which could not be excited by light were found throughout the whole nucleus. This, however, in some cases may be due to inappropriate stimuli (see above) or due to other effects (Schmielau, 1979). The receptive field location of nucleus reticularis thalami cells within the contralateral visual hemifield was in each case close to that of relay cells of the underlying geniculate layers. This demonstrates clearly that projection columns continue from LGN to nucleus reticularis thalami, as has been noted by Sanderson (1971) and Dubin and Cleland (1977) for neurons of the perigeniculate nucleus. In contrast to Sanderson (1971) who has described monocular (n = 83) and binocular (n = 136) cells within the nucleus perigeniculatus up to 2 mm above LGN, Dubin and Cleland (1977) mentioned only binocular cells (n = 19) up to 500μm dorsal from lamina A.

Cells of the nucleus perigeniculatus, as well as those of the more dorsal parts of the nucleus reticularis thalami, often responded to both the onset and offset of light with a short burst (on/off cells). As will be shown elsewhere (Schmielau, 1979) the relative intensity of on and off responses may vary with the size and the site of stimuli within the receptive field.

Besides some bias concerning the distribution of ocular dominance in nucleus reticularis thalami cells with respect to the location of cells within special parts of the nucleus, the data derived from electrical stimulation of cortical and sub-cortical structures clearly demonstrate electrophysiological differences between cells of the nucleus perigeniculatus and those of the rest of the nucleus reticu-laris thalami. In Fig. 2(B), latencies of the first (or the single) spike elicited in nucleus reticularis thalami cells by optical chiasm stimulation are plotted as a function of the distance from lamina A. In contrast to the more dorsal parts of the nucleus reticularis thalami, most of the cells of the nucleus perigeniculatus respond to this stimulus with latencies between 1.0 and 1.5 msec. Comparing the situation in LGN relay cells, this distribution is clear evidence for a *monosynaptic* retinal input. By stimulation of either of the optic nerves it was demonstrated that this input is binocular, since latencies to both stimuli were similar. Only four NPG cells were found to respond with latencies longer than 1.5 msec. On the other side the overwhelming majority of nucleus reticularis thalami cells dorsal from nucleus perigeniculatus show much longer latencies to optical chiasm stimulation, i.e., up to 6 msec for the first spike.

The latency distribution for the first spike (or the only one) elicited by the visual cortical stimulus in nucleus reticularis thalami cells, as a function of the distance from LGN is shown in Fig. 2(C). As has been mentioned, latencies range between 0.8 and 4.6 msec. The separation of cells between nucleus peri-geniculatus and the more dorsal parts of nucleus reticularis thalami is evident, as it is for the optical chiasm latency distribution (Fig. 2(B)). In both nucleus perigeniculatus and nucleus reticularis thalami latencies to visual cortical stimu-lation are similarly distributed. The clustering of long-latency cells at $h = 100$ μm and at $h = 1400$ μm, however, has to be discussed in more detail.

In Fig. 2(D) the latencies of the first or the unique spike of nucleus reticu-laris thalami cells to superior colliculus stimulation are plotted. It is shown that throughout the entire nucleus, cells which were driven by a superior colliculus stimulus with latencies from 1.4 to 10 msec were found. In contrast to LGN relay cells in which the latencies for a given cell to optic chiasm and superior colliculus stimulation show nearly a 1:1 relation, the situation is different in the nucleus reticularis thalami. Here optic chiasm latencies are usually shorter than those with superior colliculus stimulation.

As has been mentioned above, the primary excitation of nucleus reticularis thalami neurons due to electrical stimulation of the optic chiasm, visual cortex, and superior colliculus is usually directly followed by an inhibitory interval. Only in rare cases ($n = 3$) did one of these stimuli cause a long excitation (see Fig. 3). In contrast, mesencephalic reticular formation stimulation either excites or inhibits a neuron of the nucleus reticularis thalami; the vast majority of these cells, however, is inhibited.

In Fig. 3 the duration of inhibition and excitation of the nucleus reticularis thalami neurons is plotted as a function of the distance from geniculate lamina A. From the second and third columns in Fig. 3, it may be seen that the dura-tion of inhibition due to optic chiasm and visual cortical stimulation does not change very much within the vertical extent of the nucleus reticularis thalami. The duration of visual cortex induced inhibition, however, is generally longer.

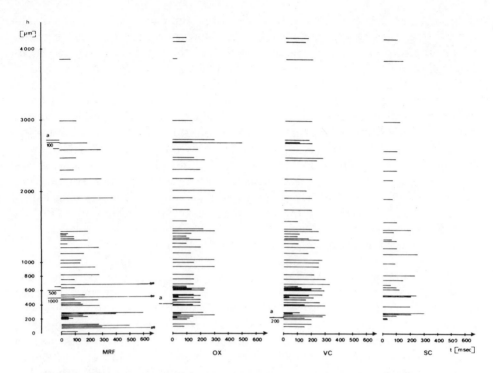

FIGURE 3 Inhibition and activation of nucleus reticularis thalami cells due to electrical stimulation as a function of the height above LGN lamina A. Data derived from the same sample of neurons as Fig. 2. See text for details.

Some clusters of neurons which are only shortly inhibited by both stimuli are found within a part of the nucleus reticularis thalami close to the LGN (h ⩽ 500 μm). This is especially true for the optic chiasm induced inhibition. From these data, as well as from those of Fig. 2, the separation between the nucleus perigeniculatus (h ⩽ 300 μm) and the rest of the nucleus reticularis thalami stands out well. Within the lower 100 μm of the nucleus perigeniculatus, however, no cells were found which were inhibited by optic chiasm or visual cortical stimulation. As might be seen from the fourth column in Fig. 3, the duration of postexcitatory inhibition in nucleus reticularis thalami cells due to superior colliculus stimulation is generally even shorter than that elicited by an optic chiasm stimulus. In the lower two-thirds of the nucleus perigeniculatus (h ⩽ 200 μm) no cells were found which were inhibited by stimulation of the superior colliculus. The lack of all three types of inhibition in the very deep part of the nucleus reticularis thalami is a striking feature contrasted by the abundance of short latency primary excitation especially demonstrated in Fig. 2. Neurons with long latency inhibitory intervals caused by repetitive stimulation of the mesencephalic reticular formation, however, were predominantly found in the perigeniculate nucleus and the adjacent more dorsal parts of the nucleus reticularis thalami up to h = 700 μm. In some neurons, spontaneous activity was inhibited by the mesencephalic reticular formation stimulus for more than 2.5 sec.

Neurons activated by mesencephalic reticular formation stimulation Two regions were found within the nucleus reticularis thalami above the nucleus perigeniculatus, one at 500-700 μm above the lamina A and another at 2.600-2.750 μm in which neurons were activated by repetitive stimulation of the mesencephalic reticular formation. Occasionally, a single stimulus was applied to the mesencephalic reticular formation. By that method five neurons were found within the nucleus reticular thalami outside of the nucleus perigeniculatus which could be activated orthodromically. Their response consisted of up to four spikes of which the first one had a latency of 0.5-3.4 msec. The latency scatter was similar to that of spikes elicited by other stimulation sites. However, in these cells the latencies due to optic chiasm stimulation were very long; with one exception the very first of several spikes was triggered only after 4-5 msec. Two cells were found to be excitable by the contralateral eye. The visual cortical stimulus elicited grouped discharges typical for nucleus reticularis thalami cells at normal latencies (> 2.0 msec). After optic chiasm, visual cortical, and superior colliculus stimulation these neurons showed postexcitatory inhibition as most of the nucleus reticularis thalami neurons did.

Cortical inactivation and change of nucleus reticularis thalami cell properties Cortical cooling, which has been shown to be an efficient tool for the investigation of feedback loops to subcortical structures was used in order to demonstrate the influence of corticofugal activity onto neurons of the nucleus reticularis thalami. From anatomical observations it was tempting to assume a participation of visual cortex in controlling the nucleus reticularis thalami. From neurophysiological data, however, this influence was, at least, a point of controversial discussion.

The data presented here are regarded by the author as a first step in the direction of understanding the high complexity and interdependency of LGN, mesencephalic reticular formation, visual cortex, and nucleus reticularis thalami activity at the single-unit level. Reversible inactivation of structures within the brain is, on the other hand, a powerful technique for the understanding of deficits caused by natural lesions. Withdrawing one major influence may severely distort the equilibrium obtained under normal working conditions and thus reduce the efficiency of the brain machinery.

By cooling the visual cortex the response properties of nucleus reticularis thalami neurons were changed significantly. An example of this is given in Fig. 4. When stimulating optical chiasm the response of a nucleus reticularis thalami cell located 600 μm above lamina A responded with the usual burst of spikes (latency of the first spike, 2.5 msec). By cortical cooling the latency of the first spike was not changed; however, the cell responded with much more prolonged primary excitation (100 msec). For the initial interval this is demonstrated in Fig. 4(E). When visual cortex had been allowed to re-warm passively (48 min after the offset of cooling), the response pattern to optic chiasm stimulation was identical to that of the "pre-cool" situation. Thus, as could be demonstrated in other neurons, the effect of cortical cooling upon nucleus reticularis thalami cell properties is reversible. This has been shown to be true for LGN relay cells before (Schmielau and Singer, 1977a). As might be seen from

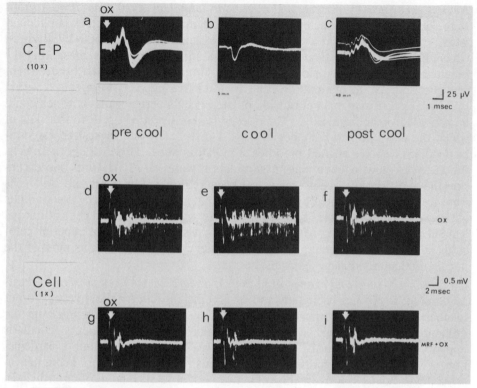

FIGURE 4 Effects of cortical cooling upon cortical evoked potential and optical chiasm elicited activation and mesencephalic reticular formation elicited inhibition of a single nucleus reticularis thalami cell. See text for details.

the cortical evoked potential to optic chiasm stimulation, after 48 min of rewarming, in this cell the cortical evoked potential had nearly the same pattern as in the pre-cool state of the visual cortex. However, the changes observed in nucleus reticularis thalami cells had been restored about 15 min after the offset of cooling. Besides the prolongation of primary excitation seen in nucleus reticularis thalami cells with inactivated visual cortex ("cool"), the strength of the postexcitatory inhibition is altered too. When stimulating the mesencephalic reticular formation 100 msec before the optic chiasm stimulus in the "pre-cool" state, the primary excitation (burst) elicited by the optic chiasm stimulus was cut down to a single spike. This may be seen from Fig. 4(G). With inactivated visual cortex, the same effect was observed (Fig. 4(F)). This is due to the inhibitory influence of the mesencephalic reticular formation demonstrated above, seen in most of the nucleus reticularis thalami cells. Thus, the mesencephalic reticular formation is still effective, even when visual cortex has been inactivated. In LGN relay cells, in contrast, the disinhibitory effect of the mesencephalic reticular formation stimulation was not abolished by cortical cooling (Schmielau and Singer, 1977b). When the mesencephalic reticular formation alone was stimulated, the strength of inhibition due to this stimulus was increased as compared to the "pre-cool" and "post-cool" situation. In addition, cortical cooling reversibly changed the spontaneous discharge rate. As has been observed before for LGN relay cells (Schmielau and Singer, 1975) in most of the cells, the spontaneous activity of nucleus reticularis thalami cells was reduced as compared to the pre- and post-cool situations. As, however, there surely are different populations (at least two) of nucleus reticular thalami cells, further investigations are needed to demonstrate the differential influence of visual cortex upon these subtypes of nucleus reticularis thalami cells.

Discussion

The data obtained by electrical stimulation of the optical chiasm demonstrate clearly that the properties of perigeniculate cells differ from those of cells in more dorsal parts of the nucleus reticularis thalami. Nucleus perigeniculatus cells receive a monosynaptic input from retinal ganglion cells of both eyes. As has been shown by the latency distribution of nucleus perigeniculatus cells to optical chiasm and optical nerve stimulation (the latter has not been shown here), this retinal input is clearly of the Y-type. These findings are contrary to those of Dubin and Cleland (1977) insofar as they mentioned a polysynaptic innervation of nucleus perigeniculatus cells. However, they stressed, too, that the input is a Y-type one. Polysynaptic retinal input, probably due to innervation of nucleus reticularis thalami cells by axon collaterals of LGN relay cells, has been shown predominantly in the more dorsal parts of the nucleus reticularis thalami *outside* of the nucleus perigeniculatus. The Y-type input to nucleus reticularis thalami cells in general could be demonstrated by recording the light responses. They were of the phasic type (illustrations will be given elsewhere).

  The identification of nucleus reticularis thalami cells, including nucleus perigeniculatus, as interneurons is similar to those given by Dubin and Cleland (1977). They respond to visual cortex stimulation with a transsynaptically elicited burst of spikes. Each spike, and especially the late ones, show a high degree of scatter. The results demonstrated above differ from those of Dubin and Cleland (1977) insofar as a large number of nucleus reticularis thalami cells is activated by visual cortical stimulation with a single spike, even at high stimulus intensities. On the basis of the findings mentioned above and previous findings (Schmielau and Singer, 1977a; Schmielau and Singer, 1977b; Singer and Schmielau, 1975), the following model is suggested (Fig. 5): cells of the perigeniculate nucleus of the nucleus reticularis thalami are innervated by monosynaptic input from both eyes. As these cells have been shown to be interneurons in the "conventional" sense and as it has been demonstrated by Ahlsen et al. (1978) that they send axons down to both main laminae of the LGN, binocular inhibition in LGN may be explained by these neurons. The visual cortex which sends axons down to the nucleus reticularis thalami and

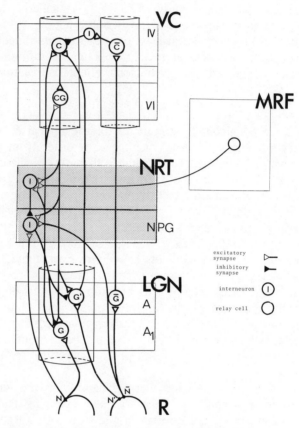

FIGURE 5 Simplified wiring diagram of the nucleus reticularis thalami with retinal, geniculate, mesencephalic reticular formation, and visual cortex afferents. No intrinsic interneurons are included. See text for details.

LGN in a retinotopical organized way has been shown by anatomical and neurophysiological findings to be excitatory with respect to *any* postsynaptic structure (Tömböl, 1966a,b; Szentágothai et al., 1972; Schmielau and Singer, 1975; Schmielau and Singer, 1977a,b). As binocular facilitation in LGN relay cells is no longer present when visual cortex is inactivated, the conclusion was drawn (Schmielau and Singer, 1977a) that layer VI pyramidal cells (complex cells) mediate binocular facilitation down to geniculate relay cells. The broad tuning of LGN relay cells by binocular inhibition that has been demonstrated by many authors is understandable on the basis of the broad axonal sprouting seen recently (Ahlsen et al., 1978). The precise binocular overlap which is needed to produce binocular facilitation in relay cells of the lateral geniculate nucleus is secured by the precise corticofugal backprojection to LGN. As corticofugal axons make synaptic contacts with nucleus reticularis thalami neurons by axon collaterals (neuropil field) (Scheibel and Scheibel, 1966; Tömböl, 1978), the influence of visual cortex onto nucleus reticularis thalami cells both in the nucleus perigeniculatus and the more dorsal parts of the nucleus reticularis thalami, described above, are more understandable. Assuming the separation of nucleus reticularis thalami in at least two groups with neurons of functionally different properties and postulating an inhibitory influence from "dorsal" nucleus reticularis thalami cells onto perigeniculate cells (Fig. 5), disinhibitory influences from cortex in some nucleus reticularis thalami cells may be understood. The inhibitory influence of "perigeniculate" cells postulated by Dubin and Cleland and the results mentioned above are the basis for effects of mesencephalic reticular formation stimulation seen in relay cells of both geniculate layers A and A_1. The short latency excitatory input from neurons of the mesencephalic reticular formation onto the "dorsal" nucleus reticularis thalami cells explains both effects of mesencephalic reticular formation stimulation in the nucleus reticular thalami. One cell type of the nucleus reticularis thalami receiving direct input is activated, whereas the other cell type is inhibited. The inhibitory influence of the nucleus reticular thalami cells of the second type, which is mediated to LGN relay cells, is blocked by the mesencephalic reticular formation stimulus. As, however, all structures mentioned here ("dorsal cells" of the nucleus reticularis thalami, perigeniculate neurons, and geniculate relay cells) are influenced by the visual cortex, the results obtained from cortical cooling can be understood on the basis of excitatory corticofugal activity. This activity influences the equilibrium set in subcortical structures as the LGN and the nucleus reticularis thalami. Differential effects to the mesencephalic reticular formation stimulation which were seen recently in LGN (Schmielau and Singer, 1977b) are due to the withdrawal of corticofugal activity in the nucleus reticularis thalami cells. The perigeniculate nucleus is a specialized substructure of the nucleus reticularis thalami mediating mainly visual input. In contrast, cells of the more dorsal part of the nucleus reticularis thalami receive visual, acoustic, somato-sensory, vestibular, oculomotor, and, possibly, various cortical influences. Thus, the findings reported by Sanderson (1971) and Dubin and Cleland (1977) that cells of the more dorsal part of the "perigeniculate nucleus" are only slightly excitable by light stimulation can now be understood on the basis of the high convergence of afferents seen in the anatomic picture (Scheibel and Scheibel, 1966).

Acknowledgements   The author thanks Professor E. Pöppel for his help and advice during the course of this investigation. I am grateful for the stimulating discussion with Professor J. Szentágothai and Professor T. Tömböl and the skilled assistance of B.Sc. D. Carr concerning the computer programming. I thank Mrs. P. Mitterhusen and Mrs. G. Lorenz for much secretarial assistance.

References

Ahlsen, G., and S. Lindström (1978). Axonal branching of functionally identified neurones in the lateral geniculate body of the cat. NELED Suppl. 1:156.

Ahlsen, G., and S. Lindström (1978). Projection of perigeniculate neurones to the lateral geniculate body in cat. NELED Suppl. 1:367.

Akimoto, H., and O. Creutzfeldt (1958). Arch. Psychiat. Nervenkr. 196:539 pp.

Bowsher, D. (1970). Reticular projections to the lateral geniculate in cat. Brain Res. 23:247-249.

Burke, W., and A. J. Sefton (1966). Discharche patterns of principal cells and interneurones in LGN in the rat. J. Physiol. 187:201-212.

Creutzfeldt, O., and H. Akimoto (1958). Konvergenz and gegenseitige Beeinflussung von Impulsen aus der Retina und den unspezifischen Thalamuskernen auf einzelne Neurone des optischen Kortex. Arch. Psychiat. Nervenkr. 196:520-538.

Dubin, M. W., and B. G. Cleland (1977). Organization of visual inputs to interneurons of lateral geniculate nucleus of the cat. J. Neurophysiol. 40:410-427.

Garey, L. (1978). This volume.

Guillery, R. W. (1978). Discussion: Symposium, Development and chemical specificity of neurons, 11-15 September 1978, Davos, Schweiz.

Holländer, H. (1978). Unpublished results, personal communication.

Jones, E. G. (1975). Some aspects of the organization of the thalamic reticular complex. J. Comp. Neurol. 162:285-308.

Montero, V. M., R. W. Guillery, and C. N. Woolsey (1977). Retinotopic organization within the thalamic reticular nucleus demonstrated by a double label autoradiographic technique. Brain Res. 138:407-421.

Moruzzi, G., and H. W. Magoun (1949). Brain stem reticular formation and activation of the EEG. Electroenceph. Clin. Neurophysiol. 1:455-473.

Negishi, K., E. S. Lu, and M. Verzeano (1961). Neuronal activity in the lateral geniculate body and the nucleus reticularis of the thalamus. Vis. Res. 1:343-353.

O'Leary, J. L. (1940). A structural analysis of the lateral geniculate nucleus in the cat. J. Comp. Neurol. 73:405-430.

Ranson, S. W., and S. L. Clarke (1953). Anatomy of the nervous system. Saunders, Philadelphia and London.

Rose, J. E. (1952). The cortical connections of the reticular complex of the thalamus. Res. Publ. Ass. Nerv. Ment. Dis. 30:454-479.

Sanderson, K. J. (1971). The projection of the visual field of the lateral geniculate and medial interlaminar nuclei in the cat. J. Comp. Neurol. 143:101-108.

Sasaki, K., T. Shimono, H. Oka, T. Yamamoto, and Y. Matsuda (1976). Effects of stimulation of the midbrain reticular formation upon thalamocortical neurones responsible for cortical recruiting responses. Exp. Brain Res. 26:261 pp.

Satinsky, D. (1966). Reticular influences on lateral geniculate neuron activity. Electroenceph. Clin. Neurophysiol. 25:543-549.

Scheibel, M. E., and A. B. Scheibel (1966). The organization of the nucleus reticularis thalami: a Golgi study. Brain Res. 1:43-62.

Schmielau, F. (1975). Corticale Beeinflussung der Verarbeitung visueller Information im Corpus geniculatum laterale. Dissertation, Fachbereich Physik der Ludwig-Maximilians- Universität München.

Schmielau, F. (1978). Cortical influence of the nucleus reticularis thalami in cat: a combined neurophysiological and neuroanatomical study. J. Comp. Neurol. (in preparation).

Schmielau, F., and W. Singer (1975). Corticofugal control of the cat lateral geniculate nucleus. Exp. Brain Res. (Suppl.) 23:363.

Schmielau, F., and W. Singer (1977a). The role of visual cortex for binocular interactions in the cat lateral geniculate nucleus. Brain Res. 120:354-361.

Schmielau, F., and W. Singer (1977b). The importance of visual cortex in reticular disinhibition in cat LGN. Proc. Int. U. Physiol. Sc. 13:1994.

Singer, W. (1973). The effect of mesencephalic reticular stimulation on intracellular potentials of cat lateral geniculate neurons. Brain Res. 61:35-54.

Singer, W. (1977). Control of thalamic transmission by corticofugal and ascending reticular pathways in the visual system. Physiol. Rev. 57:386-420.

Singer, W., and N. Bedworth (1973). Inhibitory interaction between X and Y units in the cat lateral geniculate nucleus. Brain Res. 49:291-307.

Singer, W., and U. Dräger (1972). Postsynaptic potentials in relay neurons of the cat lateral geniculate nucleus after stimulation of the mesencephalic reticular formation. Brain Res. 41:214-220.

Singer, W., F. Tretter, and M. Cynader (1976). The effect of reticular stimulation on spontaneous and evoked activity in the cat visual cortex. Brain Res. 102:71-90.

Snider, R. S., and W. T. Niemer (1970). A stereotaxic atlas of the cat brain. Univ. of Chicago Press, Chicago, Ill., 3d ed.

Sumimoto, I., M. Nakamura, and K. Iwama (1976). Location and function of the so-called interneurons of rat lateral geniculate body. Exp. Neurol. 51:110-123.

Sumimoto, I., M. Sugitani, and K. Iwama (1977). Disinhibition of perigeniculate reticular neurons following chronic ablation of the visual cortex in rats. Tohoku J. exp. Med. 122:321-329.

Suzuki, H., and E. Kato (1966). Binocular interaction at cat's lateral geniculate body. J. Neurophysiol. 29:909-920.

Szentágothai, J. (1972). Lateral geniculate body structure and eye movement. In: Cerebral Control of Eye Movements and Motion Perception. S. Karger, Basel.

Szentágothai, J., J. Hamori, and T. Tömböl (1966). Degeneration and electron microscope analysis of the synaptic glomeruli in the lateral geniculate body. Exp. Brain Res. 2:283-301.

Thuma, B. D. (1928). Studies on the diencephalon of the cat. I. The cytoarchitecture of the corpus geniculatum laterale. J. Comp. Neurol. 46:173-199.

Todd, N. B. (1977). Cats and commerce. Sci. Am. 237:100-107.

Tömböl, T. (1967). Short axon neurons and their synaptic relations in the specific thalamic nuclei. Brain Res. 3:307-326.

Tömböl, T. (1969). Two types of short axon (Golgi 2nd) interneurones in the specific thalamic nuclei. Acta morph. Acad. Sci. hung. 17:285-297.

Tömböl, T. (1978). Unpublished results, personal communication.

Vastola, E. F. (1960). Binocular inhibition in the lateral geniculate body. Exp. Neurol. 2:221-231.

Yingling, C. D., and J. E. Skinner (1975). Regulation of unit activity in nucleus reticularis thalami by the mesencephalic reticular formation and the frontal granular cortex. Electroencephal. and clin. Neurophys. 39:635-642.

A Hypothesis on the Efferent System from the Visual Cortex

GIORGIO M. INNOCENTI

Institute of Anatomy
University of Lausanne
Lausanne, Switzerland

Abstract There is evidence suggesting that the radial and tangential distributions of neurons within the sensory cortices may be main factors determining their functional properties. Cortical neurons projecting to different structures have, in general, different radial location and can have different tangential location; thus probably the cortex sends to different structures messages of different format. In the case of neurons efferent to the contralateral cortex, via the corpus callosum, the restricted radial and tangential distributions typical of the adult are acquired postnatally in somatosensory and visual areas of the cat, through a process of tangential and radial reductions from a widespread neonatal distribution. In the visual cortex, this maturational process can be affected by strabismus or by monocular deprivation. These manipulations of the visual experience result in adults having a more widespread distribution of callosal neurons in area 17 than normally observed.

A sensory cortex is capable of sending simultaneous messages over separate channels to different target structures. The *format* of each message consists of the set of functional properties (mainly receptive field properties, including location) of the cortical neurons projecting to the target structure to which the message is addressed.

There appears to be a high degree of order in the functional architecture of the cortex. Some functional properties of cortical neurons vary across the tangential (parallel to the pial surface) dimension of the cortex in either a continuous or discrete way, while they remain invariant along the radial (normal to the pial surface) dimension of the cortex. This particular distribution of functional properties (usually referred to as "columnar"; cf. Mountcastle, 1978) is found in the somatosensory cortex for receptive field (RF) position, responsiveness to sensory modality and type of adaptation (Mountcastle, 1957); in the visual cortex for orientation specificity and to some extent for ocular dominance

227

(Hubel and Wiesel, 1962, 1968, 1974); and in the auditory cortex for responsiveness to tone frequency and for binaural interaction (Abeles and Goldstein, 1970; Imig and Adrian, 1977).

Differently from the somatosensory cortex, in the visual cortex RF position has not been clearly interpreted as defining columns. However, implicit to the existence of cortical retinotopic maps is the notion that RF position of visual cortical neurons does vary systematically across the tangential cortical dimension but not across the radial one. A non-negligible degree of scatter is found, however, along a radial line across the cortex (Hubel and Wiesel, 1962; Creutzfeldt et al., 1974) and is also superimposed on the trend of variation in the RF position across the tangential cortical dimension (Albus, 1975).

Other functional properties of cortical neurons in the visual cortex, such as the degree of binocularity, RF type, and others (Hubel and Wiesel, 1962; Gilbert, 1977; Poggio, Doty, and Talbot, 1977; Leventhal and Hirsch, 1978) appear to vary in a consistent way along the radial, but not along the tangential cortical dimension. Finally, some functional properties appear to vary both along the radial and the tangential dimensions; for instance, in the visual cortex, RF size (Albus, 1975; Gilbert, 1977; Leventhal and Hirsch, 1978).

Structures that we call *efferent cortical zones* (Innocenti et al., 1977; Innocenti, 1978, 1979) are defined within the cortex by the radial and tangential location of neurons of origin of different corticofugal tracts. Different efferent zones have (with a few exceptions) different radial locations (Gilbert and Kelly, 1975; Lund et al., 1975; Wise and Jones, 1977) and can have different tangential locations (cf. below). The cortical efferent zones participate in two related operations: by means of the initial axon collaterals of each corticofugal neuron they contribute to the intracortical processing of afferent information; by means of the radial and tangential distribution of the efferent neurons and their dendrites they determine the particular format of each efferent message. The latter operation is probably a consequence of the fact that certain morphological types of cortical neurons receiving only a limited set of intrinsic and extrinsic afferents participate in each corticofugal system.

In normal adult cats (Innocenti and Fiore, 1976; Innocenti, in preparation) after multiple injections of horseradish peroxidase (HRP) into the postlateral and lateral gyri, which fill most of areas 17, 18 (visual areas V1 and V2) and part of area 19, neurons which project through the corpus callosum (callosal neurons) are found in the contralateral V1 and V2. In the tangential direction they are limited to a restricted region on each side of the boundary between the two areas. In particular, very few, if any, callosal neurons are found in the medial bank of area 17, and these only near the crest of the postlateral and lateral gyri. In some animals, however, at the rostralmost end of area 17 they can be found as medially as the fundus of the suprasplenial sulcus. Laterally, in area 18, they commonly extend as far as the medial bank and sometimes up to the fundus of the lateral and postlateral sulci. Rostrocaudal and individual differences in the location of this lateral boundary exist. They may be related to the heterogeneity and individual variability of the retinotopic representation in area 18 (Donaldson and Whitteridge, 1977). Radially, the callosal neurons are restricted to layers 3 and 4, the latter mainly in its upper part and in area 18. A few callosal neurons are found also in layer 6, but they are very rare or absent in layers 1, 2 and 5. The most common morphological type found among callosal neurons is the pyramidal (layers 3, 6); the stellate type (layers 3, 4) is also very common.

There appears to be the expected type of correlation between the functional properties of callosal neurons and the tangential and radial location and morphology of their cell bodies. The callosal neurons have RFs restricted to near the vertical meridian of the visual field (Choudhury et al., 1965; Berlucchi et al., 1967; Hubel and Wiesel, 1967; Shatz, 1977); this is represented near the areas 17/18 boundary (Bilge et al., 1967; Tusa et al., 1978). Callosal neurons have small RFs (Innocenti, in preparation) of the simple, complex, or hypercomplex type (Berlucchi et al., 1967; Hubel and Wiesel, 1967; Shatz, 1977; Innocenti, in preparation). Hypercomplex RFs are particularly common in the supragranular layers in area 17 (Kelly and Van Essen, 1974; Camarda and Rizzolatti, 1976); simple RFs are typical for area 17 where they are found mainly in layers 3, 4, and 6 (Hubel and Wiesel, 1962, 1965; Kelly and Van Essen, 1974); and in 3 and 4 especially among neurons with stellate morphology (Kelly and Van Essen, 1974).

Contrary to the callosal neurons, the cortico-collicular neurons and the cortico-pontine neurons, both located in layer 5, have RFs of the complex type (Palmer and Rosenquist, 1974; Gilbert and Kelly, 1975; Albus and Donate-Oliver, 1977).

There is both direct and inferential evidence that not only the callosal efferent zone (CZ) but also other efferent zones in area 17 (and 18) may perform a parcellation of the visual field representation in those two areas. In the cat, the cortico-pontine projection originates only from those parts of areas 17 and 18 where the periphery of the visual field is represented (Sanides et al., 1978). Since partial representations of the visual field exist in areas 18 and lateral suprasylvian (Donaldson and Whitteridge, 1977; Palmer et al., 1978) and these areas receive from the ipsilateral area 17 (Hubel and Wiesel, 1965; Gilbert and Kelly, 1975), it seems likely that the respective efferent zones must also occupy part of the total visual field representation in area 17. In the monkey, parts of V1 representing foveal and extrafoveal portions of the retina differ in their connections to other cortical visual areas (Zeki, 1978). This difference is taken as a suggestion that "among other functions, the striate cortex acts as a distribution centre for the information coming over the retino-geniculo cortical pathways parcelling this information out to different visual areas of the prestriate cortex" (Zeki, 1978).

It is possible that the parcellation of the cortical representation of the visual field performed by the efferent zones mentioned above may be accompanied by a rescaling of the portion of the visual field represented by each efferent system. In the CZ this seems to happen, since the packing density of callosal neurons appears to vary through it (Innocenti and Fiore, 1976; Innocenti, in preparation). This may indicate that in the efferent message through the corpus callosum (CC) the visual field is magnified in its various parts differently than in the cortex.

This particular type of structural to functional relation does not appear to be limited to the visual system. An interesting transformation of the cortical sensory representation takes place also in the efferent message to the CC from somatosensory area SI (Caminiti et al., 1979). Here, two partially overlapping CZs projecting to contralateral S1 and S2 exist. In the CZ projecting to S1 only the trunk is represented. In the CZ projecting to S2 both the trunk and part of the distal forepaw are represented. The highest tangential density of callosal neurons is within the portion of the CZ representing the forepaw.

The tangential distribution of efferent neurons within CZ may achieve more than parcellation and rescaling of the cortical maps of the sensory periphery since RF properties, other than location, change tangentially across the cortex. However, thus far, in the visual system of the cat we have not been able to identify variations in the tangential density of callosal neurons that we could refer to the tangential variations of either RF orientations or ocular dominance. In the auditory cortex of the cat, however, the tangential density of callosal neurons is highest in correspondence with bands of cortex where neurons can be activated by binaural stimulation and lowest along bands of cortex in which ipsilateral ear stimulation suppresses responses to stimulation of the contralateral ear (Imig and Brugge, 1976). It is not known to what functional modality the clusters of callosal neurons existing in S1 of the monkey (Jones et al.,1975) or in S2 of the cat (Caminiti et al., 1979) correspond.

Thus, the so-called Adrian-Mountcastle rule, according to which (along primary sensory pathways) the volumes of the central representations of the different parts of the sensory periphery are roughly proportional to the density of peripheral innervation, breaks down at the output from primary sensory cortices.

One wonders what the new rules are. Two factors are likely to determine the format of each message: (i) the type of information needed by the structure that will receive it and (ii) the ontogenetic sequence that generates the neuronal structure responsible for the message. In both the visual (Innocenti et al., 1977) and somatosensory (Caminiti et al., 1978) cortices of the cat, the callosal efferent neurons are, at birth, much more widespread than in the adult.

In the visual cortex, during the first postnatal week, the CZ in areas 17 and 18 differs from the adult in three major ways (Innocenti et al., 1977): (i) The packing density of the callosal neurons is higher. It has not yet been calculated whether this corresponds to a higher total number of callosal neurons at birth or whether it simply reflects the overall crowding of cortical cells due to an incomplete development of the neuropile. (ii) Near the 17/18 boundary, neurons in layer 6 make a much larger contribution to the callosum. Layer 5 is free of callosal neurons and the CZ in this region appears, at this age, clearly bilaminar. (iii) The callosal neurons extend throughout the mediolateral extent of areas 17 and 18 and continue without interruption to areas 19, 21, and to the lateral suprasylvian areas.

At the end of the second week (i.e., after about one week of visual experience), the number of neurons outside the adult boundaries of the CZ in areas 17 and 18 has decreased (Innocenti, 1978). Quite a few neurons can still be seen in the medial bank of area 17, below the suprasplenial sulcus. Laterally, there are now clear-cut strips of cortex free of callosal neurons between the CZ in area 18 and the CZ in area 19, as well as between this and the CZ in the lateral suprasylvian cortex.

At the end of the first postnatal month (Innocenti, 1978) the tangential decrease of the CZ has progressed further. Callosal neurons are found in area 17 only as far as the suprasplenial sulcus although they are more abundant than in the adult.

The final tangential distribution of the callosal neurons is acquired during the second and the beginning of the third postnatal month.

In parallel with the tangential diminution of the CZ, the callosal neurons in layer 6 decrease in both packing density and absolute number.

Simultaneously, a similar course of events takes place in the somatosensory cortex (Caminiti et al., 1978). In S1, the CZ for contralateral S2 extends at birth over the entire representation of the periphery and it includes, therefore, the part of the paw representation that is, in the adult, free of callosal neurons. As in the visual cortex, the callosal neurons in layer 6 are more numerous and more densely packed than in the adult. Owing to the lack of neurons in layers 5 and lower 4, the CZ is bilaminar. Gradually, mainly during the first postnatal month the CZ acquires an adult-like morphology.

Thus, it appears that the messages that the visual and somatosensory cortices send through the CC must be, at birth, very different than in adulthood. However, it is unknown where the fibers of the neonatal callosal neurons terminate and whether they form functioning synapses. The fate of the callosal neurons which disappear during development is also unknown: they may die, lose or withdraw their callosal axons.

The postnatal development of the CZs in both visual and somatosensory areas seems to require that information determining time course and order of the maturational process reach the callosal neurons. The information can be thought of as an instruction either to delete some callosal neurons (or axons) or to stabilize the adult connections, or both.

As to the source of the information used in the process there are four non-mutually exclusive possibilities: (i) sensory (or motor) experience; (ii) structures having efferent or afferent relations with the cortex; (iii) the cortex; or (iv) the callosal neurons themselves.

The results of experiments on the visual system (Innocenti and Frost, 1978) indicate that visual experience can be used in the maturational process. In cats reared since their day of eye opening (beginning of the second postnatal week) with divergent or convergent strabismus (provoked by bilateral sectioning of either the medial or the lateral recti muscles) or with one eye sutured shut, CZs extending further medially into area 17 than normally were found; however, they never extended as far as in normal cats during the first two or three postnatal weeks.

On the contrary, in cats raised with both eyes sutured, the CZs were not grossly different from those of normal animals. Thus, although visual experience can be used in the maturational process, information from other sources may be used also.

Acknowledgements This work was supported by the Swiss National Science Foundation (3.492.0.75).

References

Abeles, M. and M. H. Goldestein, Jr. (1970). Functional architecture in cat primary auditory cortex: columnar organization and organization according to depth. J. Neurophysiol. 33:172-187.

Albus, K. (1975). A quantitative study of the projection area of the central and the paracentral visual field in area 17 of the cat. The precision of the topography. Exp. Brain Res. 24:159-179.

Albus, K. and F. Donate-Oliver (1977). Cells of origin of the occipito-pontine projection in the cat: functional properties and intracortical location. Exp. Brain Res. 28:167-174.

Berlucchi, G., M. S. Gazzaniga and G. Rizzolatti (1967). Microelectrode analysis of transfer of visual information by the corpus callosum. Arch. ital. Biol. 105:583-596.

Bilge, M., A. Bingle, K. N. Seneviratne and D. Whitteridge (1967). A map of the visual cortex in the cat. J. Physiol. (Lond.) 191:116P-118P.

Camarda, R. and G. Rizzolatti (1976). Receptive fields of cells in the superficial layers of the cat's area 17. Exp. Brain Res. 24:423-427.

Caminiti, R., P. Barbaresi and G. M. Innocenti (1978). Callosal neurones in SI and SII of the kitten. Neuroscience Letters, Suppl. 1:S342.

Caminiti, R., G. M. Innocenti and T. Manzoni (1979). The anatomical substrate of callosal messages from SI and SII in the cat. Exp. Brain Res., in press.

Choudhury, B. P., D. Whitteridge and M. E. Wilson (1965). The function of the callosal connections of the visual cortex. Quart. J. exp. Physiol. 50:214-219.

Creutzfeldt, O., G. M. Innocenti and D. Brooks (1974). Vertical organization in the visual cortex (area 17) in the cat. Exp. Brain Res. 21:315-336.

Donaldson, I. M. L. and D. Whitteridge (1977). The nature of the boundary between cortical visual areas II and III in the cat. Proc. R. Soc. Lond. B 199:445-462.

Gilbert, C. D. and J. P. Kelly (1975). The projections of cells in different layers of the cat's visual cortex. J. comp. Neur. 163:81-106.

Gilbert, C. D. (1977). Laminar differences in receptive field properties of cells in cat primary visual cortex. J. Physiol. (Lond.) 268:391-421.

Hubel, D. H. and T. N. Wiesel (1962). Receptive fields, binocular interaction and functional architecture in the cat's visual cortex. J. Physiol. (Lond.) 160:106-154.

Hubel, D. H. and T. N. Wiesel (1965). Receptive fields and functional architecture in two nonstriate visual areas (18 and 19) of the cat. J. Neurophysiol. 28:229-289.

Hubel, D. H. and T. N. Wiesel (1967). Cortical and callosal connections concerned with the vertical meridian of visual fields in the cat. J. Neurophysiol. 30:1561-1573.

Hubel, D. H. and T. N. Wiesel (1968). Receptive fields and functional architecture of monkey striate cortex. J. Physiol. (Lond.) 196:215-243.

Hubel, D. H. and T. N. Wiesel (1974). Sequence regularity and geometry of orientation columns in the monkey striate cortex. J. comp. Neur. 158:267-294.

Imig, T. J. and J. F. Brugge (1976). Relationship between binaural interaction columns and commissural connections of the primary auditory cortical fields (A1) in the cat. Neurosci. Abs. 2:26.

Imig, T. J. and H. O. Adrian (1977). Binaural columns in the primary field (A1) of cat auditory cortex. Brain Res. 138:241-257.

Innocenti, G. M. (1978). Postnatal development of interhemispheric connections of the cat visual cortex. Arch. ital. Biol. 116:463-470.

Innocenti, G. M. (1979). Adult and neonatal characteristics of the callosal zone at the boundary between areas 17 and 18 in the cat. In: Structure and Function of the Cerebral Commissures. I. Steele Russel. M. W. Van Hof and G. Berlucchi, eds. Macmillan, London. In press.

Innocenti, G. M. and L. Fiore (1976). Morphological correlates of visual field transformation in the corpus callosum. Neuroscience Letters 2:245-252.

Innocenti, G. M., L. Fiore and R. Caminiti (1977). Exuberant projection into the corpus callosum from the visual cortex of newborn cats. Neuroscience Letters 4:237-242.

Innocenti, G. M. and D. Frost (1978). Visual experience and the development of the efferent system to the corpus callosum. Neurosci. Abs. 4:1513.

Jones, E. G., H. Burton and R. Porter (1975). Commissural and cortico- cortical "columns" in the somatic sensory cortex of primates. Science 190:572-574.

Kelly, J. P. and D. C. Van Essen (1974). Cell structure and function in the visual cortex of the cat. J. Physiol. (Lond.) 238:515-547.

Leventhal, A. G. and H. V. B. Hirsch (1978). Receptive-field properties of neurons in different laminae of visual cortex of the cat. J. Neurophysiol. 41:948-962.

Lund, J. S., R. D. Lund, A. E. Hendrickson, A. H. Bunt and A. F. Fuchs (1975). The origin of efferent pathways from the primary visual cortex, area 17, of the macaque monkey as shown by retrograde transport of horseradish peroxidase. J. comp. Neur. 164:287-304.

Mountcastle, V. B. (1957). Modality and topographic properties of single neurons of cat's somatic sensory cortex. J. Neurophysiol. 20:408-434.

Mountcastle, V. B. (1978). An organizing principle for cerebral function: the unit module and the distributed system. In: The Mindful Brain. G. M. Edelman and V. B. Mountcastle (eds.). M.I.T. Press, Cambridge, Massachusetts, pp. 7-50.

Palmer, L. A. and A. C. Rosenquist (1974). Visual receptive fields of single striate cortical units projecting to the superior colliculus in the cat. Brain Res. 67:27-42.

Palmer, L. A., A. C. Rosenquist and R. J. Tusa (1978). The retinotopic organization of lateral suprasylvian visual areas in the cat. J. comp. Neur. 177:237-256.

Poggio, G. F., R. W. Doty, Jr. and W. H. Talbot (1977). Foveal striate cortex of behaving monkey: single-neuron responses to square-wave gratings during fixation of gaze. J. Neurophysiol. 40:1369-1391.

Sanides, D., W. Fries and K. Albus (1978). The corticopontine projection from the visual cortex of the cat: an autoradiographic investigation. J. comp. Neur. 179:77-88.

Shatz, C. (1977). Abnormal interhemispheric connections in the visual system of Boston Siamese cats: a physiological study. J. comp. Neur. 171-229-246.

Tusa, R. J., L. A. Palmer and A. C. Rosenquist (1978). The retinotopic organization of area 17 (striate cortex) in the cat. J. comp. Neur. 177:213-236.

Wise, S. P. and E. G.Jones (1977). Cells of origin and terminal distribution of descending projections of the rat somatic sensory cortex. J. comp. Neur. 175:129-158.

Zeki, S. M. (1978). The cortical projections of foveal striate cortex in the rhesus monkey. J. Physiol. (Lond.) 277:227-244.

Sensitivity of Visual Neurons
to the Timing of Input from the Two Eyes

JILL G. GARDNER
Department of Psychology
Dalhousie University
Halifax, Nova Scotia, Canada

Abstract Psychophysical studies have shown that a sensation of depth can be elicited in the absence of both form and spatial disparity cues, if the input to the two eyes is separated in *time*. These data suggest that there is a "binocular delay" or "temporal disparity" system which can be used to localize objects in three-dimensional space.

Seeking a substrate for a time-based depth perception mechanism, we examined the sensitivity of cat area 18 cells to the *timing* of input from the two eyes. Using stimuli presented binocularly with variable interocular delays, it was possible to construct tuning curves for binocular interactions in the temporal domain, similar to those produced by others in the spatial domain. By varying space (position of stimuli on the two receptive fields) and time (interocular delay) simultaneously, we could show that *both* factors influence the binocular responses of single cells. Sensitivity to interocular delay characterized most units studied and indicated that simultaneity of input at the cortical cell is an important determinant of response strength.

The majority of units which showed binocular temporal sensitivity responded best when stimuli were presented to the two eyes at exactly the same time (zero interocular delay). Other cells responded best when stimuli were presented at a particular, nonzero interocular delay. In normally reared cats, the distribution of preferred interocular delay approximated a normal distribution with a mean of zero.

We wondered if we could shift the mean of this distribution by rearing cats in an environment in which they received asynchronous binocular input during development. If simultaneity of binocular input is important to visual neurons, we thought that a shift in preferred interocular delays might be produced to compensate for an externally imposed interocular latency difference. Kittens were raised wearing goggles which had a 1 log unit neutral density filter over one eye and a clear plastic lens over the other eye, so that the timing of input from one eye (the filter eye) was delayed relative to the timing of input from the second eye. Preliminary evidence suggests that the atypical visual exposure produced compensatory changes in the responses of visual cells. In experimental animals, a majority of the units which showed binocular temporal sensitivity responded best at the particular interocular delay which would have produced simultaneous binocular input during the kittens' early development.

The importance of timing in binocular vision was first emphasized by Hering over 100 years ago (Hering, 1864; cited in Von Békésy, 1969). From a theoretical basis, he predicted that a change in depth perception would occur when binocular stimuli were presented to the two eyes at slightly different times. Psychophysical data later substantiated the claim. It has been found that a vivid sensation of depth can be elicited with *asynchronous* binocular stimuli (Von Békésy, 1969; Ross, 1976), even when these stimuli lack monocular form and are presented on corresponding retinal coordinates. These data suggest that there is a "binocular delay" or "temporal disparity" system which can produce stereoscopic depth perceptions. They further suggest that this system might be found at a relatively early stage of visual processing (Julesz, 1960), possibly at the level of the visual cortex (as suggested by Schön, 1878; cited in Von Békésy, 1969). Seeking a substrate for a time-based depth perception mechanism, we examined the sensitivity of neurons in cat visual cortex to the timing of input from the two eyes.

We chose to record in cat area 18 (parastriate cortex) for a number of reasons. Our experiments required the use of static rather than moving stimuli and we had found that area 18 cells are very responsive to flashed stimuli. Most units give a transient burst of impulses to a prolonged flash (250 msec) and respond well to even very brief stimuli (10 msec). In addition, since area 18 represents only the binocular segment of the visual field (Tusa, Palmer, and Rosenquist, 1975) and since area 18 in monkey has been implicated in stereoscopic function (Hubel and Wiesel, 1970), we thought that strong binocular interactions might be observed in area 18 of the cat as well.

The procedures which were used involved four steps. First, the receptive field properties were determined for each eye separately by presenting optimally oriented flashed bars at seven to nine different positions across the receptive field. Stimuli were flashed on and off at each position and histograms were compiled after 32 to 256 stimulus presentations. The information obtained from this procedure was used to categorize units by cell type and ocular dominance group (on the seven-point scale of Hubel and Wiesel, 1962). Units were classified as complex cells if they responded to both the onset and offset of the stimulus at all receptive field locations, and were classified as simple cells if their receptive fields had spatially separate regions where only "on" or "off" responses were evoked. In the second step, binocular spatial interactions were examined by varying the spatial separation of stimuli on the two receptive fields. The position of the stimulus was held constant in the center of one receptive field (the dominant eye if the unit was driven unequally by the two eyes), while a second flashed bar was simultaneously presented at different locations within the other eye's receptive field. By comparison with the data obtained during monocular testing, it was possible to determine whether the response at different spatial disparities represented the sum of the individual monocular responses, or whether the binocular response was greater than (facilitation) or less than (occlusion or inhibition) the sum of the monocular responses. In the third step, binocular temporal interactions were examined by varying the timing of input to the two eyes. Stimuli were positioned where maximal spatial interactions had been elicited and presented to the two eyes with variable interocular delays. The temporal intervals were varied first over a broad range and then over successively narrower ranges. In the fourth and last step, responses to varied temporal and spatial relationships of binocular stimuli were examined by manipulating both variables during one block of trials.*

In a sample of over 100 units, we found that most cells in area 18 *were* sensitive to the timing of input from the two eyes. In some cells, strong binocular facilitation or inhibition occurred only when stimuli were presented to the two eyes within 20 msec of interocular synchrony. In other cells, the range of temporal selectivity was broader and interactions were observed with interocular delays of 50 to 75 msec. Units with strong temporal interactions were not identifiable on the basis of cell type, but they were commonly found either above cortical layer IV or at the bottom of layer VI and were almost always driven unequally by the two eyes (or responded poorly to monocular stimulation through both eyes). Figure 1 shows the response of two representative units. The cell shown in the top graph gave very poor monocular responses but was strongly facilitated when the two receptive fields were stimulated simultaneously. This facilitation was critically dependent on the timing of input to the two eyes—firing rate decreased substantially with an interocular delay of 20 msec. The bottom graph of Fig. 1 shows a cell whose response was inhibited when stimuli were presented at interocular delays of less than 35 msec. On closer inspection (Fig. 2), this unit proved to be *highly* sensitive to the timing of input from the two eyes. When tested with 2 msec intervals, firing rate increased monotonically with each additional delay, changing by a factor of 4 over the 20 msec range.

All of the temporal interactions described above were determined at one spatial location. In order to obtain a more complete characterization of a unit's response to changes in both the spatial and temporal parameters, binocular temporal interactions were measured at different spatial disparities. Since this involved the simultaneous collection of 64 to 100 histograms, the data were reduced to the form shown in Figs. 3 and 4. In these three-dimensional plots, one axis represents interocular delay (time), the other axis represents the spatial separation of stimuli on the two receptive fields (space), while the ordinate represents the firing rate of the unit. When the two variables are plotted in this manner, the data clearly show that maximal spatial interactions occur only when

---

*During all procedures, flashed stimuli were 1.0 log units above background with a duration of 10 or 250 msec. Stimulus presentation, data collection, and display were under computer control and stimuli were always presented in a randomized, interleaved fashion.

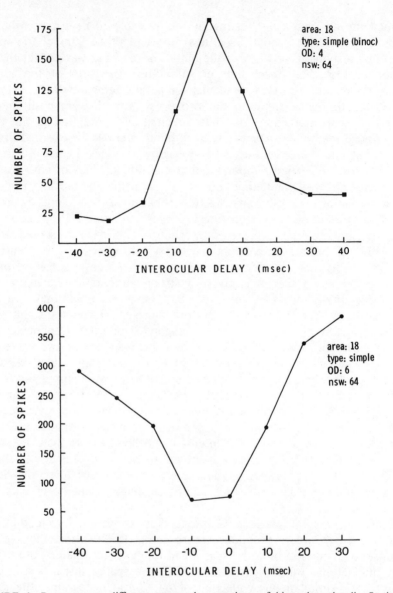

FIGURE 1 Responses to different temporal separations of binocular stimuli. Stationary, flashed stimuli (250 msec) with a constant spatial disparity were presented to the two receptive fields at a number of interocular delays. The temporal interval, in msec, between presentation of the two stimuli is shown on the abscissa. Zero delay represents simultaneous stimulation of the two receptive fields. Values to the left and right of 0, respectively, indicate that one eye was stimulated before (-) or after (+) the other eye. The number of spikes evoked with 64 stimulus presentations at each interocular delay is shown on the ordinate. The unit of the top graph (ocular dominance group 4) was most strongly facilitated when two receptive fields were simultaneously stimulated (0 interocular delay), while the unit represented in the bottom graph (ocular dominance group 6) showed maximal inhibition with 10 msec of interocular delay.

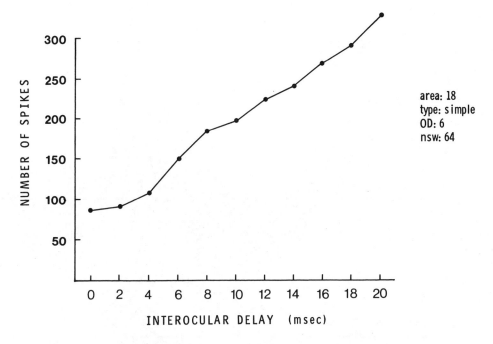

FIGURE 2 Binocular temporal tuning curve across a narrow range of temporal intervals. This unit (shown in Fig. 1, bottom graph) was highly sensitive to the timing of input from the two eyes. Conventions as in Fig. 1.

the timing of input to the two eyes is appropriate; conversely, maximal temporal interactions occur only when stimuli are presented to the two eyes at appropriate spatial disparities. Figure 3 illustrates this point. The unit represented in this graph responded very poorly to stimulation of each eye alone, but was strongly facilitated when the two receptive fields were stimulated simultaneously. This facilitation occurred only over a limited range of both spatial and temporal disparities. With interocular delays of more than 20 msec the response dropped almost to monocular levels.

The unit shown in Fig. 3 responded best to simultaneous stimulation of the two receptive fields (zero delay). Other cells, such as the one shown in the bottom graph of Fig. 1, showed maximum interactions with asynchronous binocular stimulation (nonzero delay). We have found that there is a relationship between the preferred interocular delay of a unit and its ocular dominance. In general, units which are driven with similar strength through the two eyes (ocular dominance groups 3, 4, and 5) respond best to stimuli presented at zero or near zero delay; units which are strongly dominated by one eye (ocular dominance groups 1, 2, 6, and 7) respond best to stimuli presented at a nonzero interocular delay. In most cases, the strongest response is elicited when the nondominant eye is stimulated first. This implies that the input from that eye is taking *longer* to influence the cortical cell than is input from the dominant eye. In a previous experiment, we found that units driven unequally through

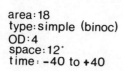

area:18
type:simple (binoc)
OD:4
space:12°
time:−40 to +40

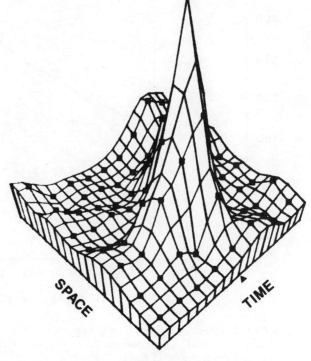

SPACE

TIME

FIGURE 3 Responses evoked with binocular stimuli of particular spatial and temporal dispari-ties. The time axis represents interocular delays of ±40 msec (0 delay is indicated by arrow), the space axis represents 12° of visual angle, and the number of spikes is shown on the ordi-nate. To measure binocular spatial interactions, the position of the stimulus was held constant in one eye and varied in the second eye. Binocular temporal interactions were studied as in Fig. 1. At each of nine spatial disparities, responses were examined at nine interocular delays. Eighty-one histograms were produced and each one is represented as a data point. Interpo-lated vectors are shown. This unit responded poorly to monocular stimulation, but gave a strong response to binocular stimuli with the appropriate spatial and temporal disparities.

the two eyes had different monocular response latencies through each eye* and that the dominant eye response was consistently faster. Units in ocular domi-nance group 4 had nearly identical latencies through each eye (mean interocular difference: 2.2 msec) while units in ocular dominance groups 2 and 6 had an average interocular latency difference of 13 msec. With the unit shown in Fig. 4, it was possible to predict the binocular preferred interocular delay from a knowledge of monocular response latencies. This cell was strongly dominated by one eye (ocular dominance group 6) and showed maximum excitation and inhibition with asynchronous stimulation of the two receptive fields. When tested monocularly, the response latency of the dominant, ipsilateral eye was 10

---

*The measure of response latency which we used was *peak* latency—the time from the onset of the stimulus to the bin with the largest number of spikes. Data were analyzed with a 1 msec bin width.

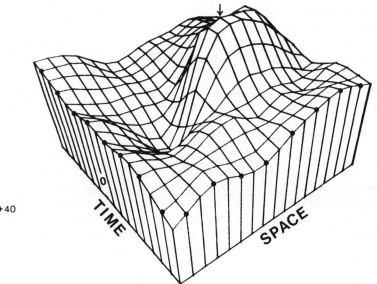

area : 18
type : simple
OD : 2
space : 8.8˚
time : −30 to +40

FIGURE 4 Responses of an area 18 simple cell to varied temporal and spatial relationships of binocular stimuli. The time axis represents interocular delays of -30 to +40 msec (0 delay indicated) and the space axis represents 8.8° of visual angle. Conventions as in Figs. 1 and 3. The number of spikes evoked by eight spatial and eight temporal disparities produced 64 data points. When tested monocularly, this unit (ocular dominance group 6) had different response latencies through each eye. With binocular stimulation, optimal responses were seen when the long latency (55 msec), nondominant eye was stimulated 10 msec before the dominant eye (latency, 45 msec). When stimuli were presented to the two eyes with the appropriate temporal relationship, responses at different spatial locations showed maximal facilitation or inhibition.

msec faster than that of the contralateral eye. Under conditions of binocular stimulation, maximal interactions were seen when the contralateral eye was stimulated 10 msec *before* stimulation of the ipsilateral eye.

In normal cats, the majority of cells which show binocular temporal interactions respond best when stimuli are presented to the two eyes with little or no interocular delay. Cells which are strongly dominated by one eye respond best when the weak eye is stimulated before the strong eye. Since the ocular dominance of the normal cat is not heavily biased toward either the ipsilateral or the contralateral eye, the distribution of preferred interocular delays we have obtained approximates a normal distribution with a mean of zero (standard deviation: 11 msec).

We wondered if we could alter the distribution of preferred interocular delays by raising cats in an environment in which they received asynchronous binocular input during development. If simultaneity of input is important to visual cells, we thought that we might see a shift in the mean of this distribution toward compensation for an externally imposed interocular latency difference.

To examine this hypothesis, kittens were raised wearing goggles which had a 1 log unit neutral density filter over one eye and a clear plastic lens over the other eye.* The ambient illumination of the rearing room was held constant at a level at which the difference in illumination of the two eyes was estimated to produce an interocular latency difference of 10 msec.

In addition to causing a latency increase, a reduction in whole field luminance can be expected to have a number of effects on the firing of visual neurons. Experiments we have conducted on normal cats have shown that a one log unit dimming filter produces the following changes in the cells' response to a visual stimulus: (1) the total number of spikes elicited is reduced, (2) peak firing rate drops, and (3) there is an increase in the duration of the response. To make sure that these differences in binocular input had not resulted in a breakdown of binocularity or left the animals with grossly abnormal cortical responses, we qualitatively studied the response characteristics of cells in the striate cortex (area 17). The data from 270 units obtained from four kittens showed a slight breakdown in binocularity but otherwise normal cortical responses. In the hemisphere contralateral to the eye which wore the filter during development, there was a clear tendency for this eye to control more units (Fig. 5, left bottom histogram). When the results were pooled across hemispheres, however, this trend was much weaker (Fig. 5, top histogram). It was apparent that the cortices of these kittens were not grossly abnormal and that they were left with a substantial proportion of binocularly driven cells.

Our quantitative studies of these animals are incomplete at this time and so the data presented must be regarded as preliminary. Nevertheless, the results obtained thus far have been extremely consistent. Of the units found in experimental cats which showed binocular temporal interactions, fewer (30%) responded best to stimuli presented at zero or a near zero interocular delay than in normal cats (50-60%). Unlike normally reared cats, where the distribution of preferred interocular delays is centered around zero, the distribution from experimental animals was centered around an interocular delay of 10 msec. The extent and direction of this apparent shift in preferred delays was appropriate to compensate for the filter-imposed interocular latency difference.

If, during the kitten's development, a cell responded best to simultaneous input from the two eyes, then during the testing procedure, when the filter was removed, that cell should respond best when the normal eye was stimulated about 10 msec *before* the previously filtered eye. Two of the units which responded optimally under these conditions are shown in Figs. 6 and 7. On the abscissa, "0" represents the onset time of the stimulus in the normal eye. Under conditions of binocular stimulation, the response of the unit shown in Fig. 6 was strongly inhibited. This inhibition was maximal when the normal eye was stimulated 10 to 20 msec before the filter eye. Similarly, the unit of Fig. 7 was most strongly facilitated when stimulation of the normal eye preceded that of the filter eye (Fig. 7, squares). The response peaked at an interocular delay of 10 msec and dropped almost to monocular levels with a 20 msec

*Kittens were reared normally until they were 25 days old and then given *continuous* exposure with the goggles on until recorded at 4 to 6 months of age.

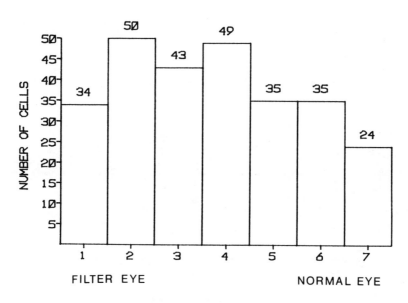

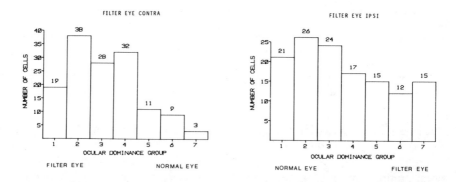

FIGURE 5 Ocular dominance of 270 units from kittens reared with an interocular delay during early development. Kittens were raised wearing goggles which had a 1 log unit neutral density filter over one eye and a clear plastic lens over the other eye. In the two lower histograms, units are categorized by ocular dominance on the seven-point scale of Hubel and Wiesel (1962). Groups 1 to 7, respectively, represent a contralateral to ipsilateral trend in ocular dominance. With monocular testing, cells in group 1 were driven only by the contralateral eye, cells in group 4 were driven equally by the two eyes and cells in group 7 were driven only through the ipsilateral eye. The two lower histograms show the ocular dominance of units recorded in the hemisphere which was contralateral (left) and ipsilateral (right) to the eye which wore the filter during development. The top distribution represents the pooled data for both hemispheres with 1-7, respectively, now representing a trend in ocular dominance from the filter eye to the normal eye. The extent of binocularity seen in these cats was similar to that of normal cats. There was a tendency, however, for the filter eye to control more units and this trend was most apparent when the filter eye was also the contralateral eye.

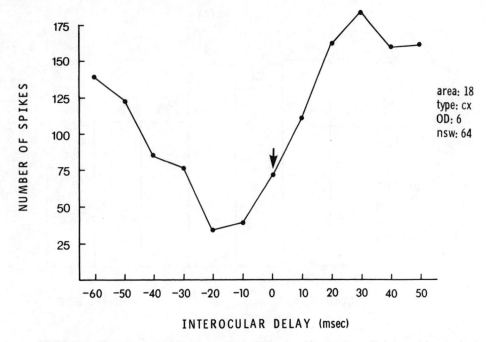

FIGURE 6 Binocular temporal interactions of an area 18 complex cell from a kitten raised with an interocular delay during development. Responses were examined at 12 interocular delays across a range of -60 to +50 msec. Conventions as in Fig. 1. Zero on the abscissa represents simultaneous stimulation of the two receptive fields. The eye which wore the filter during development was stimulated at time 0 and the time of stimulus onset to the normal eye was varied. Maximal inhibition was seen when the nondominant, normal eye (ocular dominance group 6) was stimulated 10-20 msec before the previously filtered eye.

increment in delay. To simulate the conditions in which the cat had been reared, a 1 log unit neutral density filter was placed in front of the previously filtered eye. This manipulation had the effect of shifting the peak of the temporal tuning curve back to zero (Fig. 7, circles) suggesting that during the kitten's daily exposure, this cell responded optimally at zero delay, as do the majority of cells in the normal cat.

As previously mentioned, cells in normally reared cats show a relationship between their preferred interocular delay and their ocular dominance. A unit driven unequally by the two eyes generally responds best when the weak eye is stimulated before the strong eye. Only infrequently has the reverse been found to be true. Thus, we were rather surprised to find that the usual relationship between ocular dominance and preferred interocular delay was not evident in these specially reared cats. The unit shown in Fig. 7, for instance, was dominated by the ipsilateral, normal eye (ocular dominance group 6) and would be expected to respond optimally when that eye was stimulated *after* the nondominant, previously filtered eye (see Fig. 4). However, optimal responses were observed when the normal, dominant eye was stimulated 10 msec *before* the

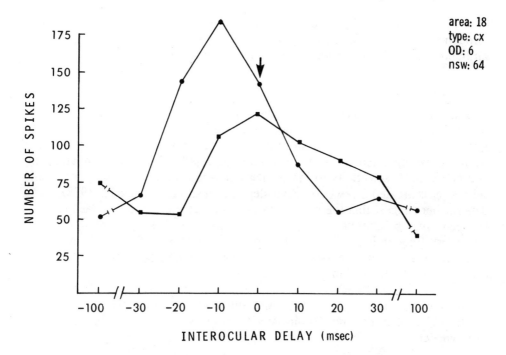

FIGURE 7 Responses to different temporal separations of binocular stimuli under two view-ing conditions. Binocular temporal interactions of an area 18 complex cell from an experimen-tal cat tested (1) when both eyes viewed normally (circles) and (2) when one eye viewed nor-mally and the other eye (the eye which wore the filter during development) viewed through a 1 log unit neutral density filter. Conventions as in Figs. 1 and 6. Under normal viewing con-ditions, this unit responded optimally when the *dominant* (ocular dominance group 6), normal eye was stimulated 10 msec before the previously filtered eye. When tested with a 1 log unit filter over the previously filtered eye, the peak of the temporal tuning curve shifted to 0 delay. This manipulation simulated the conditions in which the cat had been reared and suggested that during the kitten's early exposure, this cell responded optimally at 0 delay, as do the majority of cells in the normal cat.

filter eye. These data indicate that the normal relationship between a units-preferred interocular delay and ocular dominance can be altered by atypical visual exposure.

When a kitten is reared with a filter-imposed interocular latency difference, one way in which the system could compensate would be to shift the ocular dominance of cells toward the filter eye. If the filter eye was dominant, that eye would have a relatively shorter latency with respect to the nondominant eye and the interocular latency difference caused by the filter could be neutralized. This may, in part, explain why in experimental animals there was an ocular dominance trend toward the filter eye rather than toward the normal eye as might be expected. Nevertheless, the finding that the usual relationship between preferred interocular delay and ocular dominance was not evident in the specially reared cats suggests also that the system can compensate for the filter-induced interocular latency difference by "speeding up" transmission through the filter eye or "delaying" it through the unfiltered eye without necessarily producing a corresponding change in response strength.

Conclusion

The data presented indicate that single cells in cat visual cortex may be highly sensitive to the timing of input from the two eyes; as such, they provide a possible basis for the psychophysical findings which show that asynchronous input from the two eyes can result in a sensation of depth. The detailed relationship between a unit's sensitivity to interocular delay and the mechanisms underlying stereopsis is treated elsewhere (Cynader, Gardner, and Douglas, 1978). The data further suggest that the sensitivity of binocular units to particular interocular delays may be altered by atypical visual exposure. Whereas units in normal cats most frequently prefer simultaneous binocular stimulation, delaying the input to one eye during development results in an apparent compensatory change in the timing of binocular responses in cells of the visual cortex.

Acknowledgements This research was supported by Grant MT-5201 from M. R. C. of Canada and Grant A9939 from N. R. C. of Canada to M. Cynader.

References

Cynader, M., J. Gardner, and R. M. Douglas (1978). Neural mechanisms underlying stereoscopic depth perception in cat visual cortex. In: Frontiers of Vision Research. S. Cool and E. L. Smith, III (eds.). Springer-Verlag.

Hubel, D. H., and T. N. Wiesel (1962). Receptive fields, binocular interaction and functional architecture in the cat visual cortex. J. Physiol. (Lond.) 160:106-154.

Hubel, D. H., and T. N. Wiesel (1970). Stereoscopic vision in macaque monkey. Nature 225:41-42.

Julesz, B. (1960). Binocular depth perception of computer-generated patterns. Bell System Tech. J. 39:1125-1162.

Ross, J. (1976). The resources of binocular perception. Scientific American 234:80-86.

Tusa, R. J., L. A. Paulmer, and A. C. Rosenquist (1975). The retinotopic organization of the visual cortex in the cat. Soc. for Neurosci. Abs 1:53.

Von Békésy, G. (1969). The smallest time difference the eyes can detect with sweeping stimulation. Proc. N. A. S., Wash. 64:142-147.

STUDIES OF THE MONKEY'S VISUAL SYSTEM

Genesis of Visual Connections in the Rhesus Monkey

PASKO RAKIC

Section of Neuroanatomy
School of Medicine
Yale University
New Haven, Connecticut USA

Abstract The basic afferent connections of the visual system in the rhesus monkey are laid down before birth, although the process of segregation of terminals and synaptogenesis continue into postnatal period. Autoradiographic studies show that projections subserving each eye initially overlap in the dorsal lateral geniculate nucleus (LGd) and in the cerebral cortex of fetal monkeys. In the LGd, retinal terminals originating from each eye become segregated from each other during the middle of the 165-day gestational period. In the cortex, axons representing each eye are intermixed in layer 4 until three weeks before birth when ocular dominance stripes first begin to emerge. This process of segregation in the distribution of geniculocortical afferents is not completed until the second postnatal month. Cortical efferents also begin to develop at the end of the first half of gestation. Corticogeniculate terminals appear characteristically wedge-shaped and topographically organized by midgestation.

Considerable rearrangement of axon terminals is visible in the mature monkey if one eye is enucleated by intrauterine surgery at critical prenatal stages. Thus, when one eye is enucleated during the first third of gestation and the animal survives until the second postnatal month, the LGd is devoid of laminae and the remaining eye projects diffusely throughout the nucleus. Transneuronally transported tracers indicate that ocular dominance stripes fail to develop in the visual cortex. Thus, it appears that both the development of cellular laminae in the LGd as well as the segregation of afferent connections in both the LGd and cortex may depend on competition between projections subserving the two eyes.

Introduction

The binocular visual system of primates is an excellent model for the study of mechanisms of development of central neuronal connections. In this species, input from the two eyes is separated in the dorsal lateral geniculate nucleus

(LGd) as well as in the primary visual cortex (Fig. 1). The projections of the retina to the LGd as well as the projections of LGd neurons to primary visual cortex can be examined by injecting one eye with radioactive amino acids and/or sugars and allowing the radioactively labeled metabolites to be transported first to the LGd, and then by means of transneuronal transport to the

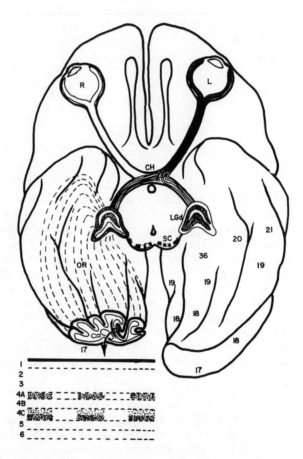

FIGURE 1 The connections underlying binocular vision in the rhesus monkey are schematically illustrated on the ventral view of the cerebrum (from Rakic, 1977b). The dorsal lateral geniculate body (LGd) and superior colliculus (SC) are slightly enlarged to render the details of binocular representation legible. Also, a region of area 17 in the depth of calcarine fissure is enlarged in the left lower corner of the diagram. The axons originating from retinal ganglion cells of each eye partially cross at the optic chiasm (CH) and become distributed in the three appropriate laminae of the LGd and in the proper territories representing each eye in the SC. Principal neurons of the LGd project to the primary visual cortex via the optic radiation (OR) and terminate mostly in sublayers 4A and 4C in the form of alternating columns that receive input from one or the other eye.

cortex. This method can be used to determine the well-established fact that in primates axons of retinal ganglion cells from each layer terminate separately in alternating laminae of the LGd (e.g., Polyak, 1957). Additionally, it shows that radioactivity is transported transneuronally to the primary visual cortex (area 17 of Brodmann, 1905) and is distributed within cortical sublayers 4A and 4C in a system of stripes approximately 350-400 μm wide (Wiesel, Hubel, and Lam, 1974). These alternate with unlabeled stripes of the same width that receive input from the uninjected eye (Fig. 1, enlarged square of area 17). These alternating stripes correspond to ocular dominance columns as physiologically defined (Hubel and Wiesel, 1968; Hubel and Wiesel, 1977).

Development of Afferent Connections
to the LGd and Primary Visual System

As part of our ongoing studies on the develoment of the primate brain, we have examined the genesis of retinogeniculate, and geniculocortical connections following unilateral eye injection of a mixture of ^{3}H—proline and ^{3}H—fucose in monkey fetuses. Each fetus was exteriorized by hysterotomy (Rakic, 1976, 1977b, 1979b) and after the injection of isotopes returned to the uterus, and either 20 hours or 14 days later returned by a second caeserean section. Their brains were fixed by intracadial perfusion and processed for autoradiography.

In fetuses sacrificed at E68 (embryonic day 68) and E77, orthogradely transported radioactivity was distributed uniformly throughout the full extent of the LGd on both sides without segregation into the laminae characteristic of this nucleus in the adult monkey (Rakic, 1976) (Fig. 2(A,B)). Our previous ^{3}H—thymidine autoradiographic analysis of the time of neuron origin of visual structures demonstrated that at this fetal age all neurons of the LGd have been generated and are already situated within the nucleus (Rakic, 1977b). However, the nucleus has still not attained its mature configuration, characteristic laminar pattern, and adult position within the diencephalon (Rakic, 1977a).

In a fetus injected at E77 and sacrificed at E91 (E77-E91), the separation of the axons, axon terminals, or both, that originate from one or the other eye is discernible at the caudal pole of the LGd as irregularly shaped areas of lower and higher silver grain densities (Fig. 2(C)). The caudal pole of the monkey LGd receives input from the central retina and is the part of the LGd where laminae can first be discerned in Nissl stained material between E90 and E95 (Hendrickson and Rakic, 1977). In an E110-E124 specimen, the projections from the two eyes are segregated in the form of a somewhat irregular six-layered pattern throughout the entire LGd; prospective laminae 2, 3, and 5 are labeled on the side ipsilateral to the injected eye and 1, 4, and 5 are labeled on the contralateral side (Fig. 2(D)). The distribution of grains over the territory of appropriate laminae assumes the typical adult pattern in the E130-E144 specimen (Fig. 2(E)). Thus, since gestation lasts 165 days in the rhesus monkey, the segregation of afferents from the two eyes at the thalamic level is completed at least three weeks before birth (Rakic, 1976, 1977b).

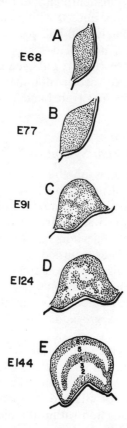

FIGURE 2 Schematic representation of the distribution of radioactive tracers over the lateral geniculate body (LGd) following injections of radioactive tracers (^{3}H-proline and ^{3}H-fucose mixture) into the contralateral eye in five monkey fetuses of various prenatal ages. After injection, each fetus was replaced into the uterus and sacrificed 20 hours (A) or 14 days (B-E) later at the embryonic (E) days indicated at the right side of each LGd. The position and shape of the LGd is outlined as it appears in coronal sections of the diencephalon aligned identically in relation to the midline in each monkey. Although all neurons of the LGd are generated before E45 (Rakic, 1977a), the nucleus changes considerably in size and shape during the course of development as it rotates from a lateral to ventral position in the thalamus. Note that between E68 and E78, radioactivity is distributed uniformly over the entire nucleus, the process of segregation of input from the two eyes occurs mainly between E91 and E124 and is completed before E144. (For details see Rakic, 1976, 1977b).

The process of segregation of retinal projections may be a general feature in mammalian species which have semidecussation at the level of the optic chiasm. Recent analysis of the formation of retinogeniculate projections during postnatal development in the opossum and golden hamster shows a similar pattern of initial overlap, although the timing and sequence of development are somewhat different (Cavalcante and Rocha-Miranda, 1978; So, Frost, and Schneider, 1978).

The same series of specimens were used to analyze the development of geniculocortical projections by means of transneuronal transport. A distinct optic radiation can be discerned in the occipital lobes of a fetus killed at E78, 14 days after unilateral eye injection (Rakic, 1976). However, axons of LGd neurons do not enter the cortical plate in any appreciable number (Rakic, 1977b). Instead, axons, their endings, or both, accumulate below the developing cortical plate (Fig. 3(A)). It is of interest that only a small fraction of neurons destined for layer 4C of the visual cortex have been generated in the proliferative ventricular zone at this stage of fetal development (Rakic, 1974). Furthermore, many neurons of layer 4C which have already undergone their last divisions have not yet reached their final position in the cortical plate (Rakic, 1975). Thus, only the prospective layers 6 and 5 and a fraction of 4C are present in the visual cortex by E78 (Fig. 3(A)).

In the slightly older specimen, E77-E91, some geniculocortical axons do invade the territory of the prospective primary visual cortex. They become uniformly distributed within layer 4, again, however, without evidence of the preferential segregation into ocular dominance stripes with layers 4A and 4C (Fig. 3(B)). After all cortical neurons destined for the primary visual cortex have been generated (Rakic, 1974) and have attained their final positions (Rakic, 1975), the number of LGd axons entering the cortex increases further so that somewhere between E110 and E124, sublayers 4A and 4C become apparent (Fig. 3(C)). It should be emphasized that alternating territories corresponding to ocular dominance stripes are not yet discernible at this age (Fig. 3(C)).

The vertical segregation of input into sublayers 4A and 4C becomes more visible in a fetus injected a few weeks later at E130 and killed at E144 (Fig. 3(D)). Simultaneously, the horizontal segregation of axons carrying input from the two eyes into incipient ocular dominance stripes begins to emerge (Fig. 3(D)). A subtle fluctuation in density of grains is difficult to discern upon simple inspection of the slides, but grain counts and measurements demonstrate clearly alternating 250-300 μm wide territories containing slightly higher and lower grain densities (Rakic, 1976). The combined width of one set of ipsilateral and contralateral ocular dominance stripes as determined by the distance between two peaks of grain concentrations is about 20-25% smaller than in the mature monkey. This indicates that there must be an additional increase in the cortical surface area between E144 and maturity (Rakic, 1977b), although the number of stripes may remain constant. The process of segregation of geniculocortical afferents into ocular dominance stripes continues in the immediate postnatal period and is completed by about 3 weeks of age (Hubel, Wiesel, and LeVay, 1977) when the final pattern, present in the adult, is attained (Fig.

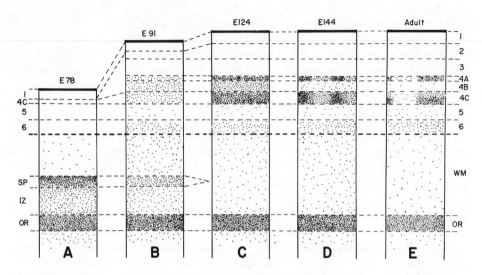

FIGURE 3 Semidiagrammatic summary of development of geniculocortical connections and ocular-dominance columns in the occipital lobe of the rhesus monkey from the end of the first half of pregnancy to adulthood (from Rakic, 1979b). The columns illustrate a portion of the lateral cerebral wall in the region of area 17 as seen in autoradiograms of animals that had received unilateral injection of a mixture of ^{3}H−proline and ^{3}H−fucose 14 days earlier. The age of animals at the moment of sacrifice is provided at the top of each column in embryonic (E) and postnatal (P) days. Cortical layers 1-6 are delineated according to Brodmann's (1905) classification. Note that at E78 the cortical plate consists of only layers 6, 5, and a portion of 4C (see Rakic, 1974). Abbreviations: OR, optic radiation; SP, deep portion of subplate layer; WM, white matter.

3(E)). It has subsequently been shown that the process of segregation of the geniculocortical projection obeys a similar progression in the visual cortex of the cat, except that corresponding stages occur later with respect to birth (LeVay, Stryker, and Shatz, 1978).

The development of an initially diffusely organized set of axons into precise laminar or columnar aggregates subserving binocular vision passes through two broad phases: in the *first phase* axons derived from each eye invade their target structures and their endings are distributed in an overlapping manner; in the *second phase* the axon terminals derived from the two eyes become segregated from each other into separate territories concerned predominantly with one or the other eye (Rakic, 1976). The mechanisms underlying this segregation are not understood, but computer simulation indicates the possibility that two rules may be sufficient to account for formation of stripes assuming an initial random intermixing (Swindale, 1979, this volume). This hypothesis is attractive because it indicates the possibility that a complex structural pattern may be achieved by relatively simple genetic information.

The process of segregation from a diffuse to a patterned termination of central neuronal connections seems not to be confined to the visual system. Thus, the cells of origin of the corpus callosum seem to be initially widespread (Innocenti, Fiore, and Caminti, 1977). Likewise in the pyriform cortex, olfactory and association input initially overlap before becoming separated into different strata of the molecular layer (Price, Maxley, and Schwob, 1976). Intricate corticocaudate projections in forms of patches and rings, described in postnatal monkeys (Goldman and Nauta, 1977), are also initially diffused (Goldman 1979). Therefore, the phenomenon of transformation from diffuse to patterned organization of neuronal connections may be a general rule of central nervous system development. It is possible that segregation develops by way of transient synaptic arrangements. Temporary synapses have been described during development of other systems (e.g., Ramón y Cajal, 1911; Giordano and Cunningham, 1978; Changeux and Dachin, 1976; Knyihar, Csillik, and Rakic, 1978; Rakic, 1979a).

Development of Efferent Projections from the Visual System

Recently, we began an investigation of the prenatal development of the cortical projections to the LGd (Shatz and Rakic, 1978). Fetuses were temporarily exteriorized from the uterus and injected with 0.1 μl of ^{3}H–proline (20-30 μCi). The site of injection was the visual cortex at the posterior pole of the developing occipital lobe. Each fetus was returned to the uterus and 24 hours later was removed, fixed by perfusion, and its brain was processed for autoradiography.

In the autoradiographs of a fetus injected at E63, no radioactive label was seen in the corticothalamic radiations or within the LGd. In fetuses injected at E69, E71, and E78, label was present mostly in the cell-poor zones surrounding the LGd. Within the LGd, label was confined to the lateral-most margin, the prospective magnocellular layers (Rakic, 1977a). The remainder of the nucleus was free of label. The large injection sites made it impossible to determine whether visuotopic order is present at these early ages. In fetuses injected at E83 and E85, the portion of the prospective region of the LGd adjacent to the white matter also contained label. In a somewhat older monkey (E95), label extended throughout both the parvo- and magnocellular layers, as in the adult. Thus, the efferent pathways from the visual cortex in primates are present by the middle of the gestation (Shatz and Rakic, 1978). Their development is in rough synchrony with that of development of the afferent pathway.

In both E83 and E95 specimens the labeling pattern in the LGd appeared characteristically wedge-shaped and appropriately located with respect to the cortical injection (Fig. 4). These wedge-shaped territories correspond to the projection lines of Bishop, Kozak, Levick, and Vakkur (1962). This finding indicates that topographic order in the corticogeniculate projection may be established already by the middle of the gestational period (Shatz and Rakic, 1978). It should be emphasized that during the same period, fibers originating from the two eyes are still intermixed within the LGd. Therefore, it appears that the cortical efferent pathway is topographically ordered prior to the lamination of the LGd and before the retinal afferents to the LGd become fully segregated.

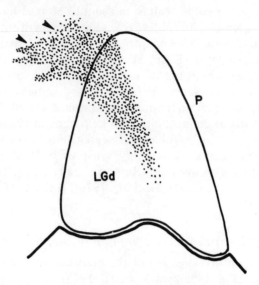

FIGURE 4 Semidiagrammatic representation of a frontal section through the LGd of a rhesus monkey that received an injection of ^{3}H—proline into the ipsilateral visual cortex at midgestation (around E83) and was sacrificed 24 hours later (based on Shatz and Rakic, 1978). The wedge-shaped projection territory marked by the presence of silver grains in photosensitive emulsion is distributed in the topographical position appropriate to the site of injection.

Manipulation of Normal Development by Unilateral Eye Enucleation Before Birth

By using prenatal surgery to enucleate one eye I was able to show considerable rearrangements of afferent axons in the primate visual system (Rakic, 1977c). Thus, in a monkey in which one eye was enucleated on E64 radioactive tracers injected in the remaining eye before sacrifice at the third postnatal month were distributed uniformly over the entire LGd in both cerebral hemispheres (Fig. 5A). Preliminary analysis shows that the number and position of neurons in the two LGd's is not substantially affected by unilateral eye removal at this age, even though the characteristic laminae fail to develop (Rakic, 1977c). A ^{3}H—thymidine autoradiographic analysis of the distribution of labeled neurons within the LGd after various short intervals following exposure to this nucleotide indicates that during the rotation and shifting of this nucleus in the course of thalamic development, the relative positions of LGd cells to each other generally do not change (Rakic, 1977a). Therefore, neurons that are situated in presumptive layers 1 and 6 at the ventral and dorsal edges of the nucleus, which normally receive input from the contralateral (enucleated) eye, now come in contact with axons originating from the ipsilateral (remaining) eye. The projections from the remaining eye occupy twice as large a territory within the mature LGd as they would occupy under normal circumstances. The

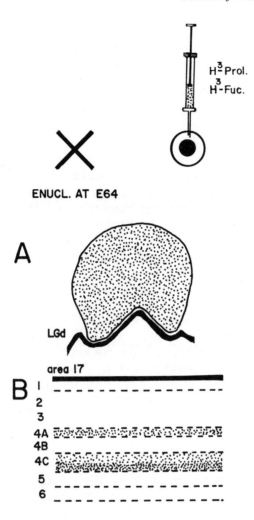

FIGURE 5 Schematic representation of the distribution of radioactive tracers in the LGd and primary visual cortex of a 2-month-old monkey 14 days following injection of a mixture of ^{3}H−proline and ^{3}H−fucose into one eye. In this animal, the other eye had been removed on the 64th embryonic day (E64). The fetus was then returned to the uterus and delivered near term. Under these circumstances, orthogradely transported radioactive label is distributed uniformly over the entire LGd and transneuronally transported label forms a uniform band over layers 4A and 4C without a trace of alternating ocular dominance columns. (Based on data from Rakic, 1977c and 1979a).

increased space occupied by the projection from the remaining eye is, however, not necessarily achieved through expansion of retinal fibers into new territories. Rather, since projections from the two eyes initially overlap in the LGd of fetal monkeys (Rakic, 1976, 1977b), terminals from the remaining eye may simply fail to retract in the absence of competition from the contralateral eye.

From light microscopic examination of autoradiograms it is not possible to determine whether or not axon terminals from the remaining eye actually form synaptic contacts with LGd neurons irrespective of their position. For example, actual synapses formed by the intact eye could be established exclusively with neurons that are originally "committed" to their eye, i.e., on neurons situated in presumptive layers 2, 3, and 4 at the same time axons from the intact eye may be intermixed with neurons "committed" to the contralateral (extirpated) eye (in prospective layers 1, 5, and 6) without making synaptic junctions. Our preliminary electron microscopic analysis indicates a uniform distribution of typical retinal synaptic terminals throughout the entire LGd, including the extreme periphery of the LGd (in prospective layers 1 and 6) which normally receive input exclusively from the remaining ipsilateral eye. Thus, not only are axons that originate from the remaining eye present in the territory which would normally be occupied by the enucleated eye, they also seem to be able to establish ultrastructurally defined synaptic junctions with neurons which they do not normally contact in the adult (Rakic, 1977c, 1979a).

Transneuronally transported radioactive label in the visual cortex of a monkey unilaterally enucleated at E64 is distributed over sublayers 4A and 4C in a continuous band (Fig. 5(B)). Since retinal terminals from the remaining eye are present within the entire volume of the LGd (Fig. 5(A)), all LGd neurons are exposed to equal amounts of radioactive label. Therefore, on technical grounds one can expect an apparent absence of ocular dominance in the cortex, whether or not there is any change in the organization of the pathway from the retina to the cortex. This poses a major problem of interpretation of results on transneuronal transport of radioactivity in the cortex (e.g., see LeVay et al., 1978). It is important to note that the geniculate input to the cortex in monkeys with early eye removal is sorted out "vertically" as indicated by denser radioactivity over sublayers 4A and 4C. Thus, competition between afferents serving the two eyes is not essential for the vertical segregation of afferent input in cortex even though it may have a significant effect on the organization of local neuronal circuits in this structure (Valverde, 1968; Rakic, 1979a).

Normal and experimentally perturbed prenatal development of visual connections in rhesus monkey should be viewed in relation to the consequence of monocular deprivation described by Hubel et al. (1977). The difference reported by them in the relative widths of ocular dominance stripes subserving functional and deprived eyes could be interpreted as arrest of a normal developmental process. If geniculocortical terminals subserving the functional eye fail to retract, they may simply retain the territory they had occupied at the time monocular deprivation began, while terminals from the deprived eye regress (Hubel et al., 1977). Although LGd neurons subserving the functional eye need not have invaded new territory, or expanded as such, the possibility that they form a larger number of synapses in the visual cortex is not ruled out. Thus, as in the case of prenatal unilateral eye enucleation (Rakic, 1977c), the effect of monocular deprivation on the ocular dominance domains in the primate visual system can be explained by hypothesis of synaptic competition for space on postsynaptic cells (Guillery, 1972; Hubel et al., 1977).

Acknowledgements This work was supported by U.S.P.H.S. Grant EY02593. It was conducted in part at the New England Regional Primate Research Center, Southborough, Massachusetts. I am thankful to Dr. P. S. Seghal for his generous help in surgery and care of animals.

References

Bishop, P. O., W. Kozak, W. R. Levick, and G. J. Vakkur (1962). The determination of the visual field on the lateral geniculate nucleus in the cat. J. Physiol. Lond. 163:503-539.

Brodmann, K. (1905). Beitrage zur histologischen lokalization der Grosshirnrinde Dritte Mitteilung: Die Rinderfelder niederen Affen. J. Psychol. Neurol. (Leipzig) 9:177-226.

Cavalcante, L. A., and C. E. Roche-Miranda (1978). Postnatal development of retinogeniculate retinopretectal and retinotectal projections in the opposum. Brain Res. 146:231-248.

Changeux, J.-P., and A. Danchin (1976). Selective stabilization of developing synapses as mechanism for the specification of neuronal networks. Nature 264:705-712.

Giordano, D. L., and T. J. Cunningham (1978). Naturally occurring neuron death in the superior colliculus of the postnatal rat. Anat. Rec. 190:402 (abstract).

Goldman, P. S. (1979). Prenatal development of cortico-striatal connections in the rhesus monkey: transformation from diffuse to patterned projections (submitted).

Goldman, P. S., and W. J. H. Nauta (1977). An intricately patterned prefronto-caudate projection in the rhesus monkey. J. Comp. Neur. 171:369-386.

Guillery, R. W. (1972). Binocular competition in the control of geniculate cell growth. J. Comp. Neur. 144:117-130.

Hendrickson, A., and P. Rakic (1977). Histogenesis and synaptogenesis in the dorsal lateral geniculate nucleus (LGd) of the fetal monkey brain. Anat. Rec. 187:602 (abstract).

Hubel, D. H., and T. N. Wiesel (1968). Receptive fields and functional architecture of monkey striate cortex. J. Physiol. 195:215-243.

Hubel, D. H., and T. N. Wiesel (1977). Functional architecture of macaque monkey visual cortex. Proc. R. Soc. Lond. B. 198:1-49.

Hubel, D. H., T. N. Wiesel, and S. LeVay (1977). Plasticity of ocular dominance in monkey striate cortex. Phil. Trans. Roy. Soc. Lond. B. 278:377.

Innocenti, G. M., L. Fiore, and R. Caminiti (1977). Exuberant projection into corpus callosum from the visual cortex of newborn cats. Neurosc. Letters 4:237-242.

Knyihar, E., B. Csillik, and P. Rakic (1978). Transient synapses in the embryonic primate spinal cord. Science 202:1206-1209.

LeVay, S., M. P. Stryker, and C. J. Shatz (1978). Ocular dominance columns and their development in layer IV of the cat's visual cortex: a quantitative study. J. Comp. Neur. 179:223-244.

Price, J. L., G. F. Moxley, and J. E. Schwob (1976). Development and plasticity of complementary afferent fiber systems in the olfactory cortex. Exp. Brain Res. Suppl. 1:148-154.

Rakic, P. (1974). Neurons in rhesus monkey visual cortex: systematic relation between time of origin and eventual disposition. Science 183:425-427.

Rakic, P. (1975). Timing of major ontogenetic events in the visual cortex of the rhesus monkey. In: Brain Mechanisms in Mental Retardation. N. A. Buchwald and M. Brazier (eds.),

Rakic, P. (1976). Prenatal genesis of connections subserving ocular dominance in the rhesus monkey. Nature 261:467-471.

Rakic, P. (1977a). Genesis of the dorsal lateral geniculate nucleus in the rhesus monkey: site and time of origin, kinetics of proliferation, routes of migration, and pattern of distribution of neurons. J. Comp. Neur. 176:23-52.

Rakic, P. (1977b). Prenatal development of the visual system in the rhesus monkey. Phil. Trans. Roy. Soc. Lond. ser B. 278:245-260.

Rakic, P. (1977c). Effects of prenatal unilateral eye enucleation on the formation of layers and retinal connections in the dorsal lateral geniculate nucleus (LGd) of the rhesus monkey. Neuroscience Absts. 3:573 (abstract).

Rakic, P. (1979a). Genetic and epigenetic determinants of local neuronal circuits in the mammalian central nervous system. In: Neurosciences. Fourth Study Program. F. O. Schmitt and F. G. Worden (eds.), MIT Press, Cambridge, Massachusetts, pp. 109-127.

Rakic, P. (1979b). Mode of genesis of central visual connections revealed by orthograde axonal flow and transneuronal transport of radioactive tracers following unilateral eye injection in temporarily exteriorized monkey fetuses (submitted).

Ramón y Cajal, S. (1911). Histologie du Système Nerveux de l'Homme et des Vertébrés Paris, Maloine Reprinted by Consejo Superior de Investigaciones Cientificus, Madrid, 1955, Vols. I and II.

Shatz, C., and P. Rakic (1978). Prenatal development of efferent projections from the visual cortex in the rhesus monkey. Neurosc. Absts. 4:654 (abstract).

So, K., G. E. Schneider, and D. O. Frost (1978). Postnatal development of retinal projections to the lateral geniculate body in Syrian hamsters. Brain Res. 142:575-583.

Swindale, N. V. (1979). How ocular dominance stripes may be formed (this volume).

Valverde, F. (1968). Structural changes in the area striata of the mouse after enucleation. Exp. Brain Res. 5:274-292.

Wiesel, T. N., D. H. Hubel, and D. M. K. Lam (1974). Autoradiographic demonstration of ocular-dominance columns in the monkey striate cortex by means of transneuronal transport. Brain Res. 79:273-279.

Reversal of the Effects
of Visual Deprivation in Monkeys

L. J. GAREY

Institut d'Anatomie
University of Lausanne
Lausanne, Switzerland

F. VITAL-DURAND

Laboratoire de Neuropsychologie Expérimentale
INSERM
Bron, France

COLIN BLAKEMORE

The Physiological Laboratory
Cambridge, England

Abstract Wiesel and Hubel (1963a,b) found that if the lids of one eye of a kitten were sutured closed soon after birth the neurons of the visual cortex, recorded by extracellular microelectrodes some weeks later, responded almost exclusively to the open eye, the deprived eye losing its ability to drive cortical units. This contrasted with normal cats in which some three-fourths of visual cortical cells were influenced binocularly (Hubel and Wiesel, 1963). Another effect of monocular lid suture was that the neurons in the lateral geniculate nucleus (LGN), innervated by the deprived eye failed to grow fully and were smaller than normally innervated cells. Hubel and Wiesel (1970) defined a "critical" or "sensitive" period from about three weeks to three months postnatally during which these effects could be elicited.

The primate visual system has also been studied with similar experimental procedures, and it has been found that monocular deprivation in the monkey also reduces cortical binocularity (Baker et al., 1974; Crawford et al., 1975; Hubel et al., 1977; Blakemore et al., 1978) and prevents normal neuronal growth in the LGN (Headon and Powell, 1973; von Noorden, 1973; Hubel et al., 1977; Vital-Durand et al., 1978).

In order to investigate the extent to which the changes due to monocular deprivation can be reversed, the technique of "reverse-suture" has been used. The animal is monocularly deprived until cortical and thalamic effects would be

261

expected, and then the closed eye is reopened and the other closed. When done before about 6 weeks of age, in the cat, reverse-suture leads to "recapture" of cortical neurons by the initially deprived eye and to a total reversal of ocular dominance within about two weeks (Blakemore and Van Sluyters, 1974; Movshon, 1976). LGN cells, which are, in the cat, 30 to 40% smaller in cross-sectional area in histological sections in deprived laminae than in non-deprived, recover their size, and are as big as normal cells within three days. Within two weeks they are 30% *larger* than the newly deprived cells (Dürsteler et al., 1976).

We report here similar experiments performed in the monkey.

Materials and Methods

Thirty-five old-world monkeys of both sexes were used; all but four were patas (*Erythrocebus patas*). Eyelid suture was performed under ketamine anaesthesia in 31 animals aged from a few hours to about 4 years. In 11, reverse-suture was performed. Sixteen underwent electrophysiological experiments, under Nembutal and Flaxedil, in which the responses of single units in area 17 to visual stimulation were recorded with glass-insulated tungsten microelectrodes.

Animals were perfused terminally with physiological saline followed by 10% formalin or 1% paraformaldehyde plus 1.25% glutaraldehyde in 0.1 M phosphate buffer. Cortical electrode tracks were reconstructed from frozen sections stained with cresyl violet. A block containing the thalamus was embedded in paraffin and coronal sections were cut and stained with cresyl violet. Neurons in the LGN were selected and measured using the following technique. The sections were examined with a microscope whose stage was driven by stepping motors. Drawings were made through a drawing tube on a digital graphics tablet interfaced with a computer. The central axis of each LGN layer was drawn at low magnification and the computer selected five sampling points spaced equally along the layer. At each sampling point the outlines of ten neurons with clear nucleoli were drawn on the graphics tablet at 1000X magnification, and the computer calculated the cross-sectional area of each cell. Six hundred or 1200 cells were measured in each brain.

Results

Normal animals In the visual cortex of normal monkeys most cells outside layer IVc were orientation-selective and binocularly driven. In layer IVc, units were non-oriented and most were monocular; alternating patches of neurons dominated by one eye or the other reflected the distribution of LGN axons in left-eye and right-eye "stripes" (Hubel et al., 1977), of equal width for each eye.

Monocular deprivation In two animals, monocularly deprived by closure of the right eye within the first two months of life, ocular dominance in area 17 was very different from normal, most cells being only driveable by the non-deprived left eye. The proportion of layer IVc dominated by the non-deprived eye was almost three times larger than that dominated by the deprived eye (cf. Hubel et al., 1977).

In two late-deprived animals, one monocularly deprived from 11 to 16 months of age and the other aged about 4 years and monocularly deprived for more than 6 months, there was no evidence of a shift in ocular dominance within layer IVc. However, in the animal deprived at around the age of a year, there was some change in ocular dominance outside layer IVc. Thus, the sensitive period seems different for cells in layer IVc and those outside IVc.

We measured the cross-sectional areas of LGN cells in 16 monocularly deprived monkeys. After monocular deprivation during the first six weeks of life, cells in laminae innervated by the deprived eye were smaller, on average, than the corresponding cells innervated by the non-deprived eye (Headon and Powell, 1973; von Noorden, 1973; Hubel et al., 1977). However, we found that the effects were less than in the kitten, where deprived cells are 30-40% smaller than non-deprived (Guillery and Stelzner, 1970; Garey et al., 1973). In the monkey the difference was about 15-20%. Deprivation for a few days in the first week of life caused a change in the size of LGN cells, but deprivation at about one year or more had no effect.

Thus, the sensitive period for effects on LGN cell growth seems similar to that for functional modification of layer IVc of area 17.

Reverse-suture In eight monkeys, the initially closed right eye was reopened and the other eye closed at ages between 4 and 38½ weeks. In four cases the animal survived for four months with the initially deprived eye open.

In one, reverse-sutured at 5½ weeks, the ocular dominance distribution of cortical cells was a mirror image of that in a monkey simply monocularly deprived until 5½ weeks: most cells were monocularly driven by the originally deprived right eye. Animals deprived until 8 or 9 weeks before reverse-suture showed less effective recapture of the cortex by the newly opened eye. Very late reverse suturing (at 38½ weeks) caused no apparent reversal of ocular dominance.

In layer IVc the right-eye stripes were roughly the same width as the left-eye stripes in the animals reverse-sutured at 9 weeks or before. This contrasts with a ratio of about 1:3 after simple monocular deprivation.

The other four reverse-sutured monkeys were initially deprived until 4 weeks of age and then reversed for between 3 days and 18 weeks. The shift in ocular dominance and the physiological re-expansion of ocular dominance stripes in layer IVc was already detectable after three days and virtually complete in two weeks.

When reverse-suture was performed before 12 weeks the difference in cell size between LGN laminae was reduced or even slightly reversed. LGN cell size showed a detectable recovery after only three days with complete recovery within two weeks. After reverse-suture at 38½ weeks the LGN showed no sign of reversal. The sizes of LGN neurons correlate well with the widths of ocular dominance stripes in layer IVc of the visual cortex. We are now looking at the distribution of afferent axons to layer IVc, using the method of transneuronal transport of radioactive label injected into the eye (Hubel et al., 1977).

Discussion

The anatomical and physiological consequences of monocular visual deprivation in the primate brain can be reversed by forced use of the deprived eye. The major effects of monocular deprivation and reverse-suture seem to be over within about the first two months of life, rather as in the cat (Hubel and Wiesel, 1970), although cortical ocular dominance outside layer IVc is modifiable for up to about a year or so. The morphological changes in the LGN seem parallel to the cortical effects and support the concept of a relation between soma size and axonal arborization (cf. Guillery, 1972).

Summary

Monocular visual deprivation in kittens or baby monkeys causes most visual cortical neurons to become unresponsive to the deprived eye. In both kittens and monkeys, "reverse-suture" (opening the deprived eye and closing the other) leads to "recapture" of cortical cells by the initially deprived eye if done within about 6 weeks. In the monkey the ocular dominance "columns" in layer IVc of area 17 related to the newly opened eye expand. Late reversal is ineffective and intermediate reversal (8 to 9 weeks) causes partial recapture.

Monocular deprivation also inhibits normal neuronal growth in the deprived laminae of the lateral geniculate nucleus (LGN) of monkeys, an effect which can be reversed with reverse-suture with a time course similar to that of the cortical changes in layer IVc.

Acknowledgements The monkeys in this study were kindly donated by Institut Mérieux and IFFA Mérieux, Lyon, to whom we are very grateful. We thank M. C. Cruz, C. Frenois, and M. Gaillard for technical help. The work was supported by INSERM, by the European Training Programme in Brain and Behaviour Research and by grants from the Swiss National Science Foundation (No. 3.2460.74) and the Medical Research Council (No. G972/463/B).

References

Baker, F. H., P. Grigg, and G. K. Von Noorden (1974). Effects of visual deprivation and strabismus on the response of neurons in the visual cortex of the monkey, including studies on the striate and prestriate cortex in the normal animal. Brain Res. 66:185-208.

Blakemore, C., L. J. Garey, and F. Vital-Durand (1978). The physiological effects of monocular deprivation and their reversal in the monkey's visual cortex. J. Physiol. (Lond.) 283:223-262.

Blakemore, C., and R. C. Van Sluyters (1974). Reversal of the physiological effects of monocular deprivation in kittens: further evidence for a sensitive period. J. Physiol. (Lond.) 237:195-216.

Crawford, M. L. J., R. Blake, S. J. Cool, and G. K. Von Noorden (1975). Physiological consequences of unilateral and bilateral eye closure in macaque monkeys: some further observations. Brain Res. 84:150-154.

Dürsteler, M. R., L. J. Garey, and J. A. Movshon (1976). Reversal of the morphological effects of monocular deprivation in the kitten's lateral geniculate nucleus. J. Physiol. (Lond.) 261:189-210.

Garey, L. J., R. A. Fisken, and T. P. S. Powell (1973). Effects of experimental deafferentation on cells in the lateral geniculate nucleus of the cat. Brain Res. 52:363-369.

Guillery, R. W. (1972). Binocular competition in the control of geniculate cell growth. J. comp. Neurol. 144:117-130.

Guillery, R. W., and D. J. Stelzner (1970). The differential effects of unilateral lid closure upon the monocular and binocular segments of the dorsal lateral geniculate nucleus in the cat. J. comp. Neurol. 139:413-422.

Headon, M. P., and T. P. S. Powell (1973). Cellular changes in the lateral geniculate nucleus of infant monkeys after suture of the eyelids. J. Anat. (Lond.) 116:135-145.

Hubel, D. H., and T. N. Wiesel (1963). Receptive fields of cells in striate cortex of very young, visually inexperienced kittens. J. Neurophysiol. 26:994-1002.

Hubel, D. H., and T. N. Wiesel (1970). The period of susceptibility to the physiological effects of unilateral eye closure in kittens. J. Physiol. (Lond.) 206:419-436.

Hubel, D. H., T. N. Wiesel, and S. LeVay (1977). Plasticity of ocular dominance columns in monkey striate cortex. Phil. Trans. R. Soc. B 278:377-409.

Movshon, J. A. (1976). Reversal of the physiological effects of monocular deprivation in the kitten's visual cortex. J. Physiol. (Lond.) 261:125-174.

Vital-Durand, F., L. J. Garey, and C. Blakemore (1978). Monocular and binocular deprivation in the monkey: morphological effects and reversibility. Brain Res. 158:45-64.

Von Noorden, G. K. (1973). Histological studies of the visual system in monkeys with experimental amblyopia. Invest. Ophthal. 12:727-738.

Wiesel, T. N., and D. H. Hubel (1963a). Effects of visual deprivation on morphology and physiology of cells in the cat's lateral geniculate body. J. Neurophysiol. 26:978-993.

Wiesel, T. N., and D. H. Hubel (1963b). Single-cell responses in striate cortex of kittens deprived of vision in one eye. J. Neurophysiol. 26:1003-1017.

How Ocular Dominance Stripes May Be Formed

N. V. SWINDALE

Physiological Laboratory
Cambridge, England

Abstract On the basis of a computer simulation, I propose that two rules are sufficient to account for the formation of ocular dominance stripes in monkey striate cortex, assuming an initial random intermixing of left- and right-eye terminals: (1) if synapses of one eye type predominate locally around a point in the cortex (i.e., within a distance of about 200 μm) those synapses will increase in number at that point, at the expense of synapses of the other type; however, (2) this increase will be reduced or reversed if, within a surrounding annular region 200-600 μm distant, synapses of the same type predominate. This leads to a segregation of inputs into a stable pattern of non-overlapping stripes, which branch and intersect the boundaries of the modeled region at right angles.

The laminae of the lateral geniculate nucleus relay input from corresponding hemi-retinae of both eyes to the striate cortex, and in primates (and probably most other mammalian species) the inputs from each eye are confined to single laminae within the nucleus. In the macaque monkey this segregation is further maintained in the projection from the lateral geniculate to the striate cortex: if the cortical terminals conveying input from one eye are selectively labeled (e.g., by injecting a radioactive tracer into one eye) the resulting pattern of label, viewed from a direction normal to the cortical surface, takes the form of uniform parallel stripes, each about 0.4 mm wide (Hubel and Wiesel, 1972; Wiesel, Hubel, and Lam, 1974; LeVay, Hubel, and Wiesel, 1975). These stripes (Fig. 1) have a remarkable morphology: they may fuse together in "Y" or "H" formations; they may end blindly or change direction abruptly, or occur as short elongated "islands." The gaps between the stripes are occupied by unlabeled input from the other eye and show exactly the same variety of morphological features. There is thus an overall segregation of inputs into a pattern of alternating bands, and it has been convincingly shown by several independent techniques (see Hubel and Wiesel, 1977) that this patterning is responsible for

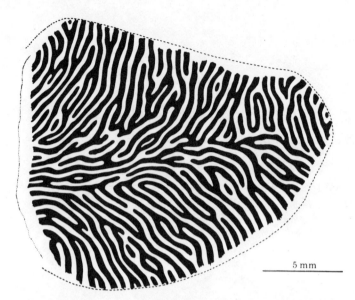

5 mm

FIGURE 1 Reconstruction of the pattern of ocular dominance stripes in area 17 of the macaque monkey, made from a series of sections tangential to the cortical surface, stained with a reduced-silver technique (LeVay et al., 1975). The stripes conveying input from one eye are black; the unshaded areas convey input from the other eye. The dashed line is the 17-18 border (taken from Hubel and Wiesel, 1977).

the organization of the visual cortex into ocular dominance columns, i.e., the tendency for cortical neurones with a similar preference for stimulation by one eye to be grouped together into vertical columns about 0.4 mm wide (Hubel and Wiesel, 1968).

It is intriguing that similar patterns can be found elsewhere in nature. The black and white stripes on zebras are a familiar example; the stripes on the body surfaces of mackerel and many other fish, and the groove and ridges of fingerprints are further examples, and others are not hard to find. Such similarities in such a diversity of systems suggest that there may exist a common set of rules governing the formation of the patterns. These rules would not lay down the precise details of the pattern, but would impose general constraints on features like the width of the stripes and the type of branching that could occur. If this is so, then it ought to be possible to arrive at an understanding of how ocular dominance stripes form without referring in detail to the underlying cortical physiology. This approach is not essentially different from one which tries to describe the behaviour of, for example, the vestibulo-ocular reflex in terms of a negative feedback loop with a variable gain, and which makes little or no reference to the underlying neural circuitry.

The theory I wish to propose makes two essential assumptions. The first is that there is initially no segregation of inputs to the cortex, but instead a nearly uniform intermixing of the inputs from the two eyes, the only deviations from uniformity being caused by naturally occurring random fluctuations. There is good experimental evidence for this: injection of labeled amino acids into one eye six weeks prenatally (which is shortly after the projection from the lateral geniculate nucleus reaches the cortex) reveals no periodicity in the resulting distribution of label in layer IV (Rakic, 1977). Later injections reveal slight fluctuations in the label from one eye, which is still distributed throughout layer IV; segregation (that is, the absence of overlap in the projections from the two eyes) is probably not complete until about six weeks after birth (Hubel et al., 1977).

The second assumption is that left- and right-eye synapses grow at rates which are not locally independent. Thus the rate of growth of synapses from one eye at a point in the cortex will be determined by the composition, in terms of relative numbers of left- and right-eye synapses, of the surrounding synaptic environment. Assuming that the two eyes are symmetrical with respect to the rules they obey, there are two types of interaction to take into account: effects on growth exerted by synapses of the same eye type, and effects exerted by synapses of opposite eye type. These two effects, which will vary with distance, can be described quantitatively by two functions, W_s and W_0, respectively. Two suitable functions are shown in Fig. 2. They are assumed to be radially symmetric and have the following interpretation: W_s is positive for distances of less than 200 μm, and this implies that synapses of the same eye type will reinforce each other's rate of growth if situated less than this distance apart (i.e., within region E of Fig. 2); for distances of 200-600 μm (i.e., within region I) the effect is an inhibition, since for these distance W_s is negative. The function W_0 is to be understood in the same way with respect to the interaction between synapses of different eye type. To find the rate of growth $dR(i,j)/dt$ of right-eye synapses, for example, at a point with coordinates (i,j) on the cortical surface, the effects of all surrounding left- and right-eye synapses have to be taken into account. This is done by summing over neighbouring points (k,l) distant from (i,j) so that

$$\frac{dR(i,j)}{dt} = \sum_{k,l} \left\{ W_s(k,l) R(i-k,j-l) + W_0(k,l) L(i-k,j-l) \right\} \tag{1}$$

where $R(i,j)$ and $L(i,j)$ represent the densities of right- and left-eye synapses, respectively, at a point (i,j). A similar summation is carried out to determine the rate of growth of left-eye synapses. The rate of growth of synapses is limited by two other factors: one is that densities cannot be negative, and the other is that there must be an upper limiting density.

When these rules are implemented in the form of a computer program, patterns like that shown in Fig. 3 develop. The figure shows the initial condition: the cortex is represented by an array of points, approximately 100 μm distant from each other, and two similar values, representing the densities of right- and left-eye synapses, have been assigned to each point. Small random fluctuations (whose amplitude is uncritical) are present in these values, and since only those points where more than 50% of the total input from one eye are displayed, a

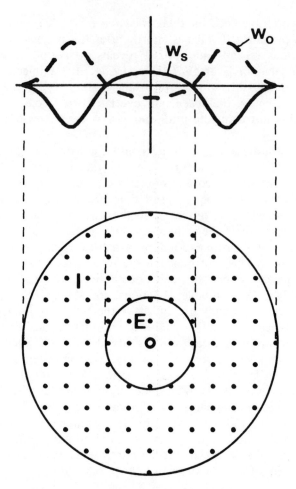

FIGURE 2 The functions W_0 (shown as a continuous line) and W_s (dotted line) graphed as a function of radial distance. The bottom half of the figure shows their projection onto a part of the array used to represent the cortex. The dots represent points approximately 100 μm apart from one another. The points shown are those whose synaptic densities have to be taken into account to find the rate of growth of synapses at the central point, represented by a circle.

random selection of half of the points in the array is visible. As the inputs are changed with successive iterations of the program, at rates given by Eq. (1) the randomness disappears; blotches and stripes develop and these eventually organize into a final pattern of branching stripes which is stable to repeated further iterations of the program.

The precise shapes of the functions W_0 and W_s, which are used to generate the stripes, appear to be uncritical. It is possible for W_s to be zero everywhere, and it is then necessary that W_0 should be negative within region E and positive within region I. It is possible for both functions to be zero within region E, provided W_0 is positive in region I and $W_0 = -W_s$ within this region. If both functions are zero within region I, then segregation occurs in the form of

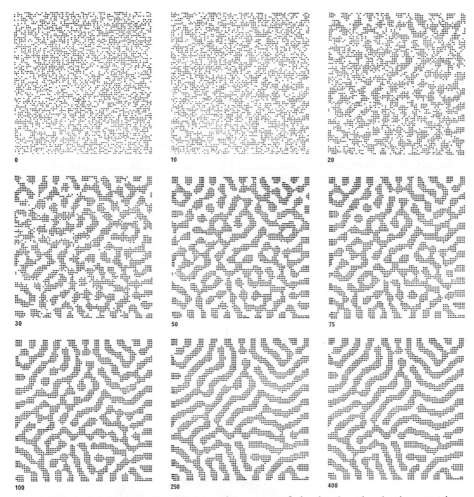

FIGURE 3 Stages in the development of a pattern of simulated ocular dominance stripes. Numbers represent the iterations needed to produce each pattern. Points are displayed wherever the input from one eye predominates over the other. In the final pattern, more than 97% of the points in the array are occupied exclusively by input from one eye; the number of such points increases most rapidly between the 50th and 100th iterations, which in the monkey would probably correspond to a period shortly after birth.

large irregular blotches. As a general rule, if the functions are large in region E, the stripes become less regular in width and branch more frequently. For the case where $W_s = -W_0$ (as in Fig. 2) the following simple interpretation of the model can be formulated: (1) if synapses of one type predominate locally (i.e., within region E), then that type of synapse will increase in number at the expense of the other type; however, (2) this increase will be reduced or reversed if within a surrounding annular region (region I in Fig. 2), synapses of the same type predominate.

Departures from radial symmetry in W_0 and W_s alter the directions in which the stripes run. If region E is elongated, there is a marked tendency for the stripes to run in the direction of elongation; if I is elongated, then the stripes run perpendicular to the direction of elongation. This behaviour has an interesting consequence for points near the boundaries of the modeled region, where the parts of regions E and I which extend over the boundary no longer have any determining effect on the behaviour of the stripes. This effectively elongates region I and as a result the stripes become oriented in a direction perpendicular to the border. This is a striking feature of the stripes in the monkey (Fig. 1), as well as some other naturally occurring patterns of stripes, such as those on zebras and mackerel.

Finally, it is interesting to consider the possible effects of growth of the striate cortex, as a whole, on stripe morphology. The striate cortex enlarges by about 20% after birth (Hubel et al., 1977), and presumably also prenatally, while the columns are forming; and although it may be that the space constants of the interaction simply enlarge in parallel with the expansion of the cortex (this is suggested by the fact that the periodicity of the columns is less in prenatal monkeys than in adults (Rakic, 1976), one might suspect that a mismatch between the two, in the presence of an overall elongation of the cortex, would influence the directional properties of the stripes. This expectation was confirmed (Fig. 4) by expanding, in one direction only, the array in which the

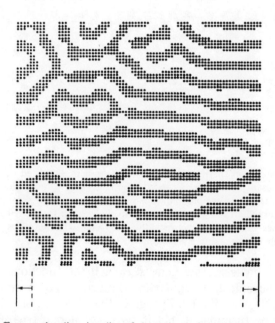

FIGURE 4 The effect on the directionality of the stripes of slowly elongating the array while they are forming. The functions W_0 and W_s remain unchanged in shape while this happens. The degree of elongation is indicated beneath the figure and was completed by the 26th iteration of the program.

stripes were forming. The array was expanded in size by 12%, and the expansion took place in the first 26 iterations of the program, long before the emergence of a final stable pattern. The effects on the directionality of the stripes were very pronounced and are clearly not due to a simple stretching of the pattern. Thus, the observed uniformities in the directional properties of the stripes in the monkey may perhaps reflect not only the influence of the boundaries of the pattern, but also the way in which the cortex grows while the stripes are forming.

Acknowledgements I am grateful to J. B. L. Bard, H. B. Barlow, C. B. Blakemore, and G. Mitchison for their advice and encouragement. The work was supported by a grant from the U.K. Medical Research Council to C. B. Blakemore.

References

Hubel, D. H., and T. N. Wiesel (1968). Receptive fields and functional architecture of monkey striate cortex. J. Physiol. Lond. 195:215-243.

Hubel, D. H., and T. N. Wiesel (1972). Laminar and columnar distribution of geniculo-cortical fibers in the macaque monkey. J. Comp. Neurol. 146:421-450.

Hubel, D. H., and T. N. Wiesel (1977). Functional architecture of macaque monkey visual cortex. Proc. R. Soc. Lond. B. 198:1-59.

Hubel, D. H., T. N. Wiesel, and S. LeVay (1977). Plasticity of ocular dominance columns in monkey striate cortex. Phil. Trans. R. Soc. Lond. B. 278:131-163.

LeVay, S., D. H. Hubel, and T. N. Wiesel (1975). The pattern of ocular dominance columns in macaque visual cortex revealed by a reduced silver stain. J. Comp. Neurol. 159:559-576.

Rakic, P. (1976). Prenatal genesis of connections subserving ocular dominance in the rhesus monkey. Nature 261:467-471.

Rakic, P. (1977). Prenatal development of the visual system in the rhesus monkey. Phil. Trans. R. Soc. Lond. B. 278: 245-260.

Wiesel, T. N., D. H. Hubel, and D. M. K. Lam (1974). Autoradiographic demonstration of ocular-dominance columns in the monkey striate cortex by means of transneuronal transport. Brain Res. 79:273-279.

STUDIES OF THE HUMAN'S VISUAL SYSTEM

Development of Optokinetic Nystagmus in the Human Infant and Monkey Infant: an Analogue to Development in Kittens

J. ATKINSON

Department of Experimental Psychology
University of Cambridge
Cambridge, England

Abstract One permanent effect of monocular deprivation on visual behaviour in kittens is that optokinetic nystagmus (OKN) in monocular viewing can only be elicited for temporal to nasal field movement; movement of the stimulus nasally to temporally elicits either no response or irregular saccades. In the present study, this condition is called "asymmetrical OKN." Asymmetrical OKN has also been previously observed in kittens under approximately 4 weeks of age.

In the present study, the development of OKN, with monocular viewing in human infants to random noise patterns, has been investigated. The results of these experiments demonstrate that infants up to about 3 months of age give asymmetrical OKN, although binocularly they give OKN in both directions. There is a gradual change-over to symmetrical OKN between 2 and 3 months of age. Preliminary results on monkey infants show a similar asymmetry over the first two to three weeks of life. Studies of OKN in human clinical cases of early binocular deprivation (strabismus, ptosis), in general, support the idea that early loss of binocularity prevents the normal development of some corticotectal connections controlling OKN.

A number of recent studies have compared the recovery of visual discriminations and visuomotor coordination in monocularly or binocularly deprived cats with the development of similar responses in newborn kittens. One response that was studied is optokinetic nystagmus (OKN). For monocular viewing this was observed only when the stimulus is moved temporally to nasally in the visual field, rather than nasally to temporally, in kittens who have had long periods of monocular deprivation (eight to ten months) followed by reverse suturing (van Hof-van Duin, 1976). (In this paper the case where OKN can be

277

driven better in one horizontal direction in the visual field will be referred to as "asymmetrical OKN," with the converse "symmetrical OKN" referring to OKN that is equally easily elicited in either direction.) These monocularly deprived (MD) kittens also showed irreversible impairment on depth discrimination on the visual cliff. A similar asymmetry of OKN was found in kittens reared in darkness from birth (BD) and in normal kittens over the first few weeks after eye opening (van Hof-van Duin, 1978). In both BD and MD cats the deficit was still present after many months. From these results it seems likely that one condition necessary for the development of symmetrical OKN is the development of normal binocular processing in the visual cortex, which is dependent on balanced binocular input early in life. The rationale behind the experiments reported here was that the time course of development of a functional binocular visual cortex (and behaviour such as stereoscopic vision which is dependent on it) may be indicated by the course of development of symmetrical OKN.

Experiment I

Human infants who had normal birth histories were observed at three different ages: 3-6 weeks, 8-9 weeks, and 12-14 weeks. A few older infants and special cases (with a history of a visual problem either within the family or the individual) were also tested. Eye movements were observed both binocularly and monocularly. The infant looked into a white translucent hemisphere (diameter 100 cm) on which was back-projected a colourful meaningless visual scene. The picture moved from left to right, or vice versa, in front of the infant over approximately 180° of visual field. The picture was made by filming an approximation to a random noise field, panning with an 8-mm cine camera through 360°. The stimulus was newspaper sheets overlaid with brightly coloured paints spots of varying density and size, the print of the newspaper also being visible on the film. The film passed in front of the infants' eyes at 36 deg/sec. The stimulus contained a wide range of spatial frequencies, individual paint spots subtending large visual angles (1-10°) at the viewing distance of 40 cm. Each infant was either seated in a modified infant seat or held sitting on a cushioned platform in front of the screen. Each trial consisted of 15 seconds of film during which time the observer noted head and eye movements. The direction of movement was randomized from trial to trial and was unknown to the observer who, from the position of the peephole (towards the bottom of the screen), could only see the infant's face. The trials were run with the room lights off, some ambient light coming from the projector. At the end of each trial, the observer had to say in which direction she judged the film to have moved. This decision was based on observation of the infant's head and eye movements and on the form the eye movements had taken. The total time spent following the direction of movement during the trial was recorded by the observer using a stopwatch. This included the following three different types of eye movements (on some occasions all three types were observed in a single trial):

1. Bursts of classic OKN consisting of a short drift in the direction of movement followed by a rapid flyback, repeated a number of times.

2. Steady, smooth pursuit across a large proportion of the visual field (probably tracking a single target). This tracking was often followed immediately by a burst of classic OKN.

3. Irregular saccades, in the approximate direction of movement, sometimes followed by a series of saccades in the reverse direction or often followed by a long stare.

At the start of testing, binocular eye movements (including OKN and smooth tracking) were observed for both directions of movement. Infants who either did not respond binocularly or fretted were not tested further. This binocular screening was followed by five or six trials with one eye covered, followed by five or six trials with the other eye covered. The eye that was initially covered was varied randomly from infant to infant. After a litle initial observation the observer only very rarely made an error in judging the direction of movement. This was because it became apparent that either eye movements were easily elicited in both directions or that when no eye movements were observed, the direction was nasal to temporal in the visual field. Consequently, a guess on such a trial was invariably correct. Sometimes, with older infants, head movements rather than eye movements indicated the direction.

The mean total time per 15-second trial is shown in Fig. 1 for each infant in each age group. From this figure it can be seen that there is a clear asymmetry in the mean time spent following the direction of movement for the two younger age groups, but not for the 3-month-olds. For the 1-month-olds much longer times are spent following field movement temporal to nasal with either eye than vice versa. If nasal to temporal movement was followed at all, it was with a number of irregular saccades. Nearly all the 3-month-olds showed the same amount of time following the movement in either direction. The 2-month-olds seem, from their results, to be at an intermediate point of development between the 1- and 3-month-olds; many of them show eye movements of mixed type for nasal to temporal field movement. Some of the 2-month-olds showed symmetrical OKN on some trials and not on others.

Table 1 shows the results for a group of individuals not falling in the 1-3 month age range, who were tested because of an individual history or family history of abnormal visual development. Their results on a subjective test of stereoscopic vision using random dot stereograms is also included, if known and possible to execute. One individual (I.A.) was tested on the moving stimulus on two occasions (approximately five months apart). It was noticed that for either eye viewing, the eye movements were more irregular when field movements were nasal to temporal and there was a large asymmetry in the mean following times per 15-second trial. These means are included in Table 1. As can be seen from this table, there does seem to be a relationship between asymmetric OKN and balanced binocular inputs early in life for some cases, but as yet there is not enough clear clinical data to strongly support this

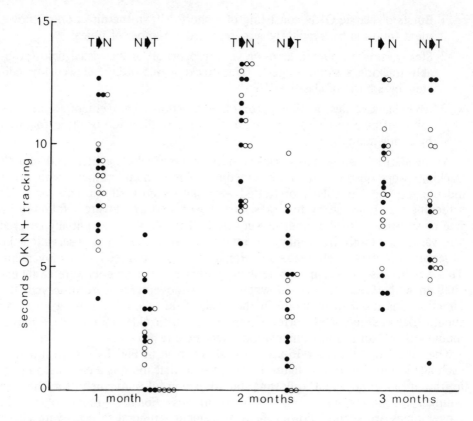

FIGURE 1 Mean total time of following eye movements per 15-second trial time for human infants in three age groups in Experiment I. T → N = temporal to nasal movement of the stimulus field; N → T = nasal to temporal movement of the stimulus field. Filled circles: right eye; open circles: left eye.

claim. All adults, children, and infants over 3 months of age, but with apparently normal vision, showed symmetrical OKN when tested monocularly.

Experiment II

The stimulus pattern used in Experiment I made it possible for the infant to track individual spots rather than the overall movement of the entire field. In Experiment II a film loop of fine random texture (Letratone) was back-projected onto a flat, wide screen (filling approximately 120° of the visual field). In adults this stimulus was a powerful elicitor of OKN when viewed either monocularly or binocularly, and it was impossible to track individual elements of the pattern over any distance. The movement was at 30 deg/sec with the infant seated 40 cm from the screen.

Three groups of infants were tested in age groups 4-6 weeks, 7-9 weeks, and 12-16 weeks, with mean ages close to 1, 2, and 3 months. Three of the infants

Table 1

SUBJECT	AGE	VISION AND HISTORY	OKN RE T → N	OKN RE N → T	OKN LE T → N	OKN LE N → T	Stereo
1. E.B.	3 mos	Bilateral convergent squint since birth; anisometropic	OKN	irr	OKN	irr	?
2. M.M.	4 mos	Normal	OKN	OKN	OKN	OKN	?
3. B.W.	4 mos	Normal	OKN	OKN	OKN	OKN	?
4. J.T.	4 mos	Left convergent squint; hypermetropic	OKN	irr	OKN	irr	?
5. R.B.	4.5 mos	Left convergent squint	OKN	OKN	OKN	irr	?
6. D.B.	5 mos (2 mos prem.)	Obstruction LE	OKN	irr	OKN	irr	?
7. R.S.	6 mos	Family history of squint, but no obvious deviation	OKN	OKN	OKN	OKN	?
8. R.M.	7 mos	Family history of squint, but no obvious deviation	OKN	OKN	OKN	OKN	?
9. B.A.	9 mos	Family history of squint, but no obvious deviation	OKN	OKN	OKN	OKN	?
10. M.M.	9 mos	Small occlusion on RE from birth; amblyopic	OKN	OKN	OKN	OKN	?
11. L.B.	9 mos	Normal	OKN	OKN	OKN	OKN	?
12. S.T.	2.5 yrs	L squint from 3-6 mos	OKN	OKN	OKN	irr	?
13. A.R.	4.0 yrs	L squint from approx. 5 mos	OKN	OKN	OKN	OKN+irr	S
14. T.M.	Adult	L squint from early life	OKN	OKN	OKN	OKN	S
15. D.R.	Adult	L squint from early life	OKN	OKN	OKN	OKN	N.S.
16. I.A.	4.5 yrs	L ptosis, vertical L deviation from birth	OKN	irr	OKN	irr	N.S.
			12	7	12	2	
			11.7	1.5	13.8	4.7	

OKN: optokinetic nystagmus; irr: irregular movements; RE: right eye; LE: left eye; T → N: temporal to nasal; N → T: nasal to temporal.

Stereo: stereoscopic vision; ?: not tested for stereoscopic vision; S: stereoscopic vision demonstrated; N.S.: sterescopic vision not demonstrated.

were tested at all three ages and so are common to all groups. Again, the mean time of following (which, with this stimulus, was nearly all classic OKN) was calculated for each eye of each infant, from a total of four to six trials using each eye. The mean for each eye separately is shown in Fig. 2. As can be seen by comparing Figs. 1 and 2, the results are very similar to those of Experiment I, namely, infants under 3 months of age show a much greater following time for a whole field movement that is temporal to nasal than for nasal to temporal. The differences in stimulus conditions did not change the apparent development of symmetrical OKN which was achieved in both samples of infants by 3 months of age.

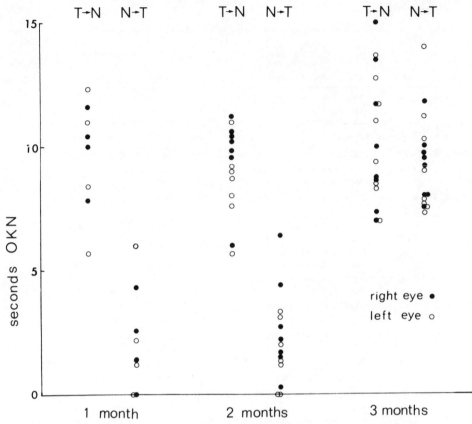

FIGURE 2 Mean total time of following eye movements per 15-second trial time for human infants in three age groups in Experiment II. T → N = temporal to nasal movement of the stimulus field; N → T = nasal to temporal movement of the stimulus field. Filled circles: right eye; open circles: left eye.

Experiment III

Monkey infants (*macaca mulatta*) between 3 to 51 days from birth were tested for symmetry of OKN responses, each monkey being tested at a number of different ages approximately a week apart. The stimulus conditions and procedure were identical to those used in Experiment II, except that the speed of the stimulus movement was 50 deg/sec rather than 30 deg/sec. It was found in pilot testing that it was easier to elicit OKN binocularly at higher speeds for monkeys than those used for the human infants. Again, at each age the mean time of following per trial of 15 seconds was calculated for each eye in both directions.

These mean times for all monkeys tested are shown in Fig. 3. The general trend which can be seen from the data is a gradual increase in time spent fol-

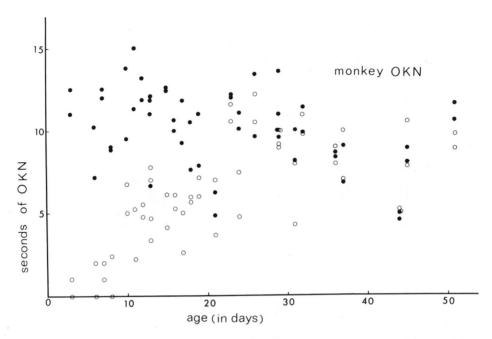

FIGURE 3 Mean total time of following eye movements per 15-second trial time for monkey infants aged 3-51 days from birth in Experiment III. Filled circles: temporal to nasal movement of the stimulus field; open circles: nasal to temporal movement of the stimulus field.

lowing for field movements nasal to temporal over the first few weeks of life, followed by symmetrical following in both directions. The changes in mean time with age are shown for individual monkeys in Figs. 4, 5, and 6. The trend seen in Fig. 3 is substantiated by individual responses shown in these graphs. In general, it would seem that symmetrical OKN is achieved in monkey infants by 2-3 weeks of age, with some individual variation in the exact time course. It is possible that some or all of the variation found is due to variation in gestational age which was not known exactly in these cases.

A clear developmental change has been observed in the ability of human infants and monkey infants to give OKN responses to a stimulus moved in the field in the nasal to temporal direction. For monkeys there is a very short period early in life (approximately two to three weeks) where OKN is asymmetrical, whereas for humans the period extends over the first two to three months of life. The asymmetry apparently persists in kittens who have had early visual deprivation and in some human cases of abnormal binocular visual input early in life (see also, Crone, 1977). It seems likely that a necessary requisite for the development of symmetrical OKN is balanced binocular input and a functioning binocular cortex, which can somehow serve to signal to subcortical levels controlling OKN. Hoffmann (1975, 1977) has found a class of neurons in the pretectum (nucleus of the optic tract, NOT) of cats that are

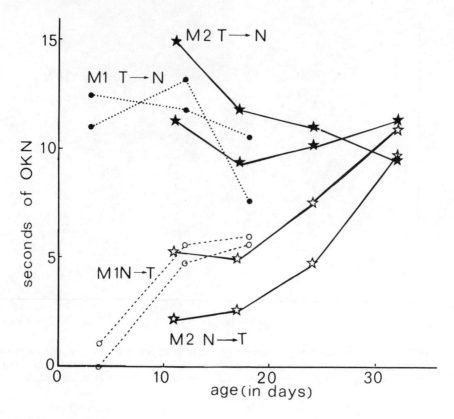

FIGURE 4 Mean total time of following eye movements per 15-second trial time for two monkeys.

excited only by movements from the periphery to the centre of the visual field. If the binocular input to the NOTs is lost due to early monocular deprivation, then only the monocular contralateral input to the NOTs can drive OKN for temporal to nasal field movements. Collewijn (1972, 1976) has found a similar set of directionally selective neurons giving rise to asymmetrical OKN responses in the NOT of the rabbit. It seems likely that a similar system controlling OKN elicited by temporal to nasal pattern movement is present in the human and monkey pretectum and that the maturation of a second system develops alongside the development of binocularity and/or differentiation of the fovea.

In humans the time at which symmetrical OKN is achieved may be related to the functional maturity of binocular cortex, necessary for stereoscopic vision. There is little evidence available concerning the time at which infants can detect disparities, but what there is suggests that detection is possible in infants

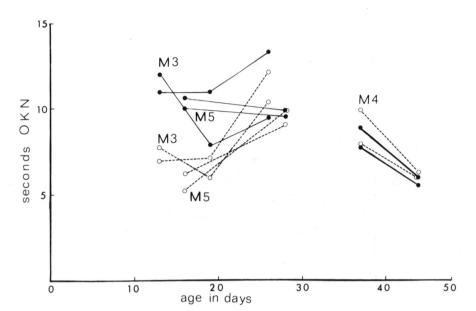

FIGURE 5 Mean total time of following eye movements per 15-second trial time for three monkeys. Filled circles: temporal to nasal movement; open circles: nasal to temporal movement.

by 2-4 months of age (Appel and Campos, 1977; Atkinson and Braddick, 1976; Fox, 1978). This is consistent with the age at which symmetrical OKN has developed.

It has recently been suggested that at least two mechanisms may be involved in driving OKN, the velocity of the movement deciding the mechanism (Cohen, Matsuo, and Raphan, 1977). One system or pathway (direct) drives the eyes to only a fraction of the stimulus velocity (gain less than 1.0). As the stimulus continues, the second slower pathway, which has a gain closer to 1.0, is activated, and can follow higher velocities. The former direct pathway is thought also to be involved with after-nystagmus (OKAN), whereas the latter is not. It is quite possible that the asymmetry of OKN found in the present study is dependent upon the velocity of the stimulus. Further research varying velocity and recording gain is needed before any answer can be given to this question.

In summary, a clear developmental change in the ease of eliciting OKN, depending on the direction of motion within the field, has been demonstrated. The exact parameters controlling the gain of OKN in these instances have not yet been demonstrated. Further neurophysiological and psychophysical research is needed before a clear understanding of the underlying mechanisms can be achieved.

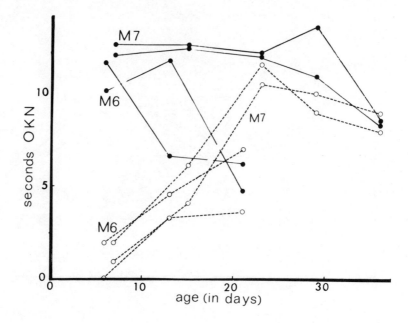

FIGURE 6 Mean total time of following eye movements per 15-second trial time for two monkeys. Symbols as in Fig. 5.

Acknowledgements My thanks are due to Professor R. Held for facilities and encouragement for Experiment II in his laboratory at MIT; Dr. T. N. Wiesel for the use of his laboratory and the staff at the NE Primate Centre, Sudbury, Massachusetts, for Experiment III; Jennifer French for assistance in Experiment I; members of Professor Held's laboratory for assistance in Experiments II and III; and, in particular, my warm thanks to Joseph Bauer for assistance and to all the parents and infants who volunteered while I was a visitor at MIT.

This work was supported in part by the Medical Research Council and in part by research grants from the National Eye Institute (NIH E01191) and CORE Grant (NIH-1-P30-EY02621-01).

References

Appel, M. A., and J. J. Campos (1977). Binocular disparity as a discriminable stimulus parameter for young infants. J. Exp. Child. Psychol. 23:47-56.

Atkinson, J., and O. Braddick (1976). Stereoscopic discrimination in infants. Perception 5:29-38.

Cohen, B., V. Matsuo, and T. Raphan (1977). Quantitative analysis of the velocity characteristics of optokinetic nystagmus and optokinetic after-nystagmus. J. Physiol. 270:321-344.

Collewijn, H. (1976). Direction-selective units in the rabbit's nucleus of the optic tract. Brain Res. 100:489-508.

Crone, R. A. (1977). Amblyopia: the pathology of motor disorders in amblyopic eyes. Docum. ophthal. Proceedings Series. Experimental and Clinical Amblyopia XIIIth I.S.C.E.R.G. Symposium 45:9-18.

Fox, R. (1978). Testing stereopsis in animals and human infants. In: Visual Deprivation Effects in Animals and Their Analogs in Human Visual Pathology. University of Rochester, Center for Visual Science, Eleventh Symposium.

Hoffmann, K.-P., and A. Schoppmann (1975). Retinal input to the direction-selective cells of the nucleus tractus opticus of the cat. Brain Res. 99:359-366.

Hoffmann, K.-P. (1977). Visual response of neurons in the nucleus of the optic tract of visually deprived cats. Soc. for Neuroscience, 7th Annual Meeting, Abstracts (vol. III).

van Hof-van Duin, J. (1976). Early and permanent effects of monocular deprivation on pattern discrimination and visuomotor behavior in cats. Brain Res. 111:261-276.

van Hof-van Duin, J. (1978). Direction preference of optokinetic responses in monocularly tested normal kittens and light-deprived cats. Arch. Ital. Biol. 116:471-477.

Accommodation and Acuity in the Human Infant

OLIVER BRADDICK
Department of Experimental Psychology
University of Cambridge
Cambridge, England
JANETTE ATKINSON
Department of Experimental Psychology
University of Cambridge
Cambridge, England

Abstract Our studies using behavioural (preferential looking) and evoked-potential techniques indicate a rapid development of acuity and contrast sensitivity over the first six months of life. The technique of "photorefraction" allows us to measure how accurately the infant's eyes are optically focused over a range of distances, and hence to investigate whether the infant's performance is limited by optical or by neural development. We find that most infants of 1 month and older accommodate accurately at distances up to 75 cm, but that accommodation at 150 cm is not accurate until later than 3 months of age. However, many newborn and 1-month infants show inconsistent fluctuations of accommodation over the whole range, indicating the physical capability to accommodate.

The errors of accommodation in young infants are not large enough to account for the low acuity; acuity appears to be determined by the development of the visual pathway.

Sixty percent of infants under 1 year are significantly astigmatic. This finding may have implications for the plasticity of the human visual system in early life.

Most of our information about the neural development of vision in human infants must necessarily come from investigations of the development of visual performance, e.g., acuity and contrast sensitivity. This performance, assessed behaviourally or by recording visual evoked potentials (VEP's) from the scalp, might reflect not only neural but also optical limitations on the transmission of visual information. This paper briefly reviews our findings on visual performance in 0-6 month-olds and then presents evidence that these results are dominated by the development of the visual pathway and not by considerations of optical image quality.

Infant Acuity and Contrast Sensitivity

Our behavioural data on contrast sensitivity have been obtained by a version of the preferential looking procedure (Teller et al., 1974). The infant is presented with a pair of screens, one of which displays a grating, while the other is uniform at the same luminance. An observer, who is "blind" as to the side displaying the grating on a particular trial, guesses which side contains the grating, based on the infant's fixations and other behaviour. A psychometric function of the observer's percent of accuracy is obtained and the interpolated 70% point taken as the infant's behavioural threshold for detecting the grating.

Figure 1 shows mean contrast sensitivity functions for groups of 2- and 3-month-old infants derived by this method (Atkinson et al., 1977a,b). The main points from these data are:

1. There is continuous improvement in visual performance over the age range, but the change from the 1- to the 2-month group is considerably more striking than that from 2 to 3 months. In terms of individual variability, there is almost no overlap between the results from the 1- and 2-month groups, but a great deal of overlap between the two older groups.

2. Acuity of even the 3-month group (mean 4 c/deg) is poor compared with adult values of at least 30 c/deg. However, vision of the levels found for infants throughout the age range would be functionally quite effective, especially over the limited range of distances which concern the infant.

3. The contrast sensitivity functions of 1-month-olds do not show any conspicuous low-frequency cut. This feature, generally attributed to the action of lateral inhibition in the visual pathway, is present for the older groups of infants. It is tempting to compare this development with the finding that kitten ganglion cells do not show inhibitory surrounds at 3-4 weeks of age (Rusoff, this meeting).

4. The use of moving stimuli yields greater sensitivity to low-frequency but not to high-spatial-frequency gratings, just as is found in the adult (Van Nes et al., 1967). Thus we have no evidence that the infant's visual system has any special differential sensitivity to moving stimuli.

In addition to these behavioural data, Fig. 1 shows a contrast sensitivity function derived from VEP recordings from a group of 97 neonates (Atkinson et al., 1979). Because of the very limited amount of data that it is possible to get from a single neonate, "threshold" in these data is defined as the stimulus for which 50% of the infants tested gave a VEP amplitude statistically significantly greater than the noise level. This leads to values of contrast sensitivity and acuity rather close to those found behaviourally for 1-month olds.

While it is debatable how directly VEP and preferential-looking estimates of infant thresholds can be compared (Dobson and Teller, 1978), we have found good agreement between the two methods when we have used them on the same individuals. This includes a study of a single 6-month-old (Harris et al., 1976) whose contrast sensitivity at low spatial frequencies was found to be very

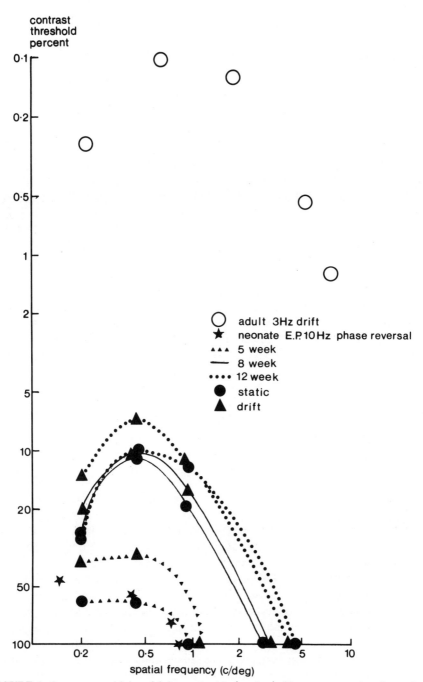

FIGURE 1 Contrast sensitivity of 0-3 month olds (see key). Neonate group data from visual evoked potentials to 10 Hz phase-reversing gratings (Atkinson et al., 1979). Mean data from 5-, 8-, and 12-week-old groups from preferential looking tests using static and 3 Hz drifting gratings (Atkinson et al., 1977b, and supplementary data).

close to adult values, with an acuity of 15-20 c/deg, by both methods. Our behavioural measures of 1-3 month-olds' contrast sensitivity are also generally in good agreement with the VEP results of Pirchio et al. (1978).

Photorefractive Study of Accommodation

To evaluate the contribution of optical factors to the development of acuity, it is necessary to assess how well focused are infant's eyes during active fixation and accommodation on a target. We have performed such a study using the method of "photorefraction" devised by Howland and Howland (1974). In this method, the eyes are photographed using a point flash source centered in the camera lens. The light reflected from the fundus returns through an arrangement of cylinder lens segments around the source, producing on the film a four-armed "star" image from each bright pupil. The measured length of the star arms indicates the dioptric defocus of the image on the subject's retina.

The method yields an instantaneous measure of the accommodative state of the subject with respect to the camera distance. There is a simultaneous record of the degree of defocus in two orthogonal meridians for each eye. It should be noted that a single photograph does not yield the sign of any defocus observed.

In our experiments infants were photorefracted with the camera at two distances, 75 cm and 150 cm. Each distance included exposures where the "target" was the experimenter's face immediately behind the camera; he or she attracted the infant's attention and operated the camera when the infant was judged to be fixating. Other exposures used targets at distances other than the camera distance; these targets were brightly coloured rattles and the exposures were made when the infant was actively tracking the target and the infant, target, and camera were momentarily in line. By these means refractions were made within 5° of the visual axis to avoid off-axis errors. Photographs with the face target at the camera distance were taken with oblique as well as horizontal-vertical orientation of the photorefractor axes, in order to estimate any component of astigmatism in each of these directions. Each of these test conditions were repeated at least twice on each infant to give a measure of the consistency of accommodative state. A photograph taken without the cylinder lens segments allowed the measurement of pupil size under the conditions of the experiment.

Using these procedures we have tested groups of infants ranging in age from newborns to 12 months. For each infant we have examined whether the exposures taken with the face target at the camera distance show accurate accommodation. The criterion of accurate accommodation is determined by an instrumental limitation: for small errors the light returning from the eyes is obscured by the flash source in the centre of the lens and so no star image appears on the film. At the 75-cm working distance, this "dead zone" corresponds to a range of 1.2 dioptres in front of the camera; at 150 cm, the criterion used corresponds to a range of 0.9 dioptres.

Figures 2 and 3 show the percentage of infants in each grooup who met this criterion of accurate accommodation at each of the two working distances.

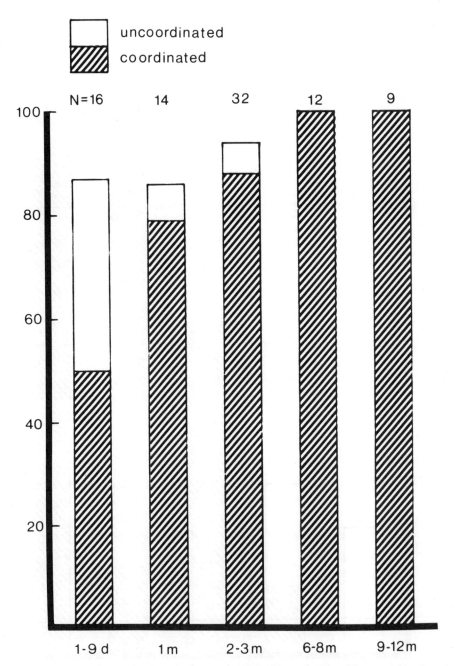

FIGURE 2 Percentage of infants in five age groups able to focus at 75 cm, within the criterion described in the text. "Uncoordinated" refers to infants who showed good focus on some but not all exposures at this distance. Numbers above each column indicate the total number of subjects in that group.

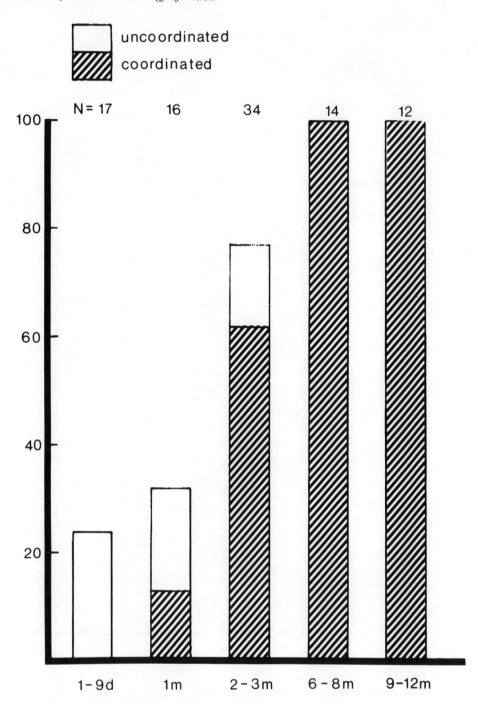

FIGURE 3 As Fig. 2, for ability to focus at 150 cm.

For inclusion as showing "coordinated" accommodation, an infant had to meet the criterion of focus in all photographs taken with the target at that working distance. Infants who were in focus for some but not all the exposures are included as "uncoordinated." These infants must have the muscular capability to focus at the distance involved, but not consistent performance in doing so.

Figure 2 shows that most infants of 1 month or older showed consistent focusing at 75 cm. Almost every infant who could focus at 75 cm also showed progressive defocus relative to the camera as the target was brought in from the camera to a 20-cm distance, indicating an accommodative response over this range. Figure 3 shows that the ability to focus at 150 cm is achieved later, but all infants over 2-3 months, and many in the 2-3 month group, were able to focus accurately and consistently at this distance. Even among the youngest groups "uncoordinated" focusing at 150 cm was often observed, indicating that these infants were capable of taking up this accommodative state.

Our results on accommodation show the same general trend as the retinoscopic data of Haynes et al. (1965), i.e., an increase with age in the ability to accommodate on relatively distant targets. However, our data do not confirm the suggestion by Haynes et al. that the 1-month-old has a fixed focus around 20 cm, and suggest that infants' focusing errors may often be due to fluctuating rather than fixed accommodation.

Infantile Astigmatism

It should be noted that, to be included as accurately focusing, an infant had to meet criterion in one meridian. In fact, many infants were found to be astigmatic (Howland et al., 1978) and so could not meet it in both meridia together. The incidence of astigmatism of 0.75 dioptre or greater (a clinically significant level) is shown for the different age groups in Fig. 4. Because the photorefractive method does not give the sign of defocus in the two meridia, these data refer to the *minimum* astigmatism consistent with the observed difference between the star arms, in the photorefractive images. Over 60% of infants aged up to 1 year showed significant astigmatism, compared with about 5% of adults. The cause of this infantile astigmatism, which has also been found using near retinoscopy (Mohindra et al., 1978), is not known; but clearly most individuals lose it some time in childhood. We have followed up a number of infants who were astigmatic when first photorefracted at 1 to 6 months, with a second set of measurements 6-14 months later. Figure 5 shows the results. In all but one of the 19 individuals, the amount of astigmatism was level or (more usually) reduced; in 14 cases it had fallen below our 0.75 dioptre cut-off. Preliminary data on a group of children aged 2-3 years indicate an approximately adult proportion of astigmats. It is intriguing to speculate whether this correction of infantile astigmatism is the result of growth processes under feedback control from retinal image quality, a possibility also suggested by the finding of Raviola and Wiesel (1978) that deprivation of pattern vision can induce myopia.

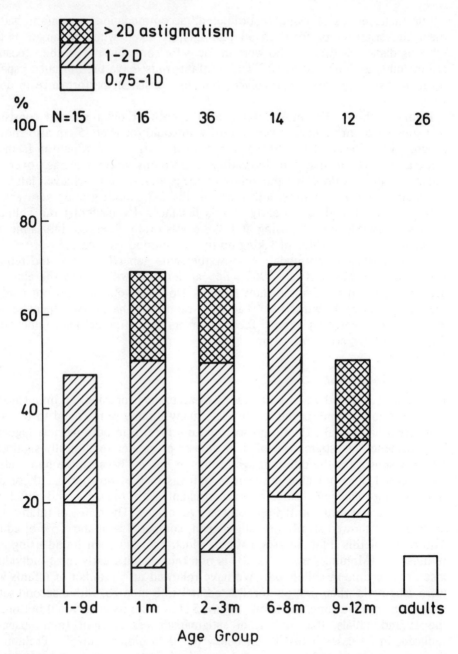

FIGURE 4 Percentage of infants in five age groups showing astigmatism greater than 0.75 dioptre. Each estimate of astigmatism is the minimum compatible with the observed meridional difference in defocus. The number above each column indicates the total number of subjects in that group. The adult group included for comparison was a sample of the infants' mothers and of laboratory personnel unselected except for exclusion of high myopes.

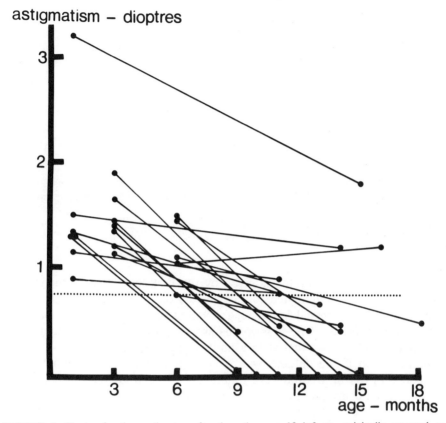

FIGURE 5 Photorefractive estimates of astigmatism on 19 infants, originally assessed as astigmatic at age 1-6 months and retested 6-14 months later. The dotted horizontal line indicates the 0.75 dioptre criterion of astigmatism used in the data of Fig. 4.

The period of infantile astigmatism falls within the likely limits of the critical period for human visual development. There is evidence that early astigmatism leads to meridional amblyopia (Mitchell et al., 1973) but we do not know whether the common astigmatism we have observed in infants has any long-term visual consequences.

Implications for Acuity

The improvement in accommodative accuracy shown in Figs. 2 and 3 takes place over a similar age range to the rapid improvement in visual acuity discussed at the beginning of this paper. Can the development of visual acuity be ascribed in whole or part to the accuracy of focusing? Quantitative considerations lead to the conclusion that it cannot.

The modulation transfer function of a defocused optical system does not show a sharp cutoff at some spatial frequency, but a reasonable estimate of the

TABLE 1 Calculated cut-off spatial frequency (cycles/degree).

PUPIL DIAMETER (mm)	DIOPTRIC DEFOCUS			
	1 D	2 D	3 D	4 D
4	5.3	2.7	1.8	1.3
6	3.6	1.8	1.2	0.9
8	2.7	1.3	0.9	0.7

These figures are calculated from the following formula derived from geometrical optics:

$$\text{MTF of defocused lens} = \frac{2J_1(x)}{x}$$

where $x = 2d(NA)f$; J_1 = first-order Bessel function; d = defocus in mm; NA = spatial frequency (c/mm). The cut-off is taken as first zero of this function for which $x = 3.83$.

highest frequency transmitted is given by the first zero of this function. (Any higher frequency is attenuated to 13% of less of its original contrast.) Table 1 gives values of this spatial frequency for various combinations of pupil diameter and dioptric defocus. One-month-olds show acuities around 1 c/deg. Inspection of the table shows that such a limit would only be imposed by inaccurate accommodation if the defocus was 3 dioptres or greater, in combination with pupil sizes of 6-8 mm. Studies of infant acuity, including ours (Atkinson et al., 1977a,b) have been undertaken at viewing distances around 40 cm. This is well within the range where the great majority of 1-month-olds are accommodating accurately (i.e. within 1 dioptre or better of the target), and pupil sizes over 6 mm would be rare at this age. The acuities of 2-5 c/deg typically found for 2-3 month olds could be limited by dioptric errors of 1-2 dioptres at normal pupil sizes, but again our data on accommodation give no reason to suppose that spherical errors of this magnitude occur for a 40-cm target distance. The only instance where refractive errors might be significant in infants is for a 2-3 month old showing (as many do) astigmatism of 1-3 dioptres; in this case, if the grating orientation did not correspond to the meridian brought into good focus, acuity could be optically limited.

Although accommodation does not limit acuity, the converse may be true. The ability to set accommodation accurately must depend on the ability to detect defocus, which would be expected to be impaired if high spatial frequencies are not transmitted by the visual system. The finding that fluctuations of accommodation over a wide range occur in young infants implies that their deficit lies in controlling, rather than executing, shifts of accommodation.

We conclude, then, that the changes in acuity observed between 1-6 months do not in general reflect improving focus of the retinal image. Instead, improved transmission of spatial information must be responsible. At present we cannot say what level or levels of the pathway are most critical in this respect. There exists evidence (Mann, 1964) that the fovea develops considerably over the first four months of life, although this is badly in need of modern histological confirmation. This is also a period when cell size and arborization increase strikingly in the visual cortex (Conel, 1939-1951). A period when many connections are formed in cortex is plausibly also a period when such connections are modifiable. Mitchell et al. (1976) have shown that in kittens, the developmental increase of visual acuity coincides with the critical period for modification of cortical connections; the same relation may also hold for human infants.

Acknowledgements Photorefractive experiments were done in collaboration with Dr. H. C. Howland. We are grateful for the assistance of Kathy Moar and Jennifer French in infant testing.

We thank Dr. N. R. C. Roberton and the staff of Cambridge Maternity Hospital for cooperation and facilities in our work with neonates.

This work was supported by a project grant from the Medical Research Council of Great Britain.

References

Atkinson, J., O. Braddick, and J. French (1979). Contrast sensitivity of the human neonate measured by the visual evoked potential. Invest. Ophthal. Vis. Sci. 18:210-213.

Atkinson, J., O. Braddick, and K. Moar (1977a). Contrast sensitivity of the human infant for moving and static patterns. Vision Res. 17:1045-1047.

Atkinson, J., O. Braddick, and K. Moar (1977b). Development of contrast sensitivity over the first 3 months of life in the human infant. Vision Res. 17:1037-1044.

Conel, J. L. (1939-1951). The Postnatal Development of the Human Cerebral Cortex. Harvard University Press, Cambridge, Mass., Vols. 1-4.

Dobson, V., and D. Y. Teller (1978). Visual acuity in human infants: A review and comparison of behavioural and electrophysiological studies. Vision Res. 18:1469-1483.

Harris, L., J. Atkinson, and O. Braddick (1976). Visual contrast sensitivity of a 6-month-old infant measured by the evoked potential. Nature 264:570-571.

Haynes, H., B. L. White, and R. Held (1965). Visual accommodation in human infants. Science 148:528-530.

Howland, H. C., J. Atkinson, O. Braddick, and J. French (1978). Infant astigmatism measured by photorefraction. Science 202:331-333.

Howland, C. H., and B. Howland (1974). Photorefraction: A technique for study of refractive state at a distance. J. Opt. Soc. Amer. 64:240-249.

Mann, I. C. (1964). The Development of the Human Eye. London, British Medical Association.

Mitchell, D. E., R. D. Freeman, M. Millodot, and G. Haegerstrom (1973). Meridional amblyopia: evidence for modification of the human visual system by early visual experience. Vision Res. 13:535-558.

Mitchell, D. E., F. Griffin, F. Wilkinson, P. Anderson, and M. L. Smith (1976). Visual resolution in young kittens. Vision Res. 16:363-366.

Mohindra, I., R. Held, J. Gwiazda, and S. Brill (1978). Astigmatism in infants. Science 202:329-331.

Pirchio, M., D. Spinelli, A. Fiorentini, and L. Maffei (1978). Infant contrast sensitivity evaluated by evoked potentials. Brain Res. 141:179-184.

Raviola, E., and T. N. Wiesel (1978). Effect of dark-rearing on experimental myopia in monkeys. Invest. Ophthal. 17:485-488.

Teller, D. Y., R. Morse, R. Borton, and D. Regal (1974). Visual acuity for vertical and diagonal gratings in human infants. Vision Res. 14:1433-1439.

Van Nes, F. L., J. J. Koenderick, H. Nas, and M. A. Bouman (1967). Spatiotemporal modulation transfer function in the human eye. J. Opt. Soc. Amer. 57:1082-1088.

Development of Infant Contrast Sensitivity Evaluated by Evoked Potentials

M. PIRCHIO
D. SPINELLI
A. FIORENTINI
L. MAFFEI
Laboratorio di Neurofisiologia del C.N.R.
Pisa, Italy

Abstract Visually evoked cortical potentials were recorded in infants of ages ranging from 2 months to 1 year. Visual acuity, optimal spatial frequency of stimulation, and maximum contrast sensitivity are evaluated as a function of age.

We studied the functional development of the visual system of infants recording the visual potentials evoked by sinusoidal grating alternating in phase at 8 Hz. It is known (Campbell and Maffei, 1970) that when this stimulus is used the amplitude of the evoked potentials is linearly related to the log contrast of the stimulus and that the extrapolation of the regression line to the zero EP amplitude gives a value of contrast that closely approximates the psychophysical threshold.

By recording evoked potentials at various levels of contrast of the grating and extrapolating the contrast threshold, we evaluated the contrast sensitivity in a range of spatial frequencies for three infants, each of which was tested at three different ages.

Details of the procedure are given elsewhere (Pirchio, Spinelli, Fiorentini, and Maffei, 1978).

There have been a few reports on contrast sensitivity functions determined either behaviourally or with the cortical evoked potentials in infants. Data are available, however, only for a single subject (Atkinson, Braddick, and Braddick, 1974; Harris, Atkinson, and Braddick, 1976) or for a few subjects of the same age (Banks and Salapatek, 1976).

301

Figure 1 reports the results obtained from one infant: (i) There is a progressive increase with age of the contrast sensitivity for all spatial frequencies tested, but the increase is comparatively greater at higher spatial frequencies. (ii) The peak of the curve shifts progressively toward higher spatial frequencies. Note that the response to a spatial frequency of 3-4 c/deg, at which the adult is most sensitive, does not develop until about three months after birth. (iii) Visual acuity (evaluated by extrapolation of the curve as a high frequency cutoff) shows a clear increase with age. Similar results were obtained from other infants. All the data are summarized in Fig. 2.

Maximum contrast sensitivity increases progressively and reaches the adult values by the end of the first year. The peak of the curve (spatial frequency at which contrast sensitivity is highest) and the visual acuity show a rapid increase

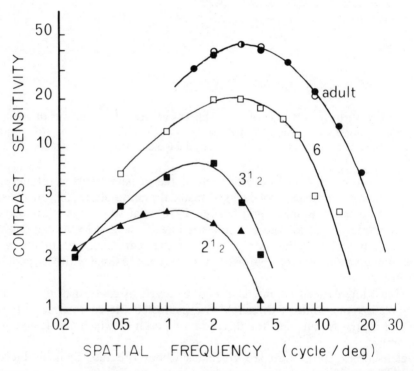

FIGURE 1 Amplitudes of EP for an infant (S.D.I.) 2½ (triangles), 3½ (filled squares), and 6-month-old (open squares) at various spatial frequencies. The points are relative amplitudes of the second harmonic of the averaged EP, obtained with a constant contrast (30%) and normalized to the contrast sensitivity at 1 c/deg. Each point is the mean of several records. Data from an adult subject are reported for comparison: full points are normalized EP amplitudes (stimulus contrast 25%), open points are extrapolated contrast sensitivities. (From Pirchio, Spinelli, Fiorentini and Maffei, 1978).

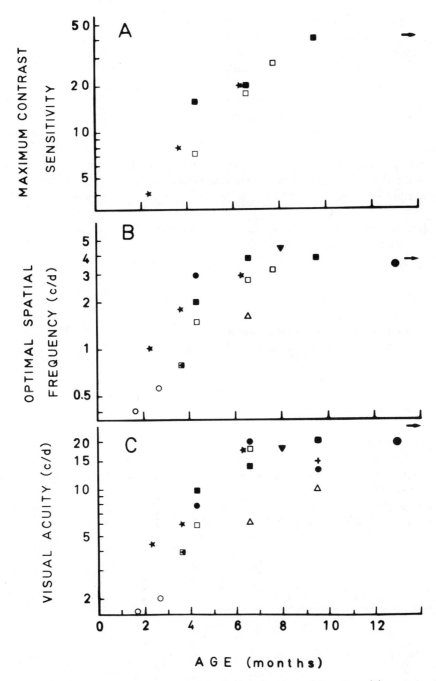

FIGURE 2 Maximum contrast sensitivity (A), optimal spatial frequency (B), and visual acuity (C) of 12 infants plotted against age in months. Different symbols have been used to indicate different infants. The arrows indicate ordinate values relative to a normal adult subject. (From Pirchio, Spinelli, Fiorentini, and Maffei, 1978).

in the first months of life and a leveling off by the fifth or sixth month. These data are in agreement with a previous report (Marg, Freeman, Peltz, and Goldstein, 1976). Even at about 1 year the visual acuity is still significantly below the average acuity of emmetropic adult subjects tested with the same apparatus.

The data show a considerable degree of variability: the difference between children of the same age was correlated with neither any obvious refractive or oculomotor anomaly, nor with the duration of gestation. To define significant deviations from normality the study should be extended to a larger infant population.

These data suggest that in the child the first year of life may be particularly critical for the development of visual acuity and for reaching a normal contrast sensitivity.

Acknowledgements This work was supported by the Italian National Research Council, Special Project on Biomedical Engineering. Dr. Pirchio was partly supported by the Scuola Normale Superiore, Pisa, Italy.

References

Atkinson, J., O. Braddick, and F. Braddick (1974). Acuity and contrast sensitivity of infant vision. Nature (Lond.) 247:403-404.

Banks, M. S., and P. Salapatek (1976). Contrast sensitivity function of the infant visual system. Vision Res. 16:867-869.

Campbell, F. W., and L. Maffei (1970). Electrophysiological evidence for the existence of orientation and size detectors in the human visual system. J. Physiol. (Lond.) 207:635-652.

Harris, L., J. Atkinson, and O. Braddick (1976). Visual contrast sensitivity of a 6-month-old infant measured by the evoked potential. Nature (Lond.) 264:570-571.

Marg, E., D. N. Freeman, P. Peltzman, and P. G. Goldstein (1976). Visual acuity development in human infants: evoked potential measurements. Invest. Ophthalm. 15:150-153.

Pirchio, M., D. Spinelli, A. Fiorentini, and L. Maffei (1978). Infant contrast sensitivity evaluated by evoked potentials. Brain Res. 141:179-184.

An Experience-dependent Aspect of Human Visual Acuity

R. D. FREEMAN

School of Optometry
University of California
Berkeley, California USA

Abstract Visual acuity tests were conducted using letters arranged in horizontal or vertical arrays. A significantly higher error rate was found for column arrangements. This effect was not evident when acuities were determined for Landolt ring targets or, with another group of subjects, for Chinese characters. Furthermore, children who knew the alphabet but did not yet read showed no effect. These results are consistent with the notion that visual experience may influence the development of visual resolution.

Introduction

The clinical determination of visual acuity is made generally with charts containing letters of different sizes. Typically, the letters are exposed in horizontal rows, but some test charts also incorporate letters in vertical columns. Although some studies suggest that these two modes of display might not yield equal acuities (Tinker, 1955; Krueger, 1970; Bryden, 1970; Mewhort and Beal, 1977), it is generally assumed in the clinic that visual resolution for letters is independent of the orientation of the test line.

This assumption was investigated in the work reported here. Subjects were tested using near-threshold-sized letters arranged pseudo-randomly with different inter-letter spacings in rows or columns. A substantially larger number of errors was made for the vertical columnar arrays. Additional tests were conducted to investigate the possibility that this effect is related to visual experience. Separate acuities were determined for letters read left to right and for those read right to left. Landolt ring targets were used on the assumption

305

that their resolution is not dependent on prior experience. Acuity was also measured using Chinese characters because in modern reading material they are displayed in both rows and columns. Finally, young children who knew the alphabet but did not yet read were tested using letters. Most of the results are in accord with the suggestion that learning patterns underlie subtle differences in visual capacity.

Methods

A standard vision chart projector was used to present rows or columns of letters, Landolt rings, or Chinese characters. Special slides which allowed spacing and size as well as type of test symbol to be varied were constructed. For the tests in which letters were used, pseudo-random combinations of E, H, R, N, D, X, O, and V were displayed in vertical or horizontal arrays. These letters were chosen arbitrarily and they may not be equally legible (Bennett, 1965). However, since the same letters were used for both row and column displays, relative comparisons may be made between these arrays.

Subjects were refracted and if any substantial visual anomaly was found, they were not included in the study. Using carefully determined ophthalmic lens corrections if required, subjects viewed monocularly test symbols presented in a row or in a column. In each case, all eight symbols were shown at each of five inter-letter separations. Ten symbols were displayed in each column or row but only the inside eight were tallied in order to avoid possible complications of edge effects at the ends of displays. Initially, charts with letters or Landolt rings subtended 4 minutes of arc at the eye and the inter-letter or inter-ring spacings were 2, 4, 8, 16, and 32 minutes. Tests with Chinese characters and letter chart measurements with children were conducted with symbols initially subtending 5 minutes and inter-symbol spacings were 2.5, 5, 10, 20, and 40 minutes. Prior to each test, the zoom system on the projector was adjusted so that isolated symbols were small enough to be identified correctly at approximately a 75% rate. The zoom setting remained constant for the entire test during which symbols were presented in columns or in rows and the subject identified them from top to bottom or from left to right, respectively. For the initial measurements with letter charts, half of the subjects were tested using columns first and for the other half, rows were presented first. No differences were found related to order of presentation.

Results

Results for the initial experiment in which letter chart acuities of 11 subjects were tested are shown in Fig. 1. In this and in subsequent figures, the mean numbers of errors are given (ordinate) for inter-letter separations (abscissa) that were used at the start of the experiment. Actual inter-letter spacings vary slightly between individuals. The data show for both vertical and horizontal arrays that the error rate increases as the inter-letter separation decreases so that there are over three times as many errors at the closest as compared to the

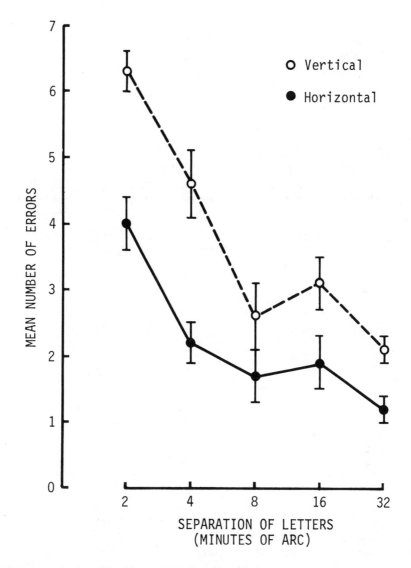

FIGURE 1 Results for visual acuity tests are given for letters presented in vertical columns (open symbols, dashed lines) or in horizontal rows (filled symbols, solid lines). Each point represents the mean number of errors out of eight (ordinate) shown at a given inter-letter separation (abscissa). Vertical bars represent ±1 standard error of the mean.

widest spacing. This finding is in accord with a number of studies showing that borders close to a target can interfere with its resolution (see, e.g., Flom, Weymouth, and Kahneman, 1963; Loomis, 1978). Figure 1 also shows that the mean error rate for letters in columns is substantially higher than that for row arrangements and the difference is highly significant ($\chi^2 = 26.12$, p = 0.0062).

There are several possible explanations for this pronounced difference between vertical and horizontal arrays. For example, it could have an optical basis. However, no subjects with substantial amounts of astigmatism were used and small astigmatic errors, if present, were not solely of the vertical axis variety. In addition, tests were conducted with fully correcting ophthalmic lenses and it would require a relatively considerable amount of axial blur to produce a difference of the magnitude found. It seems, then, that an optical explanation can be ruled out.

A second possibility is that there is a fundamental difference in the neural pathways responsible for resolution in horizontal and vertical meridians. But here again, the evidence does not favor this notion. Although large population studies apparently have not been made, no differences are found between vertical and horizontal orientations when gratings are used to test acuity (Appelle, 1972). This result also applies to laser-created interference-fringe gratings for which the effects of the eye's optics are largely bypassed (Freeman and Thibos, 1975).

Therefore, one must suppose that some aspect of reading letters in a sequence favors horizontal arrays and there is an obvious potential linkage to learned patterns of reading. Specifically, we learn to read letters in rows and it is possible that highest acuity for letters develops correspondingly. The four additional experiments described here were conducted to pursue this idea.

First, if acuity is linked to reading development, it might be expected that resolution for letters identified left to right might be better than that for the reverse order. To examine this possibility, a second group of subjects (ten individuals), most of whom were included in the first group, was tested using the same letter chart used before. Only, in this case, letters were identified from right to left in rows or from bottom to top in columns. Results were very similar to those shown in Fig. 1 with a significantly higher mean error rate at all inter-letter separations for letters presented in vertical columns. Therefore, if learned reading patterns are associated with the acuity differences found here, the effect is orientation-specific but not direction-specific. It is possible, of course, that a more sensitive test of direction effects such as a temporal measurement might yield a different result.

The second experiment was conducted to find out if the acuity differences between vertical and horizontal arrays extend to test targets other than letters. The Landolt ring, sometimes used as an acuity target, contains a gap to the left, right, top, or bottom of the circle. The subject identifies the position of the gap, probably by detecting a local light level difference. Therefore, learned reading habits do not apply to this task and if the acuity differences found for letters are associated with these patterns, the effect should not be found for Landolt targets. Tests were conducted with the same group of subjects whose results for letters are shown in Fig. 1. The findings for Landolt rings, given in Fig. 2, show a dramatic effect of inter-ring spacing with detection relatively easy for large values and very difficult for the closest targets. However, results for columns and rows are virtually identical. It is possible, then, that acuity

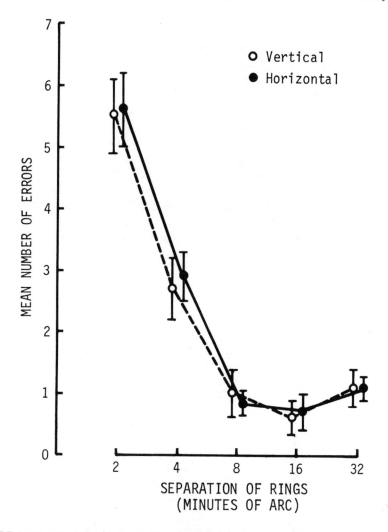

FIGURE 2 Visual acuity tests for Landolt rings yielded the data shown here. Symbols, axes, and data collection methods are the same as those given in Fig. 1.

differences with array orientation apply specifically to letters which is in accord with the proposed linkage to learned reading patterns.

The third experiment was undertaken to test this idea more directly. Modern Chinese is printed in both horizontal and vertical arrays and therefore learned reading patterns are not primarily attached to rows or columns. Based on the suggested association between acuity for symbols that are read and prior experience, resolution for Chinese characters should not be related to orientation. Although it was not possible to test Chinese who did not read or write English, subjects were found for whom English was a second language learned during teenage years. Attempts were made to conduct the tests as with English letters and eight characters, chosen for simplicity, were presented like the letter chart. Results, shown in Fig. 3, confirm that no acuity difference is evident between characters in rows and those in columns. Once again, these findings are consistent with a proposed connection between learned patterns of reading and acuity. The possible objection that racial differences complicate comparisons between Chinese and the first group of subjects may be eliminated by measuring acuity for letters using as subjects Chinese-Americans who never learned to read or write characters. Five such subjects were tested and results were very similar to those shown in Fig. 1. Thus, racial factors in the results reported here may be ruled out.

Since the native Chinese subjects were bilingual, their acuities were also determined for letters using the same chart as with the first group. A tendency for more errors for vertical arrays was found, but the difference between rows and columns, shown in Fig. 4, was not significant ($\chi^2 = 4.64$).

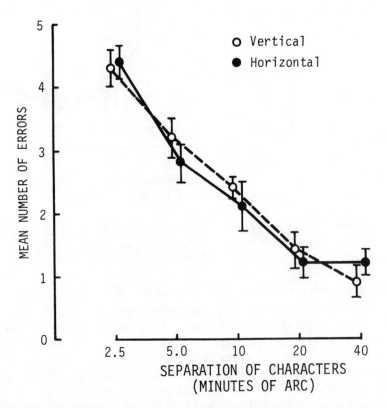

FIGURE 3 Chinese characters were used to determine the visual acuity findings presented here. Symbols and scales are as in the previous figures.

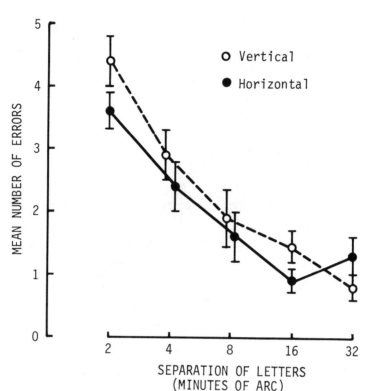

FIGURE 4 The same letter chart for which the data of Fig. 1 apply was used to measure visual acuities of the Chinese subjects whose results with characters are given in Fig. 3. Symbols and axes are as before.

The final experiment was performed to find out if children who knew the alphabet but did not read exhibited a difference in resolution for letters in rows compared to columns. If the process of training in reading results in acuity differences, they should not be expressed with this group of subjects. The results of Fig. 5 substantiate this expectation. Acuities for letters in vertical and horizontal arrays are not significantly different. As with the previous findings, these data are consistent with the proposed association between learned reading patterns and visual resolution.

The principal results are summarized in Fig. 6 which gives distributions of subjects in terms of the ratios of horizontal to vertical errors. In (A) results are shown for the initial group of subjects whose mean error rates are given in Fig. 1. Ratios for the entire group are less than 1.0 which reflects the higher rate of errors for vertical displays. The distributions of (B) and (C) which are results for the Landolt ring tests and the Chinese characters, respectively, are spread

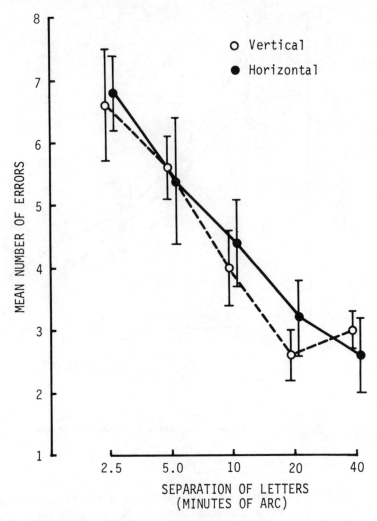

FIGURE 5 Children who knew the alphabet but could not read were tested using the letter chart. These data are presented as those of the previous figures.

around values of 1 in accord with the similarity in error rates for row and column displays. Means, indicated by arrows in Fig. 6, are 0.6, 1.08, and 0.97, respectively, for (A), (B), and (C).

Discussion

The central result of this study is that visual acuity is superior when letters are presented in horizontal arrays. It is also shown that acuities are considerably reduced when test symbols are closely spaced, and this effect is independent of

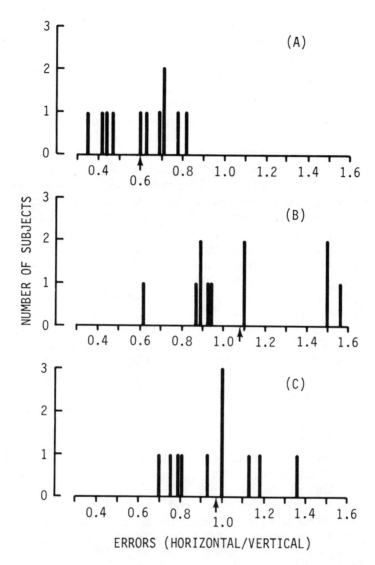

FIGURE 6 Summaries are given here of the main results. Ratios of the total number of horizontal to vertical errors are given for each subject. (A) Data for letters with corresponding mean error rates shown in Fig. 1; (B) findings using Landolt rings; (C) results for Chinese characters. Means for each distribution are indicated by arrows.

array orientation. The influence of inter-letter spacing has been alluded to in clinical treatises (Duke-Elder and Abrams, 1970) and interactions with adjacent stimuli have been investigated for a variety of tasks, including resolution (Flom, Weymouth, and Kahneman, 1963; Loomis, 1978) and increment sensitivity (Beyerstein and Freeman, 1977). However, the present results appear to be the first to show the interactive effects of rows and columns of letters or

Landolt rings. Further study is required to determine how many letters or rings of a given spacing are necessary to cause reduced acuity. It may be, as previous work suggests, that only the flanking symbols are of prime importance.

Evidence concerning the possible connection between learned patterns of reading and the difference in error rate between rows and columns of letters is supportive but not compelling. The tests with Landolt rings, Chinese subjects, and children are consistent with this proposed association. But prior reading habits cannot account for the finding that acuity is better for rows of letters even when they are read from right to left. It is possible, of course, that if a bias is established, it could develop without a directional component or that a more sensitive measure than used here might detect directional differences. In any case, the neural mechanism for the resolution differences found in this study must be quite elaborate since they seem to apply specifically to letters.

Although I have found no previous reports on visual acuity differences between rows and columns of letters, several former studies are suggestive. For example, it has been demonstrated that reading speed is slower for columns compared to rows of letters (Aulhorn, 1948; Tinker, 1955). Word or pseudo-word displays are more effectively identified when presented in rows (Bryden, 1970; Mewhorst and Beal, 1977). In addition, the reaction time for letter recognition is shorter for horizontal arrays (Krueger, 1970). There is also evidence that training in a language has a specific neural influence on word recognition capacity (Mishkin and Forgays, 1952; Orbach, 1952).

The results of the current study leave many unanswered questions. One would like to know, for example, how many letters in a given array are necessary to cause a horizontal-vertical difference. This would make it easier to relate the effects found here with previous studies of lateral masking. Would the effect be found if letters were rotated or if non-letter flanking symbols were used to cause the lateral masking? This would show the relative differences in interference effects between letters and other symbols. Would a small amount of practice eliminate the difference between horizontal and vertical arrays as seems to be the case with reading speed (Tinker, 1955)? Finally, it would be instructive to know if objectively measured eye movement patterns differ when subjects read in row as compared to column arrangements.

Acknowledgements I am grateful to Dr. Glen Kaprielian who began this study as a student project and to Drs. Doris Lin and Christopher Chan for Chinese translations and for construction of Chinese characters. This work was supported by Research Career Development Award EY00092 and grant EY01175 from the National Eye Institute, U.S. Public Health Service, National Institutes of Health.

References

Appelle, S. (1972). Perception and discrimination as a function of stimulus orientation: the oblique effect in man and animals. Psychol. Bull. 78:266-278.

Bennett, A. G. (1965). Ophthalmic test types. Brit. J. Physiol. Optics 22:238-271.

Beyerstein, B. L., and R. D. Freeman (1977). Lateral spatial interaction in humans with abnormal visual experience. Vision Res. 17:1029-1036.

Bryden, M. P. (1970). Left-right differences in tachistoscopic recognition as a function of familiarity and pattern orientation. J. exp. Psychol. 84:120-122.

Duke-Elder, S., and D. Abrams (1970). System of Ophthalmology, Vol. V. Klimpton, London.

Flom, M. C., F. W. Weymouth, and D. Kahneman (1963). Visual resolution and contour interaction. J. Opt. Soc. Am. 53:1026-1032.

Freeman, R. D., and L. N. Thibos (1975). Contrast sensitivity in humans with abnormal visual experience. J. Physiol. 247:687-710.

Krueger, L. E. (1970). Visual comparison in a redundant display. Cognitive Psychol. 1:341-357.

Loomis, J. M. (1978). Lateral masking in foveal and eccentric vision. Vision Res. 18:335-339.

Mewhort, D. J. K., and A. L. Beal (1977). Mechanisms of word identification. J. Exp. Psychol.: Human Percept. and Performance 3:629-640.

Mishkin, M., and D. G. Forgays (1952). Word recognition as a function of retinal locus. J. exp. Psychol. 43:43-48.

Orbach, J. (1952). Retinal locus as a factor in the recognition of visually perceived words. Amer. J. Psychol. 65:555-562.

Tinker, M. A. (1955). Perceptual and oculomotor efficiency in reading materials in vertical and horizontal arrangements. Amer. J. Psychol. 68:444-449.

**STUDIES OF THE VISUAL SYSTEMS
OF VARIOUS SUB-PRIMATE SPECIES**

Late LEO: A New System for the Study of Neuroplasticity in Xenopus

SCOTT E. FRASER

Jenkins Department of Biophysics
The Johns Hopkins University
Baltimore, Maryland USA

Abstract The left eye of stage 56 Xenopus tadpoles was removed. Electrophysiologic analysis of the retina to tectum projection, two to nine months later, established that the normal right eye to left (contralateral) tectum projection remained, but an additional right eye to right (ipsilateral) tectum projection was present. This anomalous ipsilateral projection was mediated by direct retina-to-tectum innervation, as shown by single-unit analysis, latency measurements, tectal lesion studies, and histological reconstruction. This anomalous projection was confined to peripheral regions of the retina that were concurrently taking part in the normal contralateral retinotectal projection. Unlike the contralateral projection from this peripheral retina, the ipsilateral projection had expanded to cover the entire dorsal surface of the right tectum. That the same region of retina can project differently to the two tecta establishes a role for fiber-fiber interaction in the development of the retinotectal map.

The connections from the eye to the primary visual center of Xenopus, the optic tectum, are characterized by a point-to-point ordering such that a "map" of the visual field of one eye is constructed over the contralateral optic tectum (Gaze, 1970; Jacobson, 1978). The ordering of the direct retinotectal projection, its independence of experiential effects (Keating, 1978), and its response to experimental manipulation (Hunt and Jacobson, 1974) have provided a wealth of indirect evidence in support of Sperry's (1950) hypothesis that each ganglion cell in the retina acquires position-dependent discriminator properties that guide it to the proper locus in the retinotectal map. However, an increasing body of literature has reported conditions under which the system "makes mistakes" and inappropriate connections are made. For example, a surgically produced half-retina spreads, with time, to cover more than its appropriate half-tectum.

319

These two classes of experiments lead to somewhat of a dilemma. Both the "specificity-type" and "plasticity-type" experiments are valid tests of the system; however, they seem to give diametrically opposed results. That there is indeed *information* of some type on the *tectum* seems unquestionable. Optic nerve fibers will follow transposed pieces of tectum (Sharma, 1975; Jacobson and Levine, 1975; Hunt, 1979a,b; Yoon, 1975; Hope, Gaze, and Hammond, 1977) even if the tectum had previously entertained only a disordered projection (Hunt, 1979b). In addition, fibers can be induced to enter the tectum by unusual pathways and still develop correctly positioned connections on the tectum (Horder, 1974; Udin, 1978; Meyer, 1978; Gaze and Grant, 1978).

The question, therefore, becomes—Is this information on the tectum that guides optic nerve fibers *absolute* (each fiber has a single matching tectal target) or only *relative* (taking fiber interaction and competition into account)? The first experiments attempting to show that tectal information was only relative were hard to interpret because they involved surgical removal of portions of the retina and tectum. The fact that the projection always compensated such that the remaining tectum and eye interconnected (violating the absolute specificity) was overshadowed by findings that such operations in other systems lead to regeneration of the missing pieces (French, Bryant, and Bryant, 1976) or rearrangement of "labels" over the remaining tissue so it became effectively a smaller whole (rev. Wolpert, 1971; Hunt and Jacobson, 1974). Recent experimentation has avoided this pitfall by elegant techniques of competing eyes for the same tectum (Schmidt, 1978; Hunt, 1977; Mayer, in press; Sharma, in press). However, what is gained in these techniques in the realm of control is nearly lost by technical ills (poor survival, few fibers to analyze, variability of time course) that limit their applicability to further experiments aimed at the mechanisms underlying neuroplasticity.

The purpose of this paper is to present a preparation in Xenopus for the study of neuroplasticity that is both simple and reliable. In the late LEO (Left Eye Out) preparation, merely removing the left eye at a late larval stage induces an anomalous direct ipsilateral projection from a circumscribed region of the remaining eye. The map the right eye makes on the right tectum is considerably different from the projection from the *same* region of retina in its normal map with the left tectum. Since the same region of retina projects to both tecta (but in different manners) and neither the experimental eye nor tecta have been surgically manipulated, arguments concerning the rearrangement of labels or regeneration do not hold. Therefore, the late LEO preparation documents a role for fiber-fiber competition in the formation of the retinotectal map.

The experimental design is derived from known facts of the Xenopus visual system. First, essentially all retinotectal fibers cross at the optic chiasm so that the *right* eye provides direct innervation to only the *left* tectum (Scalia, 1977). The representation of the *right* eye's visual field on the *right* (ipsilateral) tectum is an *indirect projection,* mediated first through the direct retinotectal connections in the *left* (contralateral) tectum, and then through a polysynaptic intertectal relay (Gaze and Jacobson, 1963; Keating, 1974). Second, removal of a left eye early in embryonic life occasionally induces bilateral innervation in which the remaining eye forms direct, complete, and topographically normal maps with both tecta (Hirsch and Jacobson, 1973; Hunt, 1975; Beazley, 1975). Third, Xenopus retina grows gradually by adding "annuli" of new cells to the retinal periphery, symmetrically at early larval stages, and later with a quantitative bias (10:1) towards ventral cell accretion during late larval and juvenile life (Hollyfield, 1971; Straznicky and Gaze, 1971; Jacobson, 1976; Beech, 1977; Hunt and Ide, 1977). The strategy, then, was to wait until the central retina had already projected to its contralateral tectum and then remove the left eye, in the hope of creating a crossing error in only the fibers yet to grow in. Since the retina grows only at the periphery (with a bias for the ventrum) only the peripheral retina should be involved in the bilateral projection.

Materials and Methods

Xenopus laevis embryos were obtained from gonadotropin-induced matings of mature animals. These were raised in a standard saline rearing solution (15% Holtfreter's solution; 5% Steinberg's solution) supplemented with ground nettle leaves as food. At Stage 56 (hindlimb stage) approximately six weeks after fertilization, the animals were lightly anaesthetized in a 1:10,000 solution of Finquel (Searle) and their left eyes were carefully removed with iridectomy scissors. Animals were raised to metamorphosis, as above, after which they were fed on live tubifex worms or ground liver. The animals were kept in clear plastic trays throughout.

Electrophysiologic analysis was performed by conventional extracellular recording with platinum-tipped, platinum-iridium micro-electrodes (F. Haer) on animals paralysed by d-tubocurarine (Lilly). A projection perimeter (aimark type) was used to search the visual field for the region that would elicit activity at the electrode tip. All electrode placements were made by means of a photograph of the dorsal surface of the animal's tecta. The points on the tecta and the visual regions that would drive these loci thereby construct the visuo-tectal map (Hunt and Jacobson, 1974). Previous work (George and Marks, 1974) has established that such a configuration records from the terminal arbors formed by the optic nerve as it branches in the tectal neuropil. A minimum of 20 recorded loci on the experimental tectum was necessary to construct the visuo-tectal map. Unit analysis was performed on the basis of amplitude with the aid of a Single-Channel Amplitude Analyser (F. Haer, New Brunswick).

Histologic reconstructions were performed by use of a drawing attachment for a compound microscope on animals fixed in Bouins fixative, dehydrated and embedded in paraffin, and cut at 10 μm. The sections were stained by the picro/panceau technique.

Tectal lesions were made by use of a fine-gauge sharpened hypodermic needle carefully drawn across the tectum in the desired pattern. These lesions, while rather deep, did not pierce the ventricle of the tectum. Electrophysiology was used to confirm the extent and completeness of the lesion. The lesion was acceptable only if it was completely impossible to obtain visually evoked potentials in the silenced region even with substantially increased stimulus strength.

Intraocular injection of ^{35}S methionine (700 Ci/m mole, New England Nuclear) was accomplished by pressure injection of .3 uCi isotope in 1 μl of saline. After 18 hours, the tecta were individually dissected out, solubilized (NCS, Amersham), and counted in a liquid scintillation counter. The vast majority of counts obtained in this manner represents isotope incorporated in proteins; the majority of these counts represents proteins synthesized in the eye and transported down the optic nerve (Szaro, Loh, and Hunt, unpublished) as shown by gel electrophoresis/autoradiography.

Results

Animals subjected to eye removal (late LEO) were recorded four to nine months after metamorphosis. As can be seen in Fig. 1, there is a normal contralateral projection in the late LEO, but an unusual ipsilateral projection. As expected from the experimental design, the ipsilateral projection is limited to the peripheral retina. In 28 of 30 animals, this projection was mediated by a temporal and ventral crescent of retina (superior and nasal visual field). In 2 of 30 animals, this crescent extended into the nasal retina (temporal visual field). The topography of this projection is quite different from the normally present intertectal relay representation as can be seen by comparison of Fig. 1(A) and (C). It also differs greatly from the map produced if the entire optic nerve is deflected to the ipsilateral tectum (Fig. 1(D)). Such a map is identical to the normal contralateral projection except for the reverse-handedness. Note that the late LEO projection has expanded to cover the entire dorsal aspect of the tectum and that the projection is ordered although considerably distorted. Figure 2 shows additional late LEO maps.

The late LEO ipsilateral projection has been shown to be due to direct retina-to-tectum innervation by three major physiologic criteria (lesion studies, unit analysis, latency of response studies). *Lesions* of the contralateral tectum that would have silenced the intertectal relay leave the late LEO ipsilateral projection unaltered. This means that the response recorded on the ipsilateral tectum are *not* due to the normal intertectal relay. Figure 3(a,b) show the lesion diagrammatically and its effect on normal and late LEO animals. The lesion was purposely designed to silence most, but not all, of the intertectal relay, documenting that the silencing was due to cutting the input to the contralateral tectum and not merely due to the trauma of the lesion itself. As can be seen in Fig. 3(a) and (b), the lesion leaves only a small patch of the ipsilateral tectum responsive in normal animals, whereas the late LEO ipsilateral projection is left unaltered.

Second, the ipsilateral tectum of late LEO frog has many *single units* present which are of *optic nerve type;* therefore, the innervation of the tectum is direct from the ipsilateral eye. These units (analyzed on the basis of amplitude) are characterized by small receptive fields (less than 15°) and simple optimal-stimulus type (Scalia, 1977).

Third, the *latency* of response after the presentation of a visual stimulus is as short to the ipsilateral tectum as it is to the contralateral tectum, indicating that

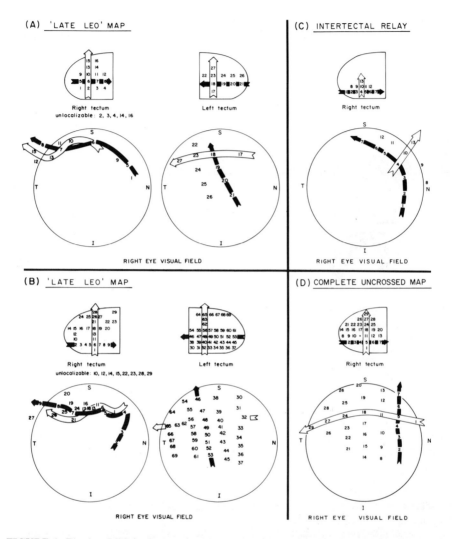

FIGURE 1 The late LEO is characterized by a normal contralateral retinotectal projection but an abnormal ipsilateral projection. As can be seen in (A) and (B), the right eye to right tectum projection is confined to superior and nasal visual field. The late LEO projection is distinctly different from the ipsilateral projection mediated by the intertectal relay in normal frogs (C). It is also quite different from the projection produced by deflecting the entire optic nerve to the ipsilateral tectum (D). Comparisons between (B) and (D) highlight some of the peculiarities of the anomalous ipsilateral projection. The normal perpendicular black and white arrows (representing orthogonal directions on the tectum) are instead intertwined in late LEO. Unlocalizable points responded to stimulation of superior visual field but had such large receptive fields that a center point could not be confidently assigned.

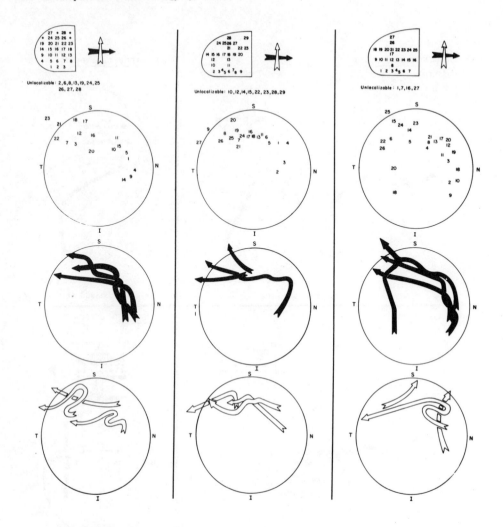

FIGURE 2 To show more LEO ipsilateral maps. The two maps on the left represent the majority result (28 of 30). The map on the right shows the temporal field involvement characteristic of the minority result (2 of 30). Note that the exact form of the projection is variable but that the ordering is similar from animal to animal. The axis corresponding to lateromedial ordering on the tectum seems to have the most regular shape (see Fig. 1(A,B)), perhaps indicating that this axis is somewhat dominant in ordering the retinotectal map.

the input to the ipsilateral tectum is direct in late LEO. As shown in Fig. 3(c), the ipsilateral and contralateral latencies of a normal frog differ significantly due to the interposed synapses in the intertectal relay (dashed lines; total, 122 measurements). This difference in latency is not found in late LEO frogs (solid lines; total, 104 measurements), making it impossible for the recorded ipsilateral activity to be due to the intertectal relay.

In addition to these physiological techniques, we have employed anatomical methods to confirm that the late LEO ipsilateral projection is direct. *Histologic reconstructions,* such as that in Fig. 3(d), indicate that a subpopulation of optic nerve fibers make a crossing error at the chiasm, coursing towards the ipsilateral as well as the contralateral tecta. Six of six late LEO frogs reconstructed to date show single optic nerve fibers making this crossing error. This result has been confirmed by autoradiographic techniques (to be published elsewhere).

Axonal transport of radiolabeled amino acids following intraocular injection of ^{35}S–Methionine also indicates a *direct* retina to tectum projection. The ratio of counts found in the ipsilateral tectum versus the counts in the contralateral tectum (18 hours post-injection) of the late LEO is .7 ± .1 in contrast to the .3 ± .1 ratio of co-reared normal frogs (ten animals). This increase in label was not due to increased systemic labeling as shown by dissection and counting of other tissues. While still preliminary, this indicates (after background subtraction) that approximately one-fourth of the normal contralateral projection is instead projecting to the ipsilateral tectum of the late LEO. This is well in line with the proportion of visual field involved in the late LEO ipsilateral projection.

Discussion

The results above indicate that a direct retinotectal projection is formed by the right eye of late LEO frogs to *both* tecta. The form of this projection is such that the same peripheral region of the same retina projects differently to the two tecta. That this cannot be due to some left-right asymmetry of the tecta is demonstrated in Fig. 1. An entire retina can make a normal (but reversed) projection on the ipsilateral tectum if all the fibers from the eye are deflected. In addition, removal of an eye at embryonic ages causes a projection from the remaining eye to both tecta in a minority of the cases. In such cases both the ipsilateral and contralateral projections are mediated by a full subset of the optic nerve fibers, resulting in normal maps to both tecta. We are, therefore, left to conclude that the causative factor of the late LEO ipsilateral projection expanding over the entire tectum is the absence of competing optic nerve fibers in the ipsilateral tectum.

The results also speak to other issues involved in the formation of the retinotectal map. The late LEO ipsilateral map demonstrates an ordered, although spatially distorted, projection. The nature of this distortion causes the normally unidirectional ordering on the tectum (corresponding to a line in visual space) to bend 90°. The fibers are still ordered on the surface of the tectum even at 90° to the normal direction of ordering. This argues against orthogonal labeling of the tectum as the *only* causative factor in ordering, as do the recent results of Schmidt and Easter (1978), showing nonlinear distortions in the map across the entire tectum following ablation of a corner of the tectum.

At the same time, however, fiber-tectum attraction is very likely to be one of the important factors in fiber ordering. The two-dimensional *ordering* on the

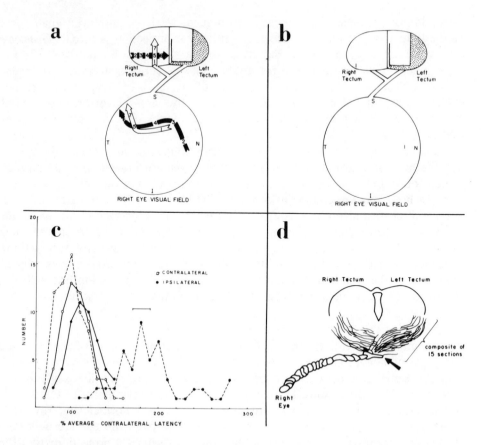

FIGURE 3 To show experiments indicating that the anomalous projection of late LEO frogs results from direct innervation from eye to ipsilateral tectum. A lesion that silences the ipsilateral projection of normal frogs by removing all but a small region of the contralateral input into the intertectal relay does not alter the LEO ipsilateral projection. A typical LEO postlesion map is shown in (a). Contrast this to the post-lesion map of a normal frog (b). Only a small region of ipsilateral response remains after the lesion in normal frogs. The responsive regions of both ipsilateral and contralateral (stippled region) tecta serve to establish that silencing was due to the lesion itself, not simply due to trauma of the tectal tissues. (c) Late LEOs do not show the difference in latencies between ipsilateral and contralateral tecta, showing that the pathway to the ipsilateral tectum is not via the multisynaptic intertectal relay. The distribution for normal animals (dashed lines) is bimodal because of the interposed synapses of the intertectal relay. The latency distributions for LEOs superimpose (solid lines). (d) Histologic reconstructions show a subpopulation of optic nerve fibers making a crossing error at the chiasm.

tectum of one dimension of the retina in late LEO argues for a role of fiber-fiber interaction in target selection. The *expansion* of the late LEO projection documents a competition or mutual repulsion between optic nerve fibers. Elsewhere, I will present a model in which selective fiber-fiber attraction and general fiber-fiber repulsion act in concert with selective fiber-tectum attraction to establish the retinotectal map. The nature of this fiber-tectum attraction is such that there is a best-fit site for each fiber (or group of fibers) at one place on the tectum. This model can generate the late LEO projection pattern in computer simulation, and many other experimental patterns, including tectal graft and retinotectal mismatch data. More important, it is very difficult to generate the late LEO pattern, and impossible to generate tectal graft data if there is no fiber-tectum attractive component in the model.

There are still many unanswered questions concerning the late LEO. The most obvious, perhaps, is why the fibers make a chiasmatic error following enucleation. It is quite possible that the main factors involved are mechanical, and that the degenerating nerve tract merely offers a rich environment for nerve growth. Also of interest is the apparent lack of an ipsilateral projection with the topography characteristic of the intertectal relay. Experimentation is under way to determine whether the intertectal relay is absent, electrophysiologically silent, or topographically altered so as to be masked by the direct ipsilateral projection. Another unanswered question is why there is such a high success rate for inducing the anomalous projection (30 cases of 32 animals). It is possible that stage 56 is in a critical period for chiasmatic error following enucleation. We have established that this period extends at least from stage 54 to stage 59. Although fiber competition seems certain in late LEO, the location of this competition is still uncertain. While the most likely candidate for the site of competition is the optic tectum, the optic nerve tract is also a possible location.

Late LEO has many facets that make it an attractive system for the study of neuroplasticity. Among these are the reliability and repeatability of the result, as well as the substantial fraction of the optic nerve involved in the projection. In addition, each animal can act as his own control (because of the normal contralateral projection) and neither the experimental eye nor tectum have been surgically altered. The ease with which embryonic procedures can be coupled with this preparation extends the type of experiments that may be done and the short time course extends the number of possible experiments. Late LEO has established a role for fiber competition in the retinotectal map and for the reasons given above promises to give insight into the underlying mechanisms.

Acknowledgements I thank K. Conway, K. Feiock, R. K. Hunt, and B. Szaro for their helpful discussions and critical reading of the manuscript; for M. Duda for her assistance with histology. This work was supported by NIH (NS-12606) (GM07231-04) and NSF (PCM77-26987).

A preliminary report of this work was presented to the Biophysical Society (Fraser and Hunt, 1978).

References

Beazley, L. (1975). Exp. Brain Res. 23:491.

Beech, D. (1977). Doctoral thesis, University of Miami, Florida.

Fraser, S. E., and R. K. Hunt (1978). Neuroplasticity in Xenopus. Biophysical J. 21:110a.

French, V., P. Bryant, and S. Bryant (1976). Pattern regulation in epimorphic fields. Science 193:969-981.

Gaze, M. (1970). The Formation of Nervous Connections. Academic Press, New York.

Gaze, M., and P. Grant (1978). The diencephalic course of regenerating retinotectal-fibers in Xenopus tadpoles. J. Embryol. Exp. Morph. 44:201-216.

Gaze, M., and M. Jacobson (1963). The path from the retina to the ipsilateral optic tectum of the frog. J. Physiol. (Lond.) 165:73-74P.

Gaze, M., and S. C. Sharma (1970). Axial differences in the reinnervation of the goldfish optic tectum by regenerating optic nerve fibers. Exp. Brain Res. 10:171-181.

George, S. A., and W. Marks (1974). Optic nerve terminal arborizations in the frog: shape and orientation inferred from electrophysiological measurements. Exp. Neurol. 42:467-482.

Hirsch, H., and M. Jacobson (1973). Development and maintenance of connectivity in the visual system of the frog. ii. The effects of eye removal. Brain Res. 49:67-74.

Hollyfield, J. (1971). Differential growth of the neural retina in Xenopus laevis larvae. Dev. Biol. 24:264-286.

Hope, R., B. J. Hammond, and R. M. Gaze (1976). The arrow model: retinotectal specificity and map formation in the goldfish visual system. Proc. R. Soc. Lond. B. 194:447.

Horder, T. (1974). Changes of fiber pathways in the goldfish optic tract following regeneration. Brain Res. 72:51-52.

Hunt, R. K. (1975). Developmental programming of retinotectal patterns. In: Cell Patterning, Ciba Foundation Symposium 29. Elsevier, New York, pp. 131-150.

Hunt, R. K. (1977). Competitive retinotectal mapping in Xenopus. Biophysical J. 17:128a.

Hunt, R. K. (1979a). Combinatorial specifiers on retinal ganglion cells for retinotectal map assembly. Nature (in press).

Hunt, R. K. (1979b). Target properties in the optic tectum for retinotectal map assembly in Xenopus. Nature (in press).

Hunt, R. K., and C. F. Ide (1977). Radial propagation of positional signals for retinotectal patterns in Xenopus. Biol. Bull. 153:431.

Hunt, R. K., and M. Jacobson (1974). Neuronal specificity revisited. In: Current Topics in Developmental Biology, vol. 8. Academic Press, New York.

Jacobson, M. (1976). Histogenesis of retina in the clawed frog with implications for the pattern of development of retinotectal connections. Brain Res. 103:541-545.

Jacobson, M. (1978). Developmental Neurobiology, 2d ed. Plenum Press, New York.

Jacobson, M., and R. L. Levine (1975). Stability of implanted duplicate tectal markers serving as targets for optic axons in adult frogs. Brain Res. 92:468-471.

Keating, M. (1974). The role of visual function in the patterning of binocular visual connexions. Br. Med. Bull. 30:145-151.

Meyer, R. (1978). Deflection of selected optic fibers into a denervated tectum in goldfish. Brain Res. 155:213-227.

Scalia, F. (1977). In: Handbook of Frog Neurobiology. Llinas and Precht (eds.). Springer-Verlag, Berlin.

Schmidt, J., and S. Easter (1978). Independent biaxial reorganization of the retinotectal projection: a reassessment. Exp. Brain Res. 31: 155-162.

Sharma, S. C. (1975). Visual projection in surgically created "compound" tectum in adult goldfish. Brain Res. 93:497-501.

Sperry, R. W. (1950). Neuronal Specificity. In: Genetic Neurology. P. Weiss (ed.). University of Chicago Press, Chicago, Ill.

Straznicky, K., and R. M. Gaze (1971). The growth of the retina in Xenopus laevis: an autoradiographic study. J. Embryol. Exp. Morph. 26:67-79.

Udin, S. (1978). Permanent disorganization of regenerating optic tract in frog. Exp. Neurol. 58:455-470.

Wolpert, L. (1971). Positional information and pattern formation. Curr. Top. Dev. Biol. 6:183-224.

Yoon, M. (1975). Readjustment of retinotectal projection following reimplantation of a rotated or inverted tectal tissue in goldfish. J. Physiol. (Lond.) 252:137-158.

The Nature of the Nerve Fibre Guidance Mechanism Responsible for the Formation of an Orderly Central Visual Projection

S. M. BUNT
T. J. HORDER
K. A. C. MARTIN
Department of Human Anatomy
South Parks Road
Oxford, England

Abstract Theories about the formation of patterned nerve connections have generally required a detailed, highly specific recognition between individual nerve fibres and their targets, or other fibres, based on unique matching cytochemical cues. A re-examination of the published data, together with new data from experiments on the retinotectal system of lower vertebrates, suggests that a considerable amount of the data can be accounted for by more simple mechanisms such as the maintenance of nearest neighbour relationships between growing fibres through the non-specific process of fasciculation and the contact guidance of fibres to their termination sites. These simple morphogenetic mechanisms are insufficient to explain the fibre selectivity in tissue exchanges between different tectal sites, or the reorientation of the optic projection after rotation of the fibre array or its target tissue. High degrees of retinal and tectal differences need not be invoked to account for the results of graft exchanges since the selectivity of the fibres is of a low order. Non-specific factors such as fibre-fibre competition, varying densities of fibre subpopulations and variable accessibility of different fibres to different tectal regions could account for the findings. The data presented suggest that fibres may respond to the same polarity cue throughout the tectum but that on occasions highly ordered projections are formed without reference to these cues. The mechanisms responsible for polarity control remain speculative but might involve generalized properties, such as tissue maturity, which could act cumulatively along the length of the optic pathway to orient the fibres, and not solely in their target tissue.

331

Some thirty-five years after R. W. Sperry first suggested that optic nerve fibres establish their extraordinarily precise and orderly connections in the midbrain tectum because of individual chemoaffinities for their corresponding postsynaptic sites, the concept still suffers from two grievous handicaps. Not only is there no direct evidence for the responsible chemical agents, but the theory seems to call for levels of cellular differentiation which make unacceptable demands on the likely scope of genetic and biochemical information storage. From an early stage Sperry has cautioned that the necessary tectal differentiation may be regional rather than truly local (Sperry, 1951; Meyer and Sperry, 1976) and many have since attempted an escape by suggesting simplified forms of tissue differentiation. However, this only clouds the essential issue. If there are not absolute distinctions between alternative termination sites in the optic tectum, how do fibres distinguish unambiguously and to a high level of refinement between sites with properties to some extent shared and therefore potentially open to other fibres? In the account which follows we wish briefly to describe evidence consistent with a quite different mechanism for the establishment of ordered fibre projections and then to consider the question of the status of major categories of evidence which can still be taken to favour the chemoaffinity view.

An Alternative to the Chemoaffinity Mechanism for the Control of Detailed Retinotopic Rank-Ordering of Optic Fibre Terminals

We have recently (Bunt and Horder, 1977; Horder and Martin, 1978) made the proposal that growing nerve fibres would be expected, by virtue of their well-known susceptibility to contact guidance, to follow their neighbours in parallel so that the order in which they leave the retina will be maintained throughout the optic pathway. Their retinotopic arrangement in the tectum would automatically follow given no further provisions than that fibres occupy vacant tectal sites on a "first come, first served" basis. In all species so far examined we have found that fibres do indeed maintain precise retinotopic ordering as they travel through the optic nerve and tract (Bunt and Horder, 1978). This includes species once thought to have random arrangements of fibres in the nerve (see Fig. 1).

We have evidence which we believe can only reasonably be interpreted to mean that a fully elaborated retinotopic map can be set up by forces intrinsic to a population of arriving optic fibres and without any assistance from possible guidance cues in the tectum. Figure 2 shows a situation in which two sets of optic fibres from entirely different retinal regions have come to form terminals simultaneously at common tectal locations: fibres have formed terminals at foreign locations, far from their usual ones, under conditions in which the "normal" state of the tectum is proved through its simultaneous innervation by its normal complement of fibres. This result, like many others now available (Horder and Martin, 1978), proves that whatever form of tectal guidance may exist, it does not have its effects by exercising an exclusive affinity for a matching fibre.

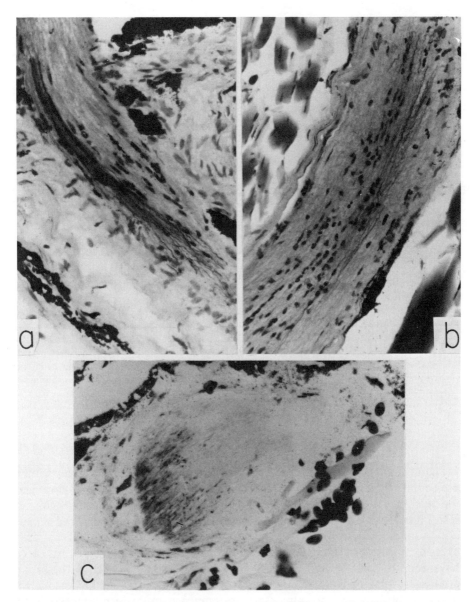

FIGURE 1 Retinotopic organization of fibres in the optic nerve of *Xenopus laevis.* (a) In a recently metamorphosed *Xenopus* frog a Gelfoam pad soaked in horseradish peroxidase had been applied to a localized cut in the retina. The animal was fixed after 24 hours. In this longitudinal section the labelled set of fibres are shown at the point at which they leave the back of the eyeball, which lies just out of view to the left. Magnification, 160X. (b) The same nerve near its point of exit from the orbit; the optic foramen lies just out of view to the left. Fewer fibres are stained at this point. Magnification, 160X. (c) In a second animal the same technique was used except that the HRP was applied directly behind the eye where the optic nerve had been partially divided. This near-transverse section, taken about 0.6 mm central to the lesion, shows that the discrete set of cut fibres taking up the HRP have remained topographically coherent as they travelled along the nerve. Magnification, 200X.

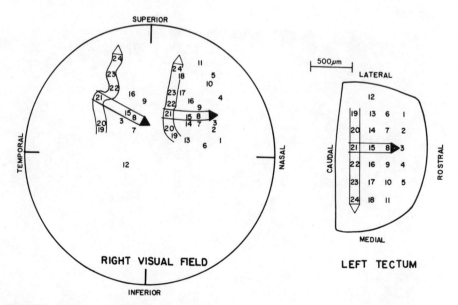

FIGURE 2 Evidence that orderly optic fibre projections can be set up under conditions in which no normal tectal guidance cues can be responsible. The figure, like those following, shows a visuotectal projection map obtained by conventional electrophysiological methods in goldfish and plotted on the usual Aimark field chart extending out to 1800°. The goldfish had had approximately two-thirds of the caudal part of the left tectum ablated 57 days previously, leaving optic pathways otherwise intact. The fish had been maintained under conditions of constant illumination, although this does not affect the result (Horder and Martin, 1977). Points 1-24 in the nasal part of the visual field are the normal, intact projection to the remaining rostral tectal fragment. Fibres excited from the temporal field have regenerated onto the rostral tectum to form a superimposed second projection to the same recording points.

Figure 3 shows a result of even greater interest since here the superimposed projection, in addition to being itself retinotopically organized, is arranged on the tectum with the reverse of the normal nasotemporal polarity. This single result demonstrates that in every respect the formation of an ordered projection can occur entirely without the benefit of tectal cues that might be used to explain normal projections. The conditions which determine polarity reversal have been to some extent defined (Horder and Martin, 1977); reversal can be predictably obtained under conditions in which regenerating fibres take previously established fibre terminals as a reference point, which suggests how the normal sequence of laying down of fibre terminals can be controlled by factors intrinsic to the fibre array itself and how one fibre might act as a cue for the next.

Figure 4 shows a result in which normal regional tectal preferences are totally, almost gratuitously, disregarded by fibres. Now lateral tectum is ablated. In some cases results comparable to those described above were obtained—duplication and polarity reversal occur in the dorsoventral axis—but

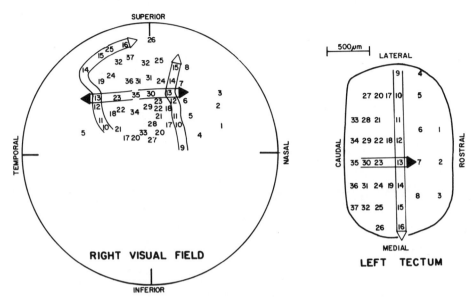

FIGURE 3 Not even polarity is controlled by the tectum. In this example of a goldfish treated similarly to that shown in Fig. 2, mapped 112 days after removing the caudal half of the tectum, the duplicated projection had a reversed nasotemporal polarity. About one-third of animals give this result. In this example the animal had been maintained under normal lighting conditions. Results are further described in Horder and Martin (1977).

here fibres have become transposed to tectal sites far from their normal ones. This is best explained by supposing that fibres cut by the ablation have been trapped by the plane of the cut edge of the tectum and during regeneration have been led by contact guidance into the caudalmost tectal tissue.

To sum up the lesson of this and much further evidence, the readiness with which fibres from particular retinal locations can come to terminate at a wide variety of tectal sites argues strongly against the existence of exclusive properties of individual sites as the basis for the formation of normal ordered projections. What is more, the formation of ordered projections under conditions in which the tectum cannot be providing any guidance cues, polarity cues included, leaves no other possibility than that detailed retinotopic ordering can be achieved by a fibre array itself. All that may be necessary is that fibres passively retain the neighbourhood relationships they have during initial growth.

Evidence Apparently Inexplicable on the New Hypothesis

In our recent review (Horder and Martin, 1978), intended as a thorough survey of the literature, we argued that many of the experimental approaches employed as evidence for selective growth by optic fibres either suffer from ambiguous results or only provide evidence for a limited form of fibre guidance. Many test no more than the ability of tectal tissue to re-orient, either in one or both axes, the array of fibres: neither selection of the group of fibres

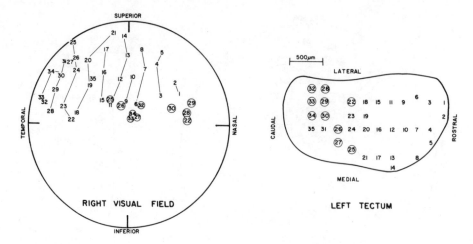

FIGURE 4 A result directly indicative of contact guidance of fibres. In this fish the lateral third of the dorsal tectum had been ablated 66 days before electrophysiological mapping, the optic pathways having otherwise been left intact. The animal was maintained under continuous illumination. Circled field positions, which would normally project to the lateral margin of *rostral* tectum, have come to project to caudal tectum. The responsible fibres form an approximately 90° rotated map superimposed on the resident normal projection; the polarity can perhaps be accounted for if, during its displacement round the caudal border of the tectum, the fibre array maintains the nasotemporal order it had originally.

involved in the re-orientation nor active control of their intrinsic rank-ordering is necessarily occurring in these experiments. Out of all the evidence we are left with a prominent exception which does appear to offer strong and direct evidence for tectal regional differentiation having a selection affinity for specific fibres, namely experiments involving exchange of grafts taken from rostral and caudal tectum. Figure 5 shows a finding of this kind, which is representative of a number of similar studies (Jacobson and Levine, 1975a, 1975b; Sharma, 1975; Hope, Hammond, and Gaze, 1976; Martin, 1978). How accurately do these grafts regain their appropriate innervations?

With regard to the caudally placed graft of rostral tectal tissue, it is unlikely that the graft receives a strictly appropriate innervation. In most cases the graft receives a diffuse collection of fibres including some representing extreme temporal field; most of them could be described as coming from local surrounding tectum, although, because the surrounding projection cannot be assumed to be in its normal position on the tectum, it is difficult to accurately judge which fibres are appropriate to the graft region. In control operations rostral tectum was left intact and the caudally placed graft (Fig. 6) was taken from the contralateral tectum. Here it is clear that the graft was precisely innervated by local fibres only. This result was not due to the contralateral graft exchange itself because a caudally originating graft similarly transposed to rostral tectum of the other side had the usual effect, in that it became innervated by fibres normally terminating in more caudal tectum. The experiment shows that rostral fibres

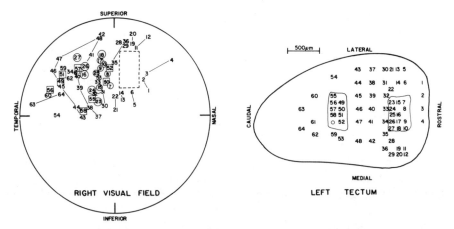

FIGURE 5 The effects of exchanging rostral and caudal fragments of dorsal tectal tissue. This visuotectal projection was mapped 189 days after crushing the right optic nerve, the exchange of grafts having been performed 71 days before crushing. A closely similar map was recorded immediately prior to interruption of the nerve. As judged on the basis of the arrangement of the projection on the tectum in normal fish, fibres normally innervating the caudally placed graft would be stimulated by visual field positions marked by dashed lines: no responses from this area were detected anywhere in the dorsal tectum. Field positions in circles project to the rostrally placed graft, those in squares to the caudally placed graft. The unfilled circle on the tectum indicates a recording position at which no responses could be obtained. The boundaries of graft tissue shown were visible at the time of mapping.

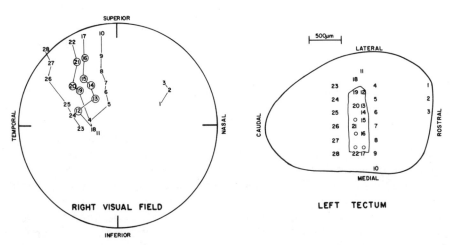

FIGURE 6 A control experiment. In this animal rostral tissue taken from the right tectum was substituted for a similarly sized fragment removed caudally in the left tectum. Optic pathways were otherwise left intact. The fish was mapped 53 days post-operatively. Fibres reinnervate the graft to form a normal projection without regard for the nature of the graft. The graft had been transplanted so that its originally medial edge lies laterally in the left tectum. Therefore, regenerating fibres disregard its mediolateral polarity which is the opposite of surrounding tectum. Circled field positions project to the graft. Similar results were obtained after the optic nerve was crushed and allowed to regenerate.

are not attracted at a distance by the caudally placed graft. It seems likely that insofar as *any* fibres normally projecting to rostral tectum reach a caudally placed graft, this is the result of the presence of the rostrally placed graft, perhaps due only to mechanical deflection of fibres caudalwards as a result of the surgical interruption to fibre paths. Finally, it is a consistent finding that there is an area of field not represented anywhere on the dorsal tectum. This corresponds closely with the area appropriate to the caudally placed graft and for this reason alone argues against accurate selective innervation of that graft.

With regard to the rostrally placed graft of caudal tectal tissue, it is a common finding (including Fig. 5 if the anomalous points 49, 51, 56-57 are disregarded) that the graft becomes innervated by fibres of more *nasal* retinal origin than the other graft. Given the way in which fibres travel across the tectum from rostral to caudal, there is no reason to doubt that *both* fibres normally destined for caudal regions *and* those normally innervating the site of the rostral implantation will have potential access to the graft. The evidence of Fig. 4 shows that fibres from rostral tectum can innervate caudal tissue. The basis on which fibres of more nasal retinal origin gain exclusive occupation of the graft is, for the present, merely a matter for speculation. For example, as a result of the different alignments of local fibre bundles and of fascicles destined for caudal tectum, the circumstances of the two sets of fibres may not be entirely comparable: some purely anatomical feature of the caudal tectal tissue may favour deflection of local fibres and facilitate entry of the other fibres. In control experiments in which grafts from the ventrolateral half of the tectum were implanted rostrally in dorsal tectum, grafts consistently became innervated by the fibres of most nasal retinal origin among those crossing the implantation site. Perhaps inappropriate tectal tissue receives, by default, fibres least coherent with fibres appropriate to the implantation site.

The failure to detect fibres appropriate to the caudally placed graft (marked by dashed lines in Fig. 5) anywhere in the tectum raises the interesting possibility that these fibres have been pushed aside or their terminal arborizations confined so that they can no longer be recorded electrophysiologically, a phenomenon which may underlie non-correspondences between anatomical and electrophysiological fibre projections reported by Levine and Jacobson (1975). Such a suppression of one set of fibres could be the result of competition for terminal space by the fibres of nasal retinal origin, which may gain the advantage as a result of being more numerous: the implantation site traverses fascicles containing a large proportion of nasal retinal fibres destined for caudal tectum while most temporal retinal fibres have terminated before reaching the site. Retinotopic coherence may ensure that one set of fibres eliminates all fibres of different retinal origin: contact guidance will ensure that fibres of related origins in the retina will assist one another in forward growth, leading eventually to the formation of increasingly large fasciculi of arriving optic fibres, which would favour exclusive occupation of the tectum nearby. In the series of animals of which Fig. 4 formed a part, it was occasionally found that the aberrantly projecting temporal retinal fibres displaced the normal fibres innervating caudal tectum. This is an interesting example, among many, of mutual exclusion of fibre populations, each of which consists of retinotopically coherent fibres, because here inappropriate fibres are excluding appropriate ones in caudal tissue similar to that used in graft exchange experiments.

To conclude, it would be dangerous to generalize from the result of such graft exchange experiments and to argue that similar distinctions apply to all tectal fragments of similar size. Corresponding results are not seen when grafts are exchanged in the mediolateral tectal axis (Martin, 1978) and even the accuracy with which rostrally placed grafts select their fibres is limited: commonly a larger than expected area of visual field is represented in it and multi-unit receptive fields are enlarged. In view of the readiness with which even these grafts can become overridden by local fibres (Jacobson and Levine, 1975a, 1975b; Sharma, 1975), it would be safer to say that all the graft can do is to tip the balance of advantage towards one population of fibres as against another. In view of the manifold uncertainties regarding the conditions under which competing fibres arrive at the graft, the actual choice may yet turn out to be the result of a purely trivial combination of circumstances. It is puzzling that the distinctions between rostral and caudal tectal tissue implied by these experiments are unlikely to underlie the control of normal events in the embryonic development of the retinotectal projection, since at the time rostral tissue is being innervated, caudal tissue has not yet been laid down.

In reviewing the evidence for and against the morphogenetic explanation for retinotopic ordering of fibres (Horder and Martin, 1978), we were careful to acknowledge that the very experiments on which Sperry founded his theory, namely those in which fibres successfully reach their normal destinations after they or their targets have been rotated with respect to one another, imply that the morphogenetic theory is insufficient. Some form of selective fibre guidance is called for by such results. In the example shown in Fig. 7, part of the tectum had been rotated *in situ* by 180° and, as comparison of the orientation of the arrows summarizing mediolateral ordering of fibres inside and outside the graft shows, fibres entering the rotated tissue have been influenced by the polarity of that tissue. This experiment reveals a significant feature of the mechanism for polarity control because the fibres are not ones which would normally innervate the graft region: in this animal, rostral tectal tissue, denervated by ablation of temporal retina, had become innervated by fibres normally destined for caudal tectum which had also been ablated. Other evidence, using different techniques (Bunt, Horder and Martin, 1978), is also available which shows polarity control by foreign tectum. It therefore follows that polarity control may not be taken as evidence for selective affinity of specific fibres for specific tectal regions. It is based on a property contained in all parts of the tectum equivalently. Taken together with the evidence that polarity cues in the tectum can be ignored and overridden (Figs. 3, 4, 6), and that, under certain conditions (Figs. 4, 7), an array of fibres can lie in any orientation on the tectal surface, the implication is that the polarity cue is a universal property of tectal tissue and of relatively low resolution. As with graft exchanges, the tissue differences appear merely to tip

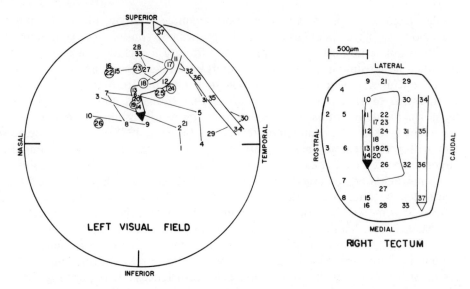

FIGURE 7 A rotated graft can control the polarity of innervating fibres even when these are foreign to the graft. A visuotectal projection mapped 387 days after crushing the left optic nerve and a total of 700 days after rotating a fragment of the right rostral tectal tissue *in situ,* together with ablating the caudal half of the right tectum and the temporal half of the left eye. The arrows summarize the polarities of typical lines of points inside and outside the graft. A very similar map had been obtained immediately prior to interrupting the optic nerve, and in eight other similarly treated animals. The abnormal angulation of projection lines outside the graft probably reflects unequal degrees of rostral transposition of fibres into the foreign tectal tissue.

the balance between two fibre arrangements: whether fibres are rotated or normally arranged in a rotated graft, all other aspects of organization within the fibre array may be internally controlled. We (Horder and Martin, 1978) have suggested that a suitable basis for the polarity cue could be an anatomical parameter directly correlated with the developmental histories of parts of the tectum: the relative maturity of tectal tissue is known, in certain instances, to exert a powerful influence on the behaviour of arriving optic fibres.

Although the question of polarity control remains possibly the most obscure and fundamental issue facing us, certain reasonable principles can be outlined. We must not be misled by the usual emphasis on possible differentiation of *targets* in the tectum as such because the essential problem is to explain how fibres, which have initially been deflected towards totally inappropriate tectal regions as a result of rotation, are able to gain access to suitable targets. It therefore seems more realistic to regard the guidance cues as operating on the choice of the pathways followed by fibres during growth, prior to their arrival at termination sites. There is direct evidence that fibres become re-oriented towards appropriate tectal regions within the optic tract. Straznicky, Gaze, and Horder (1979) show that fibres from ventral retina deflected into the ipsilateral optic tract at a point central to the chiasma become selectively segregated into the medial division of the tract. Since these fibres lay medially in their original optic tract, it would be anticipated that, as a geometrical consequence of transposition across the midline, they will occupy a relatively lateral position in the deflected tract. Therefore, between the point of optic nerve section and the division of the optic tract into medial and lateral brachia, ventral retinal fibres have been actively guided towards the appropriate half of the tectum, a phenomenon which would be a major determinant of polarity in a complete projection. Developmentally there is no fundamental distinction between the tectum and the region later occupied by the optic tract: they are initially an anatomical continuum. Perhaps the guidance cue revealed in the tract is of the same nature as the universal polarity cue in the tectum. By being subjected to a series of equivalent polar cues throughout their growth, fibres may become progressively oriented towards their targets.

A second set of considerations may also help to give fibres access to their targets. The temporal sequence in which fibres reach the tectum and the possibility that early fibres (by being less regularly arranged or more widely branching) play an exploratory role are perhaps the most neglected aspects of what we know about optic nerve growth and regeneration. Because the first growing fibres are in small numbers, it is likely that they will grow relatively irregularly and reach a potentially rather wide choice of tectal regions. But as the majority of fibres arrive later, they will be increasingly constrained by mutual contact guidance with the result that intrinsic retinotopic order will be increasingly carried through to the tectum. On the basis of weight of numbers and retinotopic coherence alone, early aberrant fibres may suffer competitive exclusion from termination sites: they may become unrecordable electrophysiologically and may later withdraw. If relatively small numbers of fibres are surgically deflected into foreign tectal regions (Bunt et al., 1978), they innervate local tissue directly with little evidence for any inclination to find their way back to more appropriate regions. But they succeed in doing this when a complete array of optic fibres has been similarly deflected (Sharma, 1973). It follows that competitive interactions between members of a fibre population play a major role in distributing fibres within the tectum. As the evidence of Cook and Horder (1974) showed, a final distribution may be the end result of a gradual and progressive series of fibre displacements.

Future Prospects

The fact that highly ordered optic projections can be formed in circumstances in which the tectum cannot be contributing any form of guidance suggests that during normal development and optic nerve regeneration detailed retinotopic fibre organization could be internally controlled. In many examples in the literature, of which Fig. 6 is an instance, fibres have been caused to form projections which do not conform with normal tectal tissue preferences and which can be readily explained as the result simply of the pattern in which fibres arrive, having traversed the optic pathways in retinotopic array. One major difficulty with the chemoaffinity theory is to see how it can readily incorporate the undeniably mechanical nature of nerve growth: "This strange tendency to follow faithfully and passively the outlines of hard and smooth organs" (Ramón y Cajal, 1928, p. 207). On the other hand, we suggest that the morphogenetic approach offers the hope of an explanatory framework which will be internally consistent and applicable at all levels of nervous organization; even such matters as polarity control can in principle be accommodated without significantly distorting its central provisions. This is undoubtedly a matter of some terminological delicacy but to retain the term "chemospecificity" to describe guidance cues underlying polarity control is likely, given its past associations, to lead to confusion.

At the heart of the matter is the question of the amount of information that must be provided during embryonic development to set up the retinotectal projection. The chemoaffinity model requires a great deal of information because, regardless of how it is applied, retinotopically neighbouring fibres must be distinguishable and therefore different. The morphogenetic approach resolves the dilemma by accounting for detailed rank-ordering of fibres in terms of the mechanics of fibre growth. We suggest that few and small tissue differences will be sufficient to account for large-scale control of optic projections. Perhaps, if we pursue this approach further, we shall be able to answer Cajal, where, writing in 1913, he said, "As for the theory of neurotropism far from being for me a dogma, it is simply a working hypothesis which I am willing to correct or even abandon in the presence of better explanations" (Ramón y Cajal, 1928, p.195).

References

Bunt, S. M., and T. J. Horder (1977). A proposal regarding the significance of simple mechanical events, such as the development of the choroid fissure, in the organization of central visual projections. J. Physiol. (Lond.) 272:10-12P.

Bunt, S. M., and T. J. Horder (1978). Evidence for an orderly arrangement of optic axons in the central pathways of vertebrates and its implications for the formation and regeneration of optic projections. Society for Neuroscience, Abstracts, Eighth Annual Meeting, p. 468.

Bunt, S. M., T. J. Horder, and K. A. C. Martin (1978). Evidence that optic fibres regenerating across the goldfish tectum may be assigned termination sites on a "first come, first served" basis. J. Physiol. (Lond.) 276:45-46P.

Cook, J. E., and T. J. Horder (1974). Interactions between optic fibres in their regeneration to specific sites in the goldfish tectum. J. Physiol (Lond.) 241:89-90P.

Hope, R. A., B. J. Hammond, and R. M. Gaze (1976). The arrow model: retinotectal specificity and map formation in the goldfish visual system. Proc. Roy. Soc. Ser. B 194:447-466.

Horder, T. J., and K. A. C. Martin (1977). Translocation of optic fibres in the tectum may be determined by their stability relative to surrounding fibre terminals. J. Physiol. (Lond.) 271:23:24P.

Horder, T. J., and K. A. C. Martin (1978). Morphogenetics as an alternative to chemospecificity in the formation of nerve connections. In: Cell-Cell Recognition. Society for Experimental Biology Symposium 32. A. S. G. Curtis (ed.). Cambridge University Press, Cambridge, pp. 275-358.

Jacobson, M., and R. L. Levine (1975a). Plasticity in the adult frog brain: filling the visual scotoma after excision or translocation of parts of the optic tectum. Brain Res. 88:339-345.

Jacobson, M., and R. L. Levine (1975b). Stability of implanted duplicate tectal positional markers serving as targets for optic axons in adult frogs. Brain Res. 92:468-471.

Levine, R. L., and M. Jacobson (1975). Discontinuous mapping of retina onto tectum innervated by both eyes. Brain Res. 98:172-176.

Martin, K. A. C. (1978). Combination of fibre-fibre competition and regional tectal differences accounting for the results of tectal graft experiments in goldfish. J. Physiol. (Lond.) 276:44-45P.

Meyer, R. L., and R. W. Sperry (1976). Retinotectal specificity: chemoaffinity theory. In: Neural and Behavioural Specificity, Vol. 3. Studies on the development of behaviour and the nervous system. G. Gottlieb (ed.). Academic Press, New York, pp. 111-149.

Ramón y Cajal, S. (1928). Degeneration and Regeneration of the Nervous System. R. M. May (translator). Hafner, New York, 1959.

Sharma, S. C. (1973). Anomalous retinal projection after removal of contralateral optic tectum in adult goldfish. Exp. Neurol. 41:661-669.

Sharma, S. C. (1975). Visual projection in surgically created "compound" tectum in adult goldfish. Brain Res. 93:497-501.

Sperry, R. W. (1951). Regulative factors in the orderly growth of neural circuits. Growth (10th Symp.) 15:63-87.

Straznicky, K., R. M. Gaze, and T. J. Horder (1979). Selection of appropriate medial brachium of the optic tract by fibres of ventral retinal origin during development and in regeneration: an autoradiographic study in *Xenopus*. J. Embryol. Exp. Morphol. (in press).

Growth and Neurogenesis
in Adult Goldfish Retina

PAMELA RAYMOND JOHNS
Department of Anatomy
The University of Michigan
Ann Arbor, Michigan USA

Abstract Growth continues in adult goldfish. Cell counts and ^{3}H−thymidine radioautography indicate that the brain and retina increase in size in part by the addition of new neurons. The retina of a large, 4-year-old fish (20 cm in length) has about 20,000,000 neurons, whereas in a small (5 cm) fish there are only about 3,000,000 retinal neurons. New cells are produced at the margins of the retina and are added appositionally at rates of up to 20,000 cells/day. Growth-related changes also occur in the older, more central regions of the retina: the eyeball expands, stretching the retina and decreasing the density of its cells. The rods alone maintain a constant density with growth, so that the proportion of rods relative to other retinal neurons increases as the fish grows. Since new rods are added only at the periphery, a shift in the position of rods with respect to their postsynaptic partners is implied. This suggests that synaptic connections may be continually broken and reformed in the functioning adult goldfish retina.

Adult goldfish continue to grow. Unlike mammals, which grow until they reach an adult body size and then stop, most teleost fish grow throughout life. Their growth is indeterminate, limited primarily by environmental restrictions (Brown, 1957). The rate of growth of individual members of a population varies tremendously in response to a complex set of social and other exogenous stimuli, but, in general, the older fish are bigger (Muller, 1952; Pfuderer et al., 1974; Johns and Easter, 1977). The brain of the fish gets larger as the body grows (Packard, 1972; Meyer, 1977; Johns, unpublished) and, what is even more remarkable, the larger brain has more neurons. While it is acknowledged that mammalian brains also increase in size somewhat during postnatal life, this growth results from glial cell production and the continued development of neuronal processes (Altman, 1970; Prestige, 1974; Watson, 1974); the production of neurons is restricted to embryonic or, in a few specific regions, early postembryonic development (Altman, 1972). Only very rarely has evidence

345

been presented for neurogenesis in the adult mammalian brain—in the olfactory bulb of the rat, for example—and the only cells produced are small, associational "microneurons" (Altman, 1966, 1969, 1970). All of the large "macroneurons" are formed prenatally (Jacobson, 1970). Thus the primary scaffolding of neuronal connections in the mammalian brain is constructed from a fixed population of neurons during embryonic development. In contrast, in the growing fish new neurons are continually being produced, and they must somehow be incorporated into an already fully active network of complex synaptic connections. The problem here is how to grow while functioning.

Let us now examine more closely the evidence for postembryonic neurogenesis in the central nervous system of fishes. Results from several different kinds of studies suggest that fish (and amphibians) maintain the capacity for neurogenesis beyond embryonic and larval stages. Localized sites of proliferating cells, called "matrix zones", have been described in the brains of fishes; these regions are associated with the ependymal layer which lines the ventricular spaces in the brain (reviewed by Kirsche, 1967). Cells derived from mitotic activity in the matrix zones are thought to contribute to brain growth (Rahmann, 1968; Richter and Kranz, 1970; Meyer, 1978) and to histotypic regeneration following brain lesions (Kirsche, 1960, 1965; Kirsche and Kirsche, 1961; Richter, 1965, 1968; Richter and Kranz, 1977). These authors suggest, but do not rigorously prove, that both neurons and glial cells are produced by the matrix zones.

The most convincing evidence for postembryonic neurogenesis in fish and amphibians comes from several studies in which the number of neurons has been quantified in animals of different ages and sizes. The neural retina has been especially useful for such studies because of the precise laminar arrangement of its cells and the relative ease with which neurons can be identified and distinguished from glia. Retinal cells (or optic nerve fibers) have been counted in at least seven species of fish or amphibians, and in all cases the numbers increase with growth (Muller, 1952; Lyall, 1957; Blaxter and Jones, 1967; Wilson, 1971; Johns and Easter, 1977; Kock and Reuter, 1978; M. A. Ali, personal communication). Our studies on growing goldfish (*Carassius auratus*) are illustrative and will be described in some detail next.

A 1-year-old goldfish is about 5 cm long; 4-year-old fish are about 20 cm in length. These values are to be regarded strictly as averages since individual fish can grow much faster or slower (Johns, 1976). The growth of the eye almost keeps pace with the body: the average diameter of the eye increases from 4 to 10 mm over four years (Johns and Easter, 1975). As the eye grows, so does the retina, and Fig. 1 shows the increase in retinal surface area (from 15 to 120 mm^2) as the fish grows. To answer this question, retinal cells were counted in 25 goldfish retinas, sectioned transversely at 5 μm parallel to the nasotemporal or the dorsoventral meridian. The density of cells was found to be quite homogeneous throughout the retina—goldfish have no specialized region adapted for increased acuity such as is found in other vertebrate retinas (Rodieck, 1973). Reasonable estimates of total cell numbers were obtained from the simple pro-

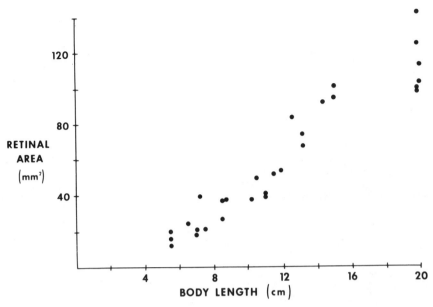

FIGURE 1 The area of the retina in goldfish of different body lengths. Retinal area was calculated from measurements on histological sections through the dorsoventral or nasotemporal meridian. Each point represents one retina; the values are corrected for an histological shrinkage of 30% (linear). Reprinted with permission from J. comp. Neurol. 176:331-342 (1977).

duct of average cell density (number/mm^2) and retinal surface area (in mm^2) , and are given in Fig. 2. The larger the retina, the more neurons it has: ganglion cells, photoreceptors (rods and cones) and cells in the inner nuclear layer all increase in number as the retina grows. The total cell number increases from about 3,000,000 to 20,000,000 in four years.

The rate of cell addition was calculated by comparing the number of cells between the two eyes from three fish, in which one eye was removed, the animal allowed to grow for several months and then the second eye removed. The average rates of cell addition for the three fish were 6,000; 9,000 and 20,000 new retinal cells added per day (Johns, 1976, 1977).

Muller (1952) has also shown that the number of cells increases in the retinas of growing fish (juvenile and adult guppies, *Lebistes reticulatus*). As we did, he counted retinal cells in all three nuclear layers and showed that all of the different retinal cell types increased in number with growth. Kock and Reuter (1978) counted ganglion cells in whole mounts of crucian carp (*Carassius carassius*) retinas. They found an increase in ganglion cell numbers as the eye diameter increased from 4 to 8 mm, but further growth of the eye up to 10 mm was not accompanied by ganglion cell addition. This is unlike our results for goldfish, in which ganglion cell numbers increased over the entire range of retinal sizes counted. The range of eye diameters that we examined was, coincidentally, the same as Kock and Reuter used—4 to 10 mm. We do not know

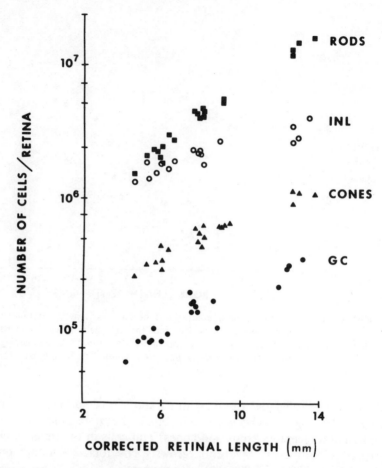

FIGURE 2 The total number of cells in goldfish retinas of different sizes. Each point represents one retina. Cells counted include rods (squares), inner nuclear layer cells (INL, open circles), cones (triangles), and ganglion cells (GC, filled circles). The retinal length is defined as the distance along the retina from one *ora terminalis* to the other in a meridional section, corrected for an histological shrinkage of 30%. Reprinted with permission from J. comp. Neurol. 176:331-342 (1977).

why our results differ from theirs. The whole mount technique which they used is less susceptible to the errors inherent in counting cells; when using sectioned material the appropriate sampling methods and correction factors must be carefully chosen (Konigsmark, 1970). But since we wanted to count retinal cells in layers other than just the ganglion cell layer, we chose to section the tissue. Being cognizant of the problems associated with the counting techniques we used, we compared the values for ganglion cell density determined from meridional sections with counts made on sections passing tangentially through the single layer of cells in the ganglion cell layer in retinas from 14 fish. Both counting methods gave similar results for ganglion cell density as a function of the size of the retina (Johns, 1976).

If not methodological differences, the discrepancy in results might be due to species differences, or it may be related to environmental variables. The carp that Kock and Reuter used were from a wild population and our goldfish were obtained commercially. The eyes of adult, wild carp which live in the cold waters of Finland apparently grow more slowly than the eyes of domestic goldfish raised in the southern United States: at 1 year of age the eyes of both species were approximately 4 mm in diameter, but after four years of growth the goldfish eyes averaged 10 mm in diameter, whereas the carp eyes required seven years to reach the same size. It is worth noting that elsewhere in the central nervous system of teleost fish the capacity to produce new neurons diminishes with age. This is shown by a decreasing level of mitotic activity in the matrix zones (Richter and Kranz, 1970) associated with a reduced ability to regenerate neural tissue as the fish gets older and bigger (Kirsche and Kirsche, 1961; Segaar, 1965). It is quite possible that if we looked at goldfish older than 4 years we might also find an age or size at which ganglion cell addition ceases.

Where do the new retinal cells come from? The best technique for demonstrating cell proliferation is radioautography following ^{3}H–thymidine injection (Sidman, 1970). When labeled thymidine is injected into goldfish, dividing cells are seen at the margin of the retina, at the *ora terminalis*. This germinal zone surrounds the retina and is a surviving remnant of the embryonic neuroepithelial germinal layer which once covered the entire presumptive neural retina. The restriction of the germinal cells to the peripheral margin in the mature fish retina is a direct consequence of the embryonic pattern of histological differentiation found in all vertebrate retinas, including those of teleost fishes (Hollyfield, 1972) and, specifically, goldfish (Sharma and Ungar, 1977). In the embryo, retinal differentiation always begins in central regions around the optic disc and then spreads in a centripetal wave toward the periphery. Peripherally placed cells continue to divide until the wave of differentiation reaches them. In mammalian retinas, all of the germinal cells are eventually exhausted (Johns et al., 1979), whereas in fishes and larval amphibians a few of them at the extreme peripheral border escape and persist in mitotic activity (Gaze and Watson, 1968; Straznicky and Gaze, 1971; Jacobson, 1976; Johns, 1977).

Postembryonic cell proliferation at the *ora terminalis* in fish and amphibian retinas results in the addition of concentric rings of new cells. In the radioautographs of retinas taken from goldfish injected with ^{3}H–thymidine at 1 year of age and allowed to survive and grow for several weeks or months afterwards, a vertical band of labeled nuclei spans the retinal layers (Fig. 3). These labeled nuclei are the differentiated progeny of the dividing germinal cells. They are no longer located at the *ora terminalis*, but are displaced centrally by the newer, unlabeled cells added peripheral to them as the retina continued to grow during the prolonged interval after injection when labeled precursor was no longer available. The band of labeled nuclei is circumferentially continuous and forms

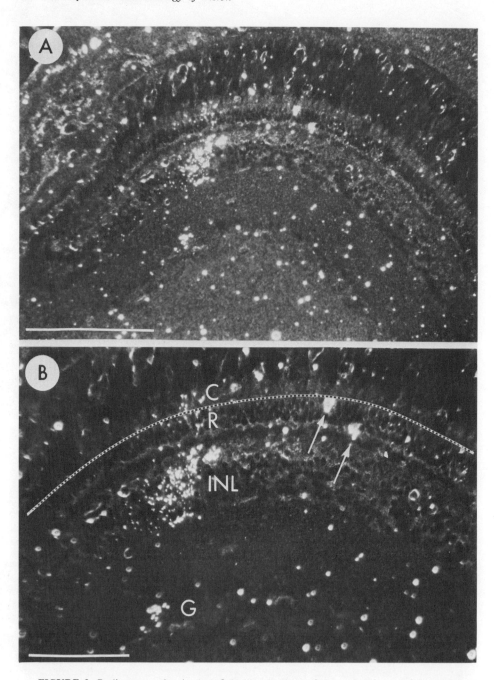

FIGURE 3 Radioautographs, in dark-field illumination, of the retina from a fish which survived for 168 days following ^{3}H–thymidine injection. (A) Low magnification of the retinal
margin. The *ora terminalis* is at the left edge of the field. A vertical band of bright silver
grains just to the left of center represents labeled retinal nuclei. Calibration bar: 100 μm. (B)
At higher magnification labeled nuclei are seen in the ganglion cell layer (G) and the inner

an annulus parallel to the *ora terminalis*. The segment of retina from the labeled annulus to the retinal margin represents the "new retina" added; it increases in length with time and continued growth. In a group of ten fish which survived from 3 to 11 months following injection, the linear increment of new retina added ranged from 80 to 570 μm (Johns, 1976).

This appositional mode of retinal growth was previously shown in larval amphibians (Gaze and Watson, 1968; Straznicky and Gaze, 1971; Jacobson, 1976) and was recently confirmed in the goldfish by Meyer (1978). Restriction of cell addition to the peripheral margin, away from the central, presumably more visually active regions may be one way of dealing with the problem of how to grow while functioning. Weiss (1949) comments that spatial segregation of proliferating and functioning compartments is common to growing tissues in general.

The central, functioning retina is not entirely free from the effects of growth, however. The retina grows not only by adding new cells but also by hypertrophy, or expansion, of the existing tissue. During development the vertebrate eyeball expands, perhaps as a result of increased intraocular pressure generated by vitreal secretion (Coulombre, 1955; Coulombre et al., 1963). The retina is stretched and the cellular density decreases as cells are pulled apart (Ali, 1964; Mann, 1969; Rusoff, 1979). This process contributes substantially to retinal growth in adult goldfish. In radioautographs prepared from fish several months following ^{3}H–thymidine injection, the expansion of the retina was measured directly. The labeled retinal cell nuclei mark the former position of the *ora terminalis* at the time of injection; they were there when they incorporated the label and they differentiated *in situ*. The cells surrounded by, that is, central to, the labeled annulus are in the "old retina" which was present at the time of injection. The size of the "old retina" approximately doubled in 11 months, and this expansion accounted for about 75 to 80% of the total increase in retinal area.

Changes even more dramatic than stretching also happen in the central retina during growth. Consider Fig. 4 in which two histological sections from central retinal regions are compared; the one on the left is from a fish 7 cm in length and that on the right is from a 20-cm fish. Note that the density of cells in the larger retina is less in the ganglion cell and inner nuclear layers and the layer of cone nuclei, scleral to the external limiting membrane. This decrease in cell density is to be expected because of stretching. (The increase in overall thickness of the retina, opposite to what would be predicted from stretching, is due largely to increases in the width of the plexiform layers presumably due to continued growth of neuronal and glial cell processes. See below for further discussion of this point.) In marked contrast to the other cells, the rod nuclei—in

nuclear layer (INL). In the outer nuclear layer, the rod nuclei (R) are on the vitreal side of the external limiting membrane, indicated by the dashed line, and the cone nuclei (C) are on the scleral side. All of the labeled nuclei except the rods lie within a circumscribed, vertical band; labeled rod nuclei (arrows) are displaced toward central retina. Calibration bar: 50 μm. Reprinted with permission from J. comp. Neurol. 176:343-358 (1977).

the outer nuclear layer vitreal to the external limiting membrane—do not change in density, nor does the outer nuclear layer change in width. This means that the proportion of rods increases relative to the other neurons with growth, a change that happens throughout the retina (Johns and Easter, 1977).

The observation that retinal cell proportions can change with growth is not new, nor is it unique to the goldfish. Differential addition of cells to certain retinal laminae has been seen in other fish and amphibians. For example, the larvae of some species of teleost fishes have only cones and no rods in their retinas; rods are added later, at metamorphosis (Lyall, 1957; Blaxter and Jones, 1967; Blaxter, 1975). And, during larval growth in *Rana* spp. (Bernard, 1900; Hollyfield, 1968) and *Xenopus laevis* (Hollyfield, 1971), cells are added differentially to the inner nuclear layer. In most of these studies dividing cells were not found near the presumed site of cell addition (but see Hollyfield, 1971), so the authors concluded that the new cells were derived from either (1) the adjacent retinal layers via transformation from another cell type, or (2) the germinal zone at the margin via centrifugal migration along the retinal laminae.

Muller (1952) observed and quantified a growth-related increase in the proportion of rods in the juvenile and adult guppy retina. His data are similar qualitatively to ours for the goldfish retina. To account for the change in cell proportions at sites far distant from the source of new cells, Muller proposed a unique and somewhat bizarre mechanism which is, nonetheless, supported by the results of our radioautographic study. Imagine that the rods are not pulled apart by the forces of expansion which act to decrease the density of cells in other layers, and that the expanding layers slide over the stationary layer of rods as the retina grows. As the other cells move out and the rods stay behind, their relative numbers would appear to increase, as we both observed. It follows that if the rest of the retina is stretched while the layer of rods is not, a gap would be created around the peripheral edge of the rod layer; this gap never actually appears because it is quickly filled by new rods derived from proliferating cells in the adjacent germinal zone.

The proposed mechanism of laminar displacement, or shear, requires that new rods should be displaced centrally relative to the new cells added to the other layers, and our radioautographs show just that (Fig. 4). Meyer (1978) and Scholes (1976) noted a similar centralward displacement of labeled rod nuclei following ^{3}H—thymidine injection and subsequent growth in goldfish and another cyprinid, the Black Molly, respectively. Meyer did not speculate on how the labeled rod nuclei got there; Scholes attributed their anomolous position to *in situ* mitotic division in the outer nuclear layer, *outside* of the marginal germinal zone. More detailed studies, needed to clarify the issue of the source of the new rods in central retinal regions, are in progress.

If slippage between retinal laminae does occur, it implies that the rods exchange postsynaptic partners as the eye grows. Neuronal connections are arranged vertically in the vertebrate retina: neurons in one layer synapse with others directly above or below (scleral or vitreal to) them in the adjacent layers (Stell, 1972). If with growth the inner nuclear layer slides laterally, the horizontal and bipolar cells nearest a given rod in the outer nuclear layer will

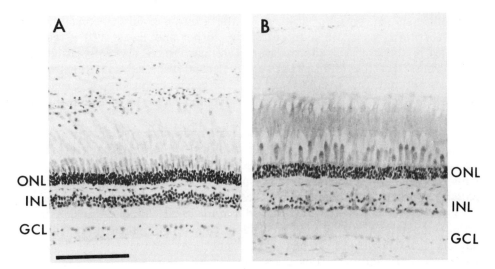

FIGURE 4 Comparison of retinas from small and large goldfish. (A) Retina from a fish 7 cm in length; (B) Retina from a fish 20 cm in length. The rod nuclei in the outer nuclear layer (ONL) remain constant in density, whereas cells in the inner nuclear layer (INL) and ganglion cell layer (GCL) decrease in density. The cones, which lie above (scleral to) the rod nuclei, also decrease in density and increase in size as the fish grows. (These sections were bleached to remove the pigment in the pigmented epithelial cells.) Calibration bar for both photomicrographs: 100 μm.

change. If synaptic connections are to be maintained, the rod axons and/or the postsynaptic cell dendrites must become skewed laterally. I calculate that the magnitude of the displacement may be up to 1 mm in 11 months, given the amount of retinal expansion observed. Stell and colleagues (1967, 1975) have provided extremely precise and detailed descriptions of rod and cone connectivity in central regions of goldfish retina, and they find a vertical, not a skewed, pattern of rod cell axons and inner nuclear layer cell dendrites. I therefore suggest that intercellular connections in the outer plexiform layer of the goldfish retina may be altered with growth as a consequence of interlaminar shear.

I have, as yet, no direct evidence for synaptic promiscuity in the outer plexiform layer of growing adult goldfish retinas. I am intrigued, however, by two recent demonstrations of instability in the ultrastructure of photoreceptor terminals. When turtle retinas are dark-adapted, the postsynaptic processes of horizontal and bipolar cells delve deeply into the cone pedicles, whereas in the light-adapted state they barely indent the surface (Schaeffer and Raviola, 1975). I wonder whether an exchange of postsynaptic partners might be facilitated when such a synapse is "unplugged." In another study, in a cichlid fish (*Nannacara*), synaptic ribbons were found in cone pedicles in light-adapted but not dark-adapted retinas (Wagner, 1975). This again suggests a cyclical building, destruction, and rebuilding of the synaptic terminal of photoreceptors. Unfortunately, both studies showed changes in cones but not rods.

Another site of synaptic plasticity in the growing goldfish retina is in the inner plexiform layer. In a recent study, Fisher and Easter (1979) counted synapses in the inner plexiform layer of retinas from 1- and 4-year-old fish. They found that the retinas from older (larger) fish had substantially more synapses per mm^2, per degree visual angle or per inner nuclear layer cell. This is direct evidence for ongoing synapse formation in the adult goldfish retina, presumably related to growth.

We do not know how the retina copes with these changes, how it integrates the new neurons and rewires synaptic pathways, while continuing to maintain visual functions. The brain, and specifically the optic tectum, has similar problems: it must accommodate the new optic fibers and itself add new cells. The pattern of tectal growth in goldfish (Meyer, 1978) and in larval amphibians (Straznicky and Gaze, 1972; Scott and Lazar, 1976) differs from that in the retina—crescents of new tissue are added instead of complete annuli. These authors postulate that the topographic connections between retinal fibers and tectal cells must shift as a consequence of this growth mismatch (but see Jacobson, 1976). The central nervous system in these animals is thus actively engaged in neurogenesis and synaptogenesis, processes normally associated with embryonic development. It is not surprising, therefore, that these brains show remarkable capacities for functional regeneration and modification in response to damage. It is likely that this neuronal plasticity is an experimentally induced exaggeration of a normal growth process.

References

Ali, M. A. (1964). Stretching of the retina during growth of salmon (*Salmo salar*). Growth 28:83-89.

Altman, J. (1966). Autoradiographic and histological studies of postnatal neurogenesis. II. A longitudinal investigation of the kinetics, migration and transformation of cells incorporating tritiated thymidine in infant rats, with special reference to postnatal neurogenesis in some brain regions. J. comp. Neurol. 128:431-474.

Altman, J. (1969). Autoradiographic and histological studies of postnatal neurogenesis. IV. Cell proliferation and migration in the anterior forebrain with special reference to persisting neurogenesis in the olfactory bulb. J. comp. Neurol. 137:433-458.

Altman, J. (1970). Postnatal neurogenesis and the problem of neural plasticity. In: Developmental Neurobiology. W. A. Himwich (ed.). Thomas, Springfield, Illinois, pp. 197-237.

Altman, J. (1972). Postnatal development of the cerebellar cortex in the rat. I. The external germinal layer and the transitional molecular layer. J. comp. Neurol. 145:353-398.

Bernard, H. M. (1900). Studies in the retina: rods and cones in the frog and in some other amphibia. Quart. J. Micros. Sci. 43:23-47.

Blaxter, J. H. S. (1975). The eyes of larval fish. In: Vision in Fishes: New Approaches in Research. M. A. Ali (ed.). Plenum Press, New York, pp. 427-444.

Blaxter, J. H. S., and M. P. Jones (1967). The development of the retina and retinomotor response in the herring. J. Mar. Biol. Assoc. UK 47:677-697.

Brown, M. E. (1957). The Physiology of Fishes, Vol. I. Metabolism, Ch. IX. Experimental studies on growth. Academic Press, New York, pp. 361-400.

Coulombre, A. J. (1955). Correlations of structural and biochemical changes in the developing retina of the chick. Amer. J. Anat. 96:153-189.

Coulombre, A. J., S. N. Steinberg, and J. L. Coulombre (1963). The role of intraocular pressure in the development of the chick eye. V. Pigmented epithelium. Invest. Ophthal. 2:83-89.

Fisher, L. J., and S. S. Easter (1979). Retinal synaptic arrays: continuing development in the adult goldfish. J. comp. Neurol. (in press).

Gaze, R. M., and W. E. Watson (1968). Cell division and migration in the brain after optic nerve lesions. In: Ciba Foundation Symposium on Growth of the Nervous System. G. E. W. Wolstenholme and M. O'Connor (eds.). Churchill Ltds., London, pp. 53-67.

Hollyfield, J. G. (1968). Differential addition of cells to the retina in *Rana pipiens* tadpoles. Devel. Biol. 18:163-179.

Hollyfield, J. G. (1971). Differential growth of the neural retina in *Xeonopus laevis* larvae. Devel. Biol. 24:264-286.

Hollyfield, J. G. (1972). Histogenesis of the retina in the killifish *Fundulus heteroclitus*. J. comp. Neurol. 144:373-380.

Jacobson, M. (1970). Developmental Neurobiology. Holt, Rinehart & Winston, New York.

Jacobson, M. (1976). Histogenesis of retina in the clawed frog with implications for the pattern of development of retinotectal connections. Brain Res. 103:541-545.

Johns, P. A. R. (1976). Growth of the adult goldfish retina. Ph.D. thesis, The University of Michigan.

Johns, P. R. (1977). Growth of the adult goldfish eye. III. Source of the new retinal cells. J. comp. Neurol. 176:343-358.

Johns, P. R., and S. S. Easter (1975). Retinal growth in adult goldfish. In: Vision in Fishes: New Approaches in Research. M. A. Ali (ed.). Plenum Press, New York, pp. 451-457.

Johns, P. R., and S. S. Easter (1977). Growth of the adult goldfish eye. II. Increase in retinal cell number. J. comp. Neurol. 176:331-342.

Johns, P. R., A. C. Rusoff, and M. W. Dubin (1979). Postnatal neurogenesis in the kitten retina. J. Comp. Neurol. (in press).

Kirsche, W. (1960). Zur Frage der Regeneration des Mittelhirnes der Teleostei. Verh. Anat. Ges. 56:259-270.

Kirsche, W. (1965). Regenerative Vorgange im Gehirn und Ruchenmark. Ergeb. Anat. Entwick. 38:143-194.

Kirsche, W. (1967). Uber postembryonale Matrixzonen im Gehirn verschiedener Vertebraten und deren Bezichung zur Hirnbauplanlehre. Z. Mikros. Anat. Forsch. 77:313-406.

Kirsche, W., and K. Kirsche (1961). Experimentelle Untersuchungen zur Frage der Regeneration und Funktion des Tectum opticum von *Carassius carassius*. L. Z. Mikros, Anat. Forsch. 67:140-182.

Kock, J.-H., and T. Reuter (1978). Retinal ganglion cells in the crucian carp (*Carassius carassius*). I. Size and number of somata in eyes of different sizes. J. comp. Neurol. 179:535-548.

Konigsmark, B. W. (1970). Methods for the counting of neurons. In: Contemporary Research Methods in Neuroanatomy. W. J. H. Nauta and S. O. E. Ebbesson (eds.). Springer-Verlag, New York, pp. 315-340.

Lyall, A. H. (1957). The growth of the trout retina. Quart. J. Micros. Sci. 98:101-110.

Mann, I. (1969). The Development of the Human Eye. Grune and Stratton, New York.

Meyer, R. L. (1977). Eye-in-water electrophysiological mapping of goldfish with and without tectal lesions. Exp. Neurol. 56:23-41.

Meyer, R. L. (1978). Evidence from thymidine labeling for continuing growth of retina and tectum in juvenile goldfish. Exp. Neurol. 59:99-111.

Muller, H. (1952). Bau und Wachstum der Netzhaut des Guppy (*Lebistes reticulatus*). Zool. Jb. 63:275-324.

Packard, A. (1972). Cephalopods and fish: The limits of convergence. Biol. Rev. 47:241-307.

Pfuderer, P., P. Williams, and A. A. Francis (1974). Partial purification of the crowding factor from *Carassius auratus* and *Cyprinus carpio*. J. Exp. Zool. 187:375-382.

Prestige, M. C. (1974). Axon and cell numbers in the developing nervous system. Brit. Med. Bull. 30:107-111.

Rahmann, H. (1968). Autoradiographische Untersuchungen zum DNS-Stoffwechsel (Mitose-Haufigkeit) im ZNS von *Brachydanio rerio* HAM. BUCH.

Richter, W. (1965). Regeneration im Tectum opticum bei *Leucaspius delineatus* (Heckel 1843). Z. Mikros. Anat. Forsch. 74:46-68.

Richter, W. (1968). Regeneration im Tectum opticum bei adulten *Lebistes reticulatus* (Peters 1859). J. Hirnforsch. 10:173-186.

Richter, W., and D. Kranz (1970). Die Abhangigkeit der DNS-Syntese in den Matrixzonen des Mesencephalons vom Lebensolter der Versuchstiere (*Lebistes reticulatus*—Teleostei). Autoradiographische Untersuchungen. Z. Mikros. Anat. Forsch. 82:76-91.

Richter, W., and D. Kranz (1977). Uber die Bedeutung der Zellproliferation fur die Hirnregeneration bei niederen Vertebraten. Autoradiographische Untersuchungen. Verh. Anat. Ges. 71:439-445.

Rodieck, R. W. (1973). The Vertebrate Retina: Principles of Structure and Function. W. H. Freeman & Co., San Francisco.

Rusoff, A. C. (1979). Development of retinal ganglion cells in kittens (this volume).

Schaeffer, S. F., and E. Raviola (1975). Ultrastructural analysis of functional changes in the synaptic endings of turtle cone cells. In: Cold Spring Harbor Symp. on Quant. Biol., Vol. XL. The Synapse. Cold Spring Harbor Laboratory, New York, pp. 521-528.

Scholes, J. H. (1976). Neuronal connections and cellular arrangement in the fish retina. In: Neural Principles in Vision. F. Zettler and R. Weiler (eds.). Springer-Verlag, New York, pp. 63-93.

Scott, T. M., and G. Lazar (1976). An investigation into the hypothesis of shifting neuronal relationships during development. J. Anat. 121:485-496.

Segaar, J. (1965). Behavioral aspects of degeneration and regeneration in fish brain: A comparison with higher vertebrates. In: Progress in Brain Research, Vol. 14. Degeneration Patterns in the Nervous System. M. Singer and J. P. S. Schade (eds.). Elsevier/North-Holland, New York, pp. 143-231.

Sharma, S. C., and F. Ungar (1977). The histogenesis of the goldfish retina. Neurosci. Abst. 3:94.

Sidman, R. L. (1970). Autoradiographic methods and principles for study of the nervous system with thymidine—H^3. In: Contemporary Research Methods in Neuroanatomy. W. J. H. Nauta and S. O. E. Ebbesson (eds.). Springer-Verlag, New York, pp. 252-274.

Stell, W. K. (1967). The structure and relationships of horizontal cells and photoreceptor-bipolar synaptic complexes in goldfish retina. Amer. J. Anat. 121:401-424.

Stell, W. K. (1972). The morphological organization of the vertebrate retina. In: Handbook of Sensory Physiology, Vol. VII/2. Physiology of Photoreceptor Organs. M. G. F. Fuortes (ed.). Springer-Verlag, New York, pp. 111-213.

Stell, W. K., and D. O. Lightfoot (1975). Color-specific interconnections of cones and horizontal cells in the retina of the goldfish. J. comp. Neurol. 159:473-502.

Straznicky, K., and R. M. Gaze (1971). The growth of the retina in *Xenopus laevis*: an autoradiographic study. J. Embryol. Exp. Morph. 26:67-79.

Straznicky, K., and R. M. Gaze (1972). The development of the tectum in *Xenopus laevis*: an autoradiographic study. J. Embryol. Exp. Morph. 26:87-115.

Wagner, H.-J. (1975). Quantitative changes of synaptic ribbons in the cone pedicles of *Nannacara*: Light dependent or governed by a circadian rhythm? In: Vision in Fishes: New Approaches in Research. M. A. Ali (ed.). Plenum Press, New York, pp. 679-686.

Watson, W. E. (1974). Physiology of neuroglia. Physiol. Rev. 54:245-271.

Weiss, P. (1949). Differential growth. In: The Chemistry and Physiology of Growth. A. K. Parpart (ed.). Princeton University Press, New Jersey, pp. 135-186.

Wilson, M. A. (1971). Optic nerve fibre counts and retinal ganglion cell counts during development of *Xenopus laevis* (Daudin). Quart. J. Exp. Physiol. 56:83-91.

The Organization of the Optic Tectum in Larval, Transforming, and Adult Sea Lamprey, Petromyzon marinus

KALMAN RUBINSON

Department of Physiology and Biophysics
New York University Medical Center
New York, N. Y. USA

MICHAEL C. KENNEDY

Department of Biology
New York University
New York, N. Y. USA

Abstract The cytoarchitecture and neuronal morphology of the optic tectum were examined in larval, transforming, and adult sea lamprey, *Petromyzon marinus*. In small larvae, the tectum exhibits a simple two-layer pattern and neurons with rudimentary dendrites of radial or tangential orientation. During growth and transformation to the adult form, the tectum becomes a larger and more prominent brain structure. Tectal neurons, in the adult, are larger than those in the larvae and exhibit more complex dendritic arrays. Radially oriented primary dendrites exhibit higher-order branches with tangential orientation and many tangential dendrites have branches with radial orientation. The appearance of distinct fiber layers is characteristic of the maturing tectum which, in the adult, is differentiated into seven layers of cells and fibers. The major growth of the optic tectum during transformation is accompanied by the maturation of the neural retina and by changes in behavioral responses to visual stimuli.

Introduction

Lampreys, together with hagfishes, are the sole extant representatives of the most primitive vertebrates, the agnathans. This group is distinguished from the remaining vertebrates, the gnathostomates, by the absence of jaws and pelvic fins. Agnathans possess cartilaginous skeletons and pore-like gill openings.

359

Larval lampreys are filter-feeders and burrow in sand or mud at the bottom of freshwater streams. Their eyes are located beneath layers of skin and connective tissue. Larval lampreys do not respond to photic stimulation of the head, although they do respond briskly to illumination of the tail (Young, 1935).

At 4 or 5 years of age, the larvae (130-170 mm in length) begin to transform into adults. The flap-like oral hood changes into a complex oral disc with numerous rows of teeth. The eyes undergo a three- to four-fold increase in size (macroophthalmia) and appear as surface features. Transforming animals respond to photic stimulation of the head (Kennedy and Rubinson, unpublished observations).

Adult lampreys swim downstream to the sea where they live for one to two years, preying upon teleosts. In early spring the adults return to freshwater streams where they spawn and die. Adult lampreys appear to rely on vision, both for direction of predatory attack (Lennon, 1954) and for nest building during spawning (Applegate, 1950).

Thus, the life cycle of the sea lamprey, *Petromyzon marinus,* consists of three relatively long stages (larval, transforming, and adult) during all of which the animals exist as free-swimming individuals (Manion and Stauffer, 1970; Hardisty and Potter, 1971). These extended and discrete stages provide an excellent opportunity for the experimental neuroanatomical investigation of a developing brain.

The present study investigates the cytoarchitecture of the optic tectum and the morphology of the tectal neurons in an attempt to clarify the changes which occur in the optic tectum during development.

Materials and Methods

Larval and transforming lampreys (*Petromyzon marinus*) were obtained from commercial sources in Maine and from the U.S. Fish and Wildlife Service, Millersburg, Michigan. Animals were kept at 15°C in a 475-liter aerated, freshwater tank, equipped with undergravel, biological filtration. Spawning adults were captured in Maine, where they were anesthetized, dissected, and fixed.

Animals were deeply anesthetized by immersion in 0.1% Finquel (tricaine methanesulfonate, Ayerst Laboratories, Inc.). For cytoarchitectural studies, brains were fixed by immersion in 4% paraformaldehyde in 0.1 M sym-collidine buffer. After at least two days in the fixative, the brains (or heads with brains *in situ*) were dehydrated and embedded in a soft Araldite mixture (Kennedy and Rubinson, 1977). The hardened blocks were sectioned in the transverse plane at 10 μm using a steel knife. Sections were stained with Toluidine blue O. Such Nissl-stained series were prepared from larval (35, 70, 105 mm), transforming (stage V), and adult (425 mm) lampreys. Transverse sections (10 μm) from paraffin-embedded brains of larval and adult lampreys were stained, using the Holmes' reduced silver method, to localize fiber tracts.

For studies of neuronal morphology, brains were impregnated by the Golgi-Kopsch technique (Colonnier, 1964). After impregnation, the brains (or heads with brains *in situ*) were dehydrated and embedded in soft Araldite. The hardened blocks were sectioned in the transverse plane at 50 μm (larvae) or 80 μm (transformers and adults). Selected sections were counterstained with Toluidine blue O or p-phenylenediamine. The brains of nine larvae (51-124 mm), one transformer (stage II), and six adults (385-787 mm) were prepared in this fashion. For diagrammatic representation, selected Golgi-impregnated neurons were enlarged and traced directly, using a microscope enlarger (Kennedy and Rubinson, 1976). Axons were rarely impregnated in our material. The few axons observed were of extremely fine caliber and traceable for only a short distance from the cell body.

Results

Cytoarchitecture of the optic tectum The optic tectum changes quite dramatically during development. The midbrain, in the larvae, is roofed by a choroidal lamina which forms a bridge between the two vertically oriented halves of the optic tectum. In small larvae the tectum is poorly developed, both in thickness and in dorsal-to-ventral extent. The lateral tectal recess, which approximates the ventral margin of the optic tectum and divides it from the torus semicircularis, is only barely apparent (Fig. 1(A,B)). In larger larvae and early transformers, this recess is deeper because the dorsal margins of the tectal halves have bent medially (Fig. 1(C)). The tectum enlarges in the dorsal-to-ventral dimension and becomes thicker. In late transformers and adults, the tectum has grown to approximately four times larger than that of the larvae. This growth occurs in both dimensions (medial-to-lateral and dorsal margin-to-ventral margin) and

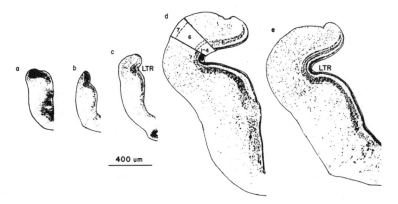

FIGURE 1 The optic tectum of the lamprey. (A) 35-mm larva. (B) 70-mm larva. (C) 105-mm larva. (D) Stage V transformer. (E) Spawning adult. These midbrain hemisections illustrate the development and growth of the optic tectum. Dorsal is at the top of the figure, medial is to the right. Tectal strata: (1) stratum ependymale, (2) stratum cellulare periventriculare, (3) stratum fibrosum periventriculare, (4) stratum cellulare et fibrosum internum, (5) stratum fibrosum centrale, (6) stratum cellulare externum, (7) stratum marginale. Toluidine blue O stain, 10 μm transverse section.

the two halves of the tectum continue to bend medially (Fig. 1(D,E)). A thin velum roofs the ventricular cavity, except at the caudal end of the tectum where the two halves are continuous with each other.

The tectum in the larva exhibits a dense cellular region adjacent to the ventricle and a superficial region of loosely scattered cells. In the smallest larva (35 mm), the deep region accounts for more than three-fourths of the tectal thickness (Fig. 1(A)). This dense cell layer reaches its greatest thickness (25-30 cells) adjacent to the lateral tectal recess and tapers as the dorsal margin is approached. As the tectum develops, the superficial region becomes more extensive, and cells appear to migrate from the periventricular region (Pfister, 1971). The superficial region can be divided into a thin, surface zone, the stratum marginale (SM), and a wider zone, the stratum cellulare externum (SCE). The SCE is a characterized by round and oval cells as well as a few triangular ones. The SM is more sparsely populated by round and oval cells, and contains the entering optic tract fibers which run parallel to the surface. As cells migrate laterally and the tectum enlarges in the dorsal-to-ventral dimension, the deep region is reduced to less than one-half of the thickness of the optic tectum (Fig. 1(B,C)). At the junction of the superficial deep cell regions, a fiber layer, the stratum fibrosum centrale (SFC), is apparent in the 70-mm larvae (Fig. 2(A)). After the appearance of the SFC, the development of an additional fiber layer, the stratum fibrosum periventriculare (SFP), divides the deep cell region into two (Figs. 1(C), 2(B)).

Transforming lampreys exhibit a well-developed tectal lamination pattern (Figs. 1(D), 2(C)), consisting of seven strata. A stratum ependymale (SE) is now distinguishable, lining the ventricular surface as a continuous sheet of columnar cells. These cells are stained much more lightly than the neurons of the other strata. Situated external to the ependymal layer, the stratum cellulare periventriculare (SCP) consists of from one to six densely packed layers of oval and pyriform cells. These neurons, as those of all other cell layers, are larger than their counterparts in the larval tectum. A thin fiber layer, the stratum fibrosum periventriculare (SFP), separates this stratum from the stratum cellulare et fibrosum internum (SCFI). The thick SCFI is composed of alternating cell and fiber layers, which we previously designated as separate layers (Kennedy and Rubinson, 1977). In late transformers, this stratum is characterized by three- to four-deep layers of round, oval, and pyriform cells, a very thin fiber layer and a superficial layer of 3-6 cells. The SCFI and the SCP become attenuated ventrally where the tectum joins the torus. All of these deep tectal strata are thickest adjacent to the lateral tectal recess and taper as they approach the dorsal margin of the tectum.

The stratum fibrosum centrale (SFC) borders the SCFI superficially, separating it from the wide stratum cellulare externum (SCE). The SCE consists of a scattered population of round, oval, pyriform, and triangular neurons which are somewhat more densely packed in the deepest third of the stratum. At the tectal surface, the stratum marginale (SM) consists of a very sparse population of round and pyriform cells. This stratum is difficult to discern in Nissl-stained

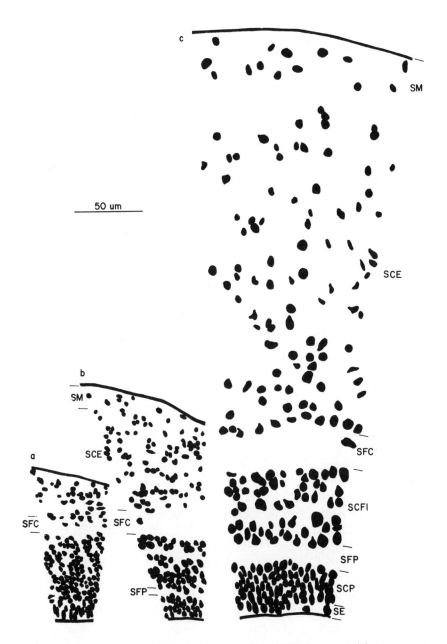

FIGURE 2 The cytoarchitecture of the optic tectum. (A) 70-mm larva. (B) 105-mm larva. (C) Stage V transformer. These segments were traced from Nissl-stained sections, and illustrate the growth in size of the tectal neurons and the differentiation of the tectal strata.

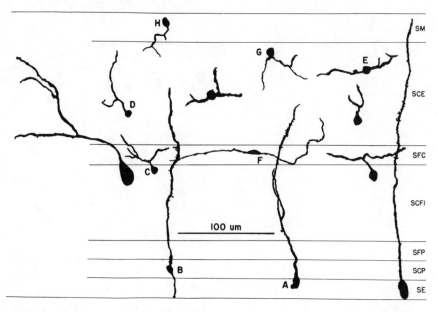

FIGURE 3 Tectal neurons in the larval sea lamprey. This composite diagram was prepared from tracings of Golgi-impregnated neurons. Although the lateral distribution of neurons in this figure is arbitrary, all neurons bear their original orientations and relationships to the tectal strata. Upper-case letters adjacent to perikarya indicate neurons referred to in text. The torus semicircularis is represented at the left.

preparations, being more obvious in fiber-stained material due to the dense aggregation of retinal efferent fibers adjacent to the tectal surface (Kennedy and Rubinson, 1977). The SCE and SM occupy slightly less than three-fourths of the tectal thickness.

The pattern observed in the late transformers is seen in the adult (Fig. 1(E)). Neurons in all strata are larger than in the transformers. The SCFI is now characterized by three cell layers (each 1-3 cells thick) alternating with two thin fiber layers. The deeper two of these cell layers merge approximately halfway to the dorsomedial margin. As in other developmental stages, the internal cell layers (SCP and SCFI) are thickest adjacent to the lateral tectal recess and taper towards the dorsomedial margin of the tectum. The SCE and SM now occupy slightly more than three-fourths of the tectal thickness.

Neuronal morphology The most characteristic feature of the tectum, seen in Golgi-impregnations, is the radial orientation of neuronal and glial processes. They constitute the great majority of the impregnated processes in larvae (Fig. 3), and a somewhat lesser majority in adults where an apparent increase in tangentially oriented processes is seen (Fig. 4). Many of the radially oriented processes in larvae are probably glial and they extend, with little branching, from the ependymal layer to the external surface. The perikarya from which

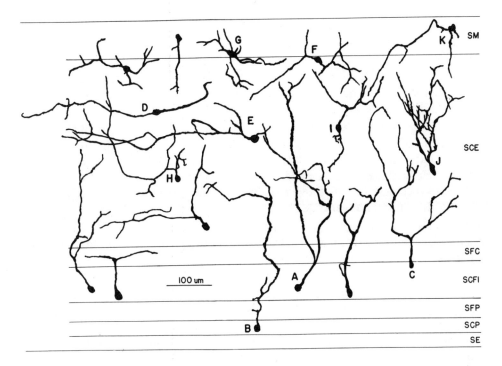

FIGURE 4 Tectal neurons in the adult sea lamprey. This composite was prepared as was Fig. 3.

these processes arise are found at varying distances from the ependymal surface (Fig. 3(B)). The radial glial processes may be utilized as guides for the lateral migration of neurons, as has been proposed, for the cerebellar cortex, by Rakic (1971).

In the deep periventricular region of the tectum of larvae, most cells exhibit a pyriform shape and a radial orientation of their processes. These processes extend into the superficial neuropil and, generally, are unbranched (Fig. 3(A)). Some cells are observed with one branch which is directed radially and a second directed ventrolaterally, toward the torus semicircularis (Fig. 3(C)). A few cells are seen with dendrites which branch extensively in the superficial neuropil while retaining their radial character.

The superficial region is characterized by pyriform and fusiform neurons with short, and often unbranched, dendrites. The pyriform cells usually have radially oriented dendrites (Fig. 3(D)), while the fusiform cell dendrites are tangentially oriented (Fig. 3(E,F)). Some of the most superficial pyriform cells give rise to dendrites which are directed away from the surface but do not enter the periventricular cell region (Fig. 3(G,H)).

Although the stage II transformer exhibits a tectal lamination pattern similar to that of the adult, its neuronal morphology seems to be more comparable to that of the larger larvae. Thus, the description of cell types given above for the larvae also applies to the stage II transformer.

In the adult, tectal dendrites exhibit more extensive branching than they do in the larvae. This change is most clearly seen in the superficial strata of the tectum, SCE and SM.

Stratum cellulare periventriculare (SCP) Neurons in this stratum have pyriform or oval perikarya and radially oriented dendrites (Fig. 4(B)). These dendrites extend their branches as far as the stratum cellulare externum.

Stratum cellulare et fibrosum internum (SCFI) The pyriform neurons of this layer have radial, often extensively branched dendrites extending into the stratum cellulare externum (Fig. 4(A,C)). Such neurons exhibit a single dendritic trunk which branches after leaving the SCFI. Some neurons, as in the larvae, have two processes, one directed radially and another directed ventrally to run in the stratum fibrosum centrale.

Stratum fibrosum centrale (SFC) Scattered pyriform and fusiform neurons are observed in the central fiber layer. These neurons are few in number and might be displaced from the cell layers on either side of the SFC, whose cells they resemble.

Stratum cellulare externum (SCE) The pyriform (Fig. 4(J)), fusiform (Fig. 4(F)), and multipolar neurons of this stratum exhibit the most complex dendritic patterns seen in the optic tectum of the lamprey. Significant radial and tangential dendritic distribution is often apparent in a single cell (Fig. 4(H)). The dendrites of fusiform neurons may extend for considerable distances in a tangential plane (Fig. 4(D)). Some fusiform neurons whose dendritic trunks are oriented tangentially exhibit higher-order dendritic branching with a radial orientation (Fig. 4(E)). Conversely, radially oriented fusiform neurons exhibit higher-order dendritic branching with a tangential orientation (Fig. 4(I)).

Stratum marginale (SM) This stratum is sparsely populated by pyriform and multipolar neurons. The branching dendrites of both cell types extend to ramify within the stratum cellulare externum (Fig. 4(G,K)).

Discussion

Tectal lamination patterns in the lamprey have been described by other investigators (Fig. 5). Tretjakoff (1909) described a medial, cell-dense region and a lateral, cell-poor region in the tectum of ammocoetes (larval lampreys) taken from the Neva River in Russia. Heier (1948) distinguished six layers in adult *Petromyzon marinus* and Leghissa (1962) identified eight layers in adult *Lampetra planeri*. Both authors divided the external cell layer (which we designate SCE) into a deeper cell layer and a more superficial cell-and-fiber layer. Such a subdivision of the SCE appears to be based on the fact that the cells in the deepest third of the stratum are more densely packed. Golgi-impregnations do not reveal any distinctions between the cells of the two regions of the SCE. Heier combined the deepest cell layer (SCP) with the ependymal layer (SE) and did not include it in his scheme. The combination of the SCP with the SE was adopted by Schober (1964) and Pfister (1971) in their studies of adult *Lampetra*

	Heier,'48	Leghissa,'62	Schober, '64 Pfister, '71
Str. marginale	Str. alb. et gris. sup.	P. degli elem. marg.	Str. superficiale
Str. cell. ext.	Str. opticum	P. fibroso esterne	Str. opticum
	Str. alb. et gris. cent.	P. fusif.-multip. est.	Str. gris. centrale
Str. fibr. centrale	Str. nerv. profund.	P. fibroso medio	
Str. cell. et fibr. int.	Str. gris. perivent.	P. pirif. e fusif. est.	Str. gris. perivent.
Str. fibr. perivent.	Str. alb. perivent.	P. fibroso int.	Str. alb. perivent.
Str. cell. perivent.	(ependyma)	P. piriforme int.	Str. ependymale
Stratum ependymale		Piano ependimale	

FIGURE 5 A comparison of the adult lamprey tectal lamination pattern described in this paper with those described by other authors.

planeri. These authors did include an ependymal layer as one of their six layers. They also divided the SCE into a deeper cell layer and a more superficial cell-and-fiber layer. Unlike Heier and Leghissa, however, Schober and Pfister did not identify a central fiber layer (SFC). They incorporated the SFC within the dense, external cell layer, which is surprising as the SFC is one of the most prominent and early appearing features of the tectum.

Tretjakoff (1909) identified two neuronal types, radial and tangential, in the optic tectum of larval lamprey, the tangential neurons being restricted to the superficial region. In his descriptive study of the adult *Petromyzon marinus* brain, Heier (1948) illustrated a few tectal neurons with radial and tangential orientations. Leghissa (1962) mentioned similar tectal neurons in adult *Lampetra planeri.* These cell types are similar to those described in the present study.

The transformation of the lamprey, *Petromyzon marinus,* is most emphatically expressed in the visual system. Studies of retinal anatomy and physiology (Dickson and Collard, 1977; Rubinson et al., 1977), visual pathways (Kennedy and Rubinson, 1977), and behavioral responses to light (Young, 1935) all suggest that the larval lamprey does not possess a functioning visual apparatus. The optic tectum, in larvae, appears to be in an arrested, immature state, growing with increasing age, but failing to further differentiate. This condition of arrested development is inferred from the observation that cells with either a restricted radial or a restricted tangential dendritic distribution are a constant feature of all larval tecta. Nonetheless, retino-tectal fibers are distributed throughout the superficial region (Kennedy and Rubinson, 1977).

The transformation from larval to adult form is accompanied by considerable growth of the eye, histological and physiological maturation of the retina, elaboration of the visual pathways and initiation of behavioral responses to visual stimuli. The tectal concomitant of this transformation is the appearance of a highly laminated structure composed of neurons with complex dendritic patterns. These dendrites afford single neurons access to input from wide expanses of the tectum, in both the radial and tangential axes. The radial distribution suggests access to inputs from different classes of ganglion cells and from different sensory modalities. The laminar separation of tectal afferents by sensory properties has not been examined in the lamprey, but has been described in other species (Maturana et al., 1960; Fite, 1969; Schroeder and Ebbesson, 1975). The tangential dendritic distribution suggests that wide receptive fields may be possessed by adult tectal neurons. We infer that these dendritic fields underlie the functional sophistication required by the adult lamprey for predation and nest-building (Applegate, 1950; Lennon, 1954).

Acknowledgements The authors gratefully acknowledge the assistance of Mr. Alexander de Nesnera in the preparation of the histological material, and of Ms. Joan D. Niemond in the preparation of this manuscript. Supported by NINCDS Grant NS 10906, NS 14156, NEI Grant EY02288, and Postdoctoral Fellowship 5 F22 EY02259.

References

Applegate, V. C. (1950). Natural history of the sea lamprey, *Petromyzon marinus,* in Michigan. Spec. scient. Rep. U.S. Fish Wildl. Serv. 55:1-237.

Colonnier, M. (1964). The tangential organization of the visual cortex. J. Anat. 98:327-343.

Dickson, D. H., and T. R. Collard (1977). The cyclostome retina: a fine structural study of photoreceptor and pigment epithelial cell development in the larval lamprey (*Petromyzon marinus*). Anat. Rec. 187:566-567 (abstract).

Fite, K. V. (1969). Single-unit analysis of binocular neurons in the frog optic tectum. Exp. Neurol. 24:475-486.

Hardisty, M. W., and I. C. Potter (1971). The Biology of Lampreys, Vols. 1 and 2. Academic Press, New York.

Heier, P. (1948). Fundamental principles in the structure of the brain. A study of the brain of *Petromyzon fluviatilus.* Acta Anat. 5(suppl. 8):7-213.

Kennedy, M. C., and K. Rubinson (1976). A microscope-enlarger for macrophotography of histological sections. Brain Res. Bull. 1:155-157.

Kennedy, M. C., and K. Rubinson (1977). Retinal projections in larval, transforming, and adult sea lamprey, *Petromyzon marinus.* J. Comp. Neurol. 171:465-480.

Leghissa, S. (1962). La struttura della corteccia mesencefalica del ciclostomi selaci ed urodeli. Atti R. Acad. Sci. Ist. Bologna, serie XI, 9:123-152.

Lennon, R. E. (1954). Feeding mechanism of the sea lamprey and its effect on host fishes. Fish. Bull. U.S. 56:247-293.

Manion, P. J., and T. M. Stauffer (1970). Metamorphosis of the landlocked sea lamprey, *Petromyzon marinus.* J. Fish. Res. Bd. Canada 27:1735-1746.

Maturana, H. R., J. Y. Lettvin, W. S. McCulloch, and W. H. Pitts (1960). Anatomy and physiology of vision in the frog *(Rana pipiens).* J. Gen. Physiol. 43:129-175.

Pfister, C. (1971). Die Matrixenwicklung in Mes- und Rhombencephalon von *Lampetra planeri* (Bloch) (Cyclostomata) in Verlaufe des Individualzyklus. J. Hirnforsch. 13:377-383.

Rakic, P. (1971). Neuron-glia relationship during granule cell migration in developing cerebellar cortex. A Golgi and electronmicroscopic study in *Macacus rhesus* J. Comp. Neurol. 141:283-312.

Rubinson, K., H. Ripps, P. Witkovsky, and M. C. Kennedy (1977). Retinal development in the lamprey, *Petromyzon marinus.* Neurosci. Abstr. 3:575 (abstract).

Schober, W. (1964). Vergleichend anatomische Untersuchungen am Gehirn der Larven und adulten Tiere von *Lampetra fluviatilis* (Linné, 1758) und *Lampetra planeri* (Bloch, 1784). J. Hirnforsch. 7:107-209.

Schroeder, D. M., and S. O. E. Ebbesson (1975). Cytoarchitecture of the optic tectum in the nurse shark. J. Comp. Neurol. 160:443-462.

Tretjakoff, D. (1909). Das Nervensystem von Ammocoetes. I. Gehirn Arch mikr. Anat. Ent. 74:636-779.

Young, J. Z. (1935). The photoreceptors of lampreys. 1. Light-sensitive fibers in the lateral line nerves. J. Exp. Biol. 12:229-238.

Connectivity of Retinal Projections in Uniocular Mice

PIERRE GODEMENT
Laboratoire de Neurophysiologie
Collège de France
Paris, France

Abstract The ipsilateral retinal projections to the dorsal lateral geniculate nucleus (dlGn) and to the superior colliculus have been traced in mice either enucleated unilaterally at birth, or bearing a congenital monophthalmy that occurs in the early embryonic life. In both cases, the uncrossed pathway spreads to deafferented regions in the dlGn and superior colliculus. In the dlGn, however, it does not reach the lateral and dorsal borders of the nucleus, and in the superior colliculus only the rostral half contains uncrossed projections. The projections to the superior colliculus are mostly found in the stratum opticum and in the upper stratum griseum superficiale, where they are frequently fragmented in patches of high density of projection.

After removal of one eye in newborn animals, the retinal axons from the remaining eye may extend to deafferented regions of the ipsilateral visual centres. The results vary greatly from one species to another (Guillery, 1972; Lund et al., 1973; Chow et al., 1973; Frost and Schneider, 1976). The variations encountered in the change in the projections after enucleation in various species and also within one species (Lund and Miller, 1975) can be related to the state of development of the retinal projections at the time of the enucleation in each species, but other factors may participate in the determination of whether abnormal growth may occur, and to what amount. This is, for example, shown by the case of the hamster (Frost and Schneider, 1976), the primary visual projections of which are at birth very immature (So et al., 1978), but where the abnormal growth after unilateral neonatal enucleation is quite limited in the superior colliculus and absent in the dorsal lateral geniculate nucleus. It would therefore seem indispensable to study in one species the degree of abnormal growth that may occur in the dorsal lateral geniculate nucleus (dlGn) and in the superior colliculus (SC) when one eye is lost at the earliest possible stage when no visual pathways have yet developed. In the following study, we have investigated this problem by using pigmented mice of the C57BL/6J strain, either *enucleated unilaterally at birth,* or bearing a *congenital monophthalmy* that is present in the embryo before the period of optic nerve growth. Ocular anomalies that are present at birth in this strain have been noted by various authors (Chase and Chase, 1941; Chase, 1942; Douglas and Russell, 1947), and there are many indications that they are analogous to those produced by a mutation effect in the "eyeless" mice, where a defect arises in the development of the eye at embryonic day 10 (Chase and Chase, 1941; Silver and Hughes, 1974; Harch et al., 1978) and produces a bilateral anophthalmy.

Materials and Methods

The mice that we used were all more than 8 weeks old at the time of sacrifice. They were obtained from an inbred stock of pigmented C57BL/6J mice. The projection fields of the uncrossed visual pathways in the dlGn and superior colliculus were traced using either the Fink-Heimer (Fink and Heimer, 1967) procedure or the orthograde transport of tritiated proline injected in the eye. A total of 18 animals were used: seven were normal, seven had been unilaterally enucleated at birth, and four had one eye missing at birth. External examination showed that the eyes of the mice that we used all had the usual degree of very dark pigmentation which is characteristic of the C57 black mice.

Fink-Heimer studies After enucleation as adults, a survival time of four days was allowed, following which the mice were perfused with 10% formalin. Procedure I of Fink and Heimer (1967) was thereafter done on frontal frozen sections.

Autoradiographic studies Nine adult mice, either normal, enucleated unilaterally at birth, or congenitally monophthalmic, were injected intraocularly with tritiated proline. For injection, the mice were maintained under Nembutal (50-60 mg/kg) anesthesia in a Horsley-Clarke apparatus and 7-8 μl of ^{3}H—proline (CEA; 2 μCi/μl; 27 Ci/mM) was injected slowly behind the vitreous humour using a 10 μl syringe with its tip inserted through a small hole at the border of the sclera. A survival time of 24 hours was allowed, followed by perfusion with 10% formalin. The brains were embedded in paraffin and serial 10 μm frontal sections were cut; an I:3 series was coated with Ilford K5 (I:2) emulsion, and exposed in the dark a 4°C in a dessicated atmosphere for 40 days. After development in Kodak D 19 developer, the slides were counterstained through the emulsion with cresyl violet or thionin.

Results

Dorsal lateral geniculate nucleus In all monocular mice, the volume of the dlGn ipsilateral to the remaining eye is decreased by about 50%; this reduction in size seems to be more important in the mice unilaterally enucleated at birth than in the congenitally monocular mice. As in other rodent species, there are no obvious laminae in the dlGn of mice. *In normal mice,* the zone of projection of the uncrossed fibres from the ipsilateral eye is limited to a patch in the medial half of the nucleus (Figs. 1, 2). In anterior sections of the nucleus, it is situated at the medio-dorsal corner of the dlGn, and in more posterior sections it is progressively displaced towards the ventro-medial corner of the nucleus; in the 200 caudal-most microns it disappears. The density of the projection is high and uniform, and its borders are sharp. Towards the end of the anterior third of the nucleus, it is separated in three distinct spots, as already noted by Hayhow et al. (1962) in the rat. The pattern of uncrossed termination in *neonatally enucleated mice* is shown in Fig. 1. In these mice, the volume of the dlGn that receives uncrossed terminals is increased compared to normal mice. The regions where the density of the projection is high are the same as in normal mice (i.e., along the medial border of the nucleus). However, surrounding this dense projection, there is a zone of abnormal projection that covers a large part of the nucleus. The density of this projection is lower, and decreases radially around the zone of high density of projection (Fig. 3). At all the antero-

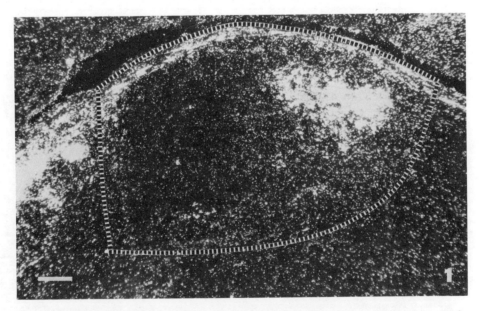

FIGURE 1 Ipsilateral innervation in the dlGn of a normal mouse. Darkfield micrograph of a counterstained proline autoradiogram. Calibration bar, 100 μm.

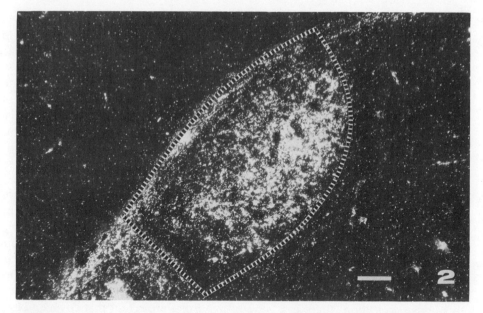

FIGURE 2 Ipsilateral innervation in the dlGn of a mouse unilaterally enucleated at birth. Darkfield micrograph of a counterstained proline autoradiogram. Calibration bar, 100 μm.

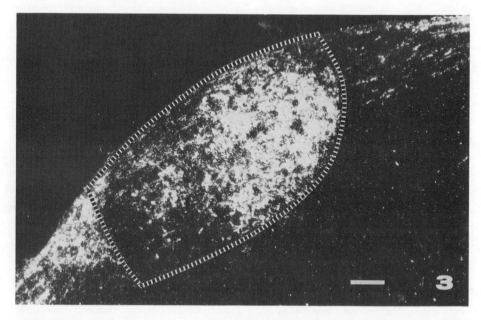

FIGURE 3 Ipsilateral innervation in the dlGn of a congenitally monophthalmic mouse. Darkfield micrograph of a counterstained proline autoradiogram. Calibration bar, 100 μm.

posterior levels of the nucleus studied, the projection fails to reach the lateral and dorsal borders of the nucleus, leaving large areas of the dlGn devoid of any retinal innervation; in these areas, the density of neurons is decreased. The same global pattern of ipsilateral innervation is observed in *congenitally mon-ophthalmic mice;* in these mice, however, the density of abnormal projection seems to be higher than in the mice enucleated unilaterally at birth. As in neonatally enucleated mice, large areas of the dlGn do not contain any retinal innervation (Figs. 1, 4).

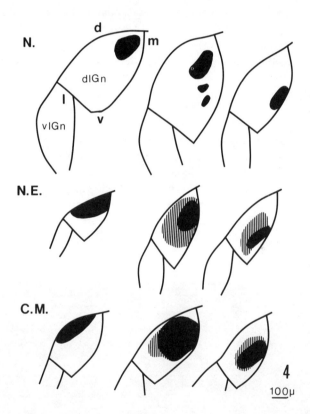

FIGURE 4 Drawings representing the ipsilateral retinal innervation in the dlGn of normal (N), neonatally enucleated (N.E.), and congenitally monophthalmic (C.M.) mice. The left column is at a rostral level of the dlGn, the middle at an intermediate level, and the right at a posterior level. Black areas indicate the zones of high density of projection, and cross-hatched areas the zones of medium to low density of projection.

Superior colliculus In monocular mice, there is a reduction in the size of the superior colliculus ipsilateral to the remaining eye. Its rostrocaudal extension is reduced from 1300 μm (as seen in paraffin-embedded normal material) to about 1000 μm, and its mediolateral width is also reduced by about 25%. The thickness of the optic layers is also decreased on the ipsilateral side.

In *normal* mice, the uncrossed projection is restricted to small patches best seen in proline preparations in the stratum opticum of the ipsilateral superior colliculus. In the anterior half, the patches are scattered along the medio-lateral extent of the stratum opticum and they are separated by areas containing no uncrossed innervation (Fig. 5). In the caudal half there is no uncrossed projection except for a medial patch in the stratum opticum that disappears before the caudal end. In *monocular* mice (i.e., unilaterally enucleated at birth, and congenitally monophthalmic) there is an augmentation of the uncrossed innervation in the ipsilateral superior colliculus. In the anterior half, the whole stratum opticum contains a higher density of terminals than in normal mice, especially near the medial border. This projection shows a tendency to be organized, as in normal mice, in patches of higher density of terminals. In addition, the uppermost part of the stratum griseum superficiale receives uncrossed inputs; the most superficial projection is often fragmented as in the stratum opticum in small patches separated by gaps containing no retinal innervation (Fig. 6). In some places, there are large areas of the stratum opticum and of the stratum griseum superficiale that contain no retinal terminals, showing that the projection is quite irregular. In the caudal half, the increase in the projection is very small compared to the projection in normal mice (Fig. 7). In one neonatally enucleated mouse, there was at some point in the anterior half of the superior colliculus an intertectal crossing of retinal fibres from the contrala-

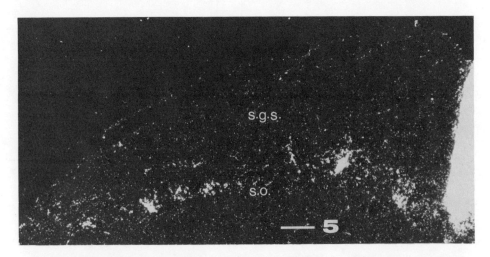

FIGURE 5 Ipsilateral innervation in the superior colliculus of a normal mouse (left side of the picture). Darkfield micrograph of a counterstained proline autoradiogram. Calibration bar, 100 μm.

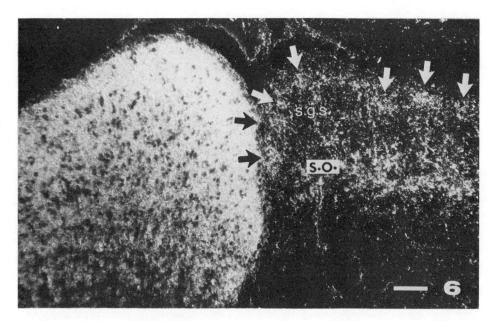

FIGURE 6 Ipsilateral innervation at an anterior level in the superior colliculus of a mouse unilaterally enucleated at birth. The contralateral superior colliculus can be seen on the left of the picture. The arrows point to patches of high density of projection in the upper stratum griseum superficiale in the ipsilateral superior colliculus. Darkfield micrograph of a counterstained proline autoradiogram. Calibration bar, 100 μm.

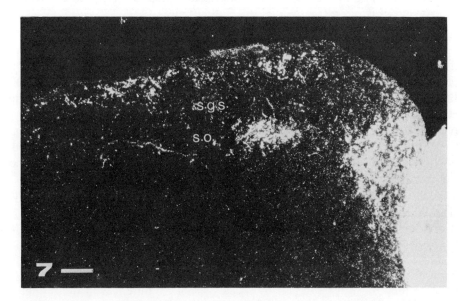

FIGURE 7 Ipsilateral innervation at an intermediate antero-posterior level in the superior colliculus of a congenitally monophthalmic mouse. Note on the lateral border (left side) the presence of a projection in the upper s.g.s., but not in the s.o. Darkfield micrograph of a counterstained proline autoradiogram. Calibration bar, 100 μm.

teral superior colliculus to the medial edge of the ipsilateral superior colliculus; the fibres that participated in this intertectal crossing appeared to do so by passing through a perforation in the pial membrane that separates both superior colliculi.

Discussion

The main findings of this study are that:

1. Prenatal or postnatal loss of one eye induces the formation of an anomalous retinal projection from the remaining eye to deafferented parts of the ipsilateral dlGn and superior colliculus.

2. There are very few differences between the pattern of uncrossed innervation in mice enucleated unilaterally at birth and in congenitally monophthalmic mice.

3. The anomalous uncrossed projection fails to fill completely both structures in both cases.

4. The density of the anomalous uncrossed projection to the dlGn is high in the zones corresponding to the emplacement of the normal ipsilateral pathway. In the deafferented regions, the density of the projection is lower, and it becomes null along the dorsal and lateral borders of the nucleus.

5. In the superior colliculus, the projection terminates mainly in the stratum opticum and in the upper stratum griseum superficiale in the anterior half, where it is often fragmented in small clumps.

This study offers no information concerning the origin of the expanded projection in the mouse. Other studies (Godement et al., in press, and unpublished observations) show that a larger than normal part of the retina participates in this expanded uncrossed projection, as in rats (Lund and Lund, 1976). Branching of the normal uncrossed axons into denervated areas of the dlGn and superior colliculus may also occur, as in the cat (Robson et al., 1978). The results obtained in the dlGn of congenitally monophthalmic mice show that a strong limit is imposed upon the size of the territory that may be innervated by uncrossed axons in the absence of binocular innervation during development. The number of retinal ganglion cells that may participate in the uncrossed pathway could be limited by some intrinsic factors; such could be the case if only a special class of retinal ganglion cells can send their axon to the ipsilateral side of the diencephalon. It is tempting to speculate that these cells are large retinal ganglion cells that effect a bilateral branching of their axon along the optic tract (Lund, 1975; Cunningham and Freeman, 1977): in this way, species differences in the amount of uncrossed projection after very early eye loss could be accounted for at least in part by species-specific differences in the retina.

The observation of a higher global density of uncrossed projection to the dlGn in congenitally monophthalmic mice, compared to neonatally enucleated mice, is difficult to appreciate, and further studies are needed in order to determine whether it is a real trend; the same observation, however, has been made in the projections of the normal eye in strongly congenitally microphthalmic mice of the same species (Godement et al., in press). The localization of the expanded uncrossed projection close to the area of normal uncrossed projection in the dlGn suggests that during development the uncrossed pathway terminates only in a definite part of the dlGn, independently from whether one or both eyes are present.

References

Chase, H. B. (1942). Studies on an anophthalmic strain of mice. III. Results of crosses with other strains. Genetics 27:339-348.

Chase, H. B., and E. B. Chase (1941). Studies on anophthalmic strain of mice. I. Embryology of the eye region. J. Morphol. 68:279-301.

Chow, K. L., L. M. Matthews, and P. D. Spear (1973). Spreading of uncrossed retinal projections in superior colliculus of neonatally enucleated rabbits. J. Comp. Neur. 151:307-322.

Cunningham, T. J., and J. A. Freeman (1977). Bilateral ganglion cell branches in the normal rat: a demonstration with electrophysiological and cobalt tracing methods. J. Comp. Neur. 172:165-175.

Douglas, P., and W. L. Russell (1947). A histological study of eye abnormalities in the C57 black strain of mice. Anat. Record. 91:414.

Fink, R. P., and L. Heimer (1967). Two methods for selective silver impregnation of degenerating axons and their synaptic endings in the central nervous system. Brain Res. 4:369-374.

Frost, D. O., and G. E. Schneider (1976). Normal and abnormal uncrossed retinal projections in Syrian hamsters as demonstrated by Fink-Heimer and autoradiographic techniques. Communication presented at the 6th Annual Meeting of the Society of Neuroscience, Toronto, November 1976.

Godement, P., P. Saillour, and M. Imbert. The ipsilateral optic pathway to the dorsal lateral geniculate nucleus and superior colliculus in mice with pre- or postnatal loss of one eye. J. Comp. Neur., in press.

Guillery, R. W. (1972). Experiments to determine whether retinogeniculate axons can form translaminar collateral sprouts in the dorsal lateral geniculate nucleus of the cat. J. Comp. Neur. 146:407-420.

Harch, C., H. B. Chase, and N. I. Gonsalves (1978). Studies on an anophthalmic strain of mice. VI. Lens and cup interaction. Dev. Biol. 63:352-357.

Hayhow, W. R., A. Sefton, and C. Webb (1962). Primary optic centers of the rat in relation to the terminal distribution of the crossed and uncrossed optic nerve fibers. J. Comp. Neur. 118:295-321.

Lund, R. D. (1975). Variations in the laterality of the central projections of retinal ganglion cells. Exp. Eye Res. 21:193-203.

Lund, R. D., T. J. Cunningham, and J. S. Lund (1973). Modified optic projections after unilateral eye removal in young rats. Brain Behav. Evol. 8:51-72.

Lund, R. D., and B. F. Miller (1975). Secondary effects of fetal eye damage in rats on intact central optic projections. Brain Res. 92:279-289.

Lund, R. D., and J. S. Lund (1976). Plasticity in the developing visual system: the effects of retinal lesions made in young rats. J. Comp. Neur. 169:133-154.

Robson, J. A., C. A. Mason, and R. W. Guillery (1978). Terminal arbors of axons that have formed abnormal connections. Science 201:635-637.

Silver, J., and A. F. W. Hugues (1974). The relationship between morphogenetic cell death and the development of congenital anophthalmia. J. Comp. Neur. 157:281-302.

So, K. F., G. E. Schneider, and D. O. Frost (1978). Postnatal development of retinal projections to the lateral geniculate body in Syrian hamsters. Brain Res. 142:343-352.

Peripheral Influences on Connectivity in the Developing Rat Trigeminal System

HERBERT P. KILLACKEY

Department of Psychobiology
University of California, Irvine
Irvine, California USA

The trigeminal system of the rat shows a high degree of discrete anatomical organization. I would like to take this opportunity to review the evidence for peripheral influences on the development of this discrete anatomical organization in both the normal animal and in ones which have suffered neonatal receptor damage. In fact, among other things, I would like to demonstrate that the loss of several whiskers can have as severe an effect on the structural organization of the brain as the loss of an eye. In this brief review I will not attempt to be all-inclusive. Rather, I will focus on the contribution that I and my several collaborators have made. The impetus for our studies of the trigeminal system was the report of Woolsey and Van Der Loos (1970) that in the portion of mouse somatosensory neocortex devoted to the vibrissae representation the cells of layer IV are arranged into discrete aggregates which they termed "barrels." Further, the number and spatial distribution of the "barrels," as well as physiological evidence (Woolsey, 1967; Welker, 1971), suggested that in both the rat and mouse there was a unique one-to-one relationship between a particular cytoarchitectonic unit, or "barrel," and a mystacial vibrissae. Given this high degree of functional and structural organization in the neocortical layer where thalamocortical projections terminate, a natural question to ask was whether or not this discrete organization was also reflected in the afferent projections to this cortical layer. Utilizing anterograde degeneration techniques, I was able to determine that in the rat the terminations of the thalamocortical projections to the vibrissae representation portion of somatosensory cortex are indeed arranged into discrete clusters (Killackey, 1973). Further, the spatial distribution of these clusters replicated the distribution of sinus hair and vibrissae follicles on the muzzle of the rat (Killackey and Leshin, 1975). The

381

discrete nature of these thalamocortical projections is relatively unique and contrasts with the continuous distribution of thalamocortical projections to the face region of cortex in such generalized mammalian species as the opossum and hedgehog (Killackey, 1973). Furthermore, thalamocortical projections are continuous in a rodent species which has developed visual specializations and inhabits an arboreal niche, the Eastern grey squirrel (Killackey, 1973). However, other terrestrial rodents in which the vibrissae are well developed, such as the gerbil, also exhibit discontinuous thalamocortical projections. The available evidence would suggest that this is a specialization of the somatosensory system which has chiefly evolved in small rodents (Woolsey, Welker and Schwartz, 1975). Perhaps the somatosensory system is able to provide the most reliable information about the external world to small rodents which are normally very close to the earth's surface, spend time burrowing beneath the ground, and generally inhabit a niche where both visual and auditory information is limited. In this regard it should be noted that rats and other rodents have evolved stereotypic patterns of whisker movements, termed "whisking," which seem to play an extremely important role in their environmental exploration (Vincent, 1912; Welker, 1964).

As an historical aside, it should be pointed out that the discrete nature of the thalamocortical projections in the mouse and the cellular aggregates in layer IV were previously noted by Lorente de N'o in his Golgi preparations of mouse parietal cortex in the early part of this century. In fact, what Woolsey and Van Der Loos termed a "barrel," Lorente de N'o had previously termed a "glomerulos." It was this work of Lorente de N'o which formed the basis of his ideas on specific and unspecific thalamocortical projections which was presented so lucidly in Fulton's "Physiology of the Nervous System," and has so widely influenced thought on cortical structure. In hindsight, Lorente de N'o's choice of the mouse parietal cortex as an area to study was a most serendipitous one.

The thalamocortical projections are the third link in the trigeminal pathway between the periphery and the primary somatosensory cortex (see Fig. 1). If these fibers exhibit a discrete organization replicating the distribution of mystacial vibrissae, the next logical question to ask is if the trigeminothalamic and primary trigeminal afferents exhibit similar organization. The results of several previous anatomical investigations of these pathways provided no evidence for a morphologically detectable discrete organization in the adult rat (Torvik, 1956; Smith, 1973), although physiological investigations have suggested a high degree of topographical organization in the cell groups in which these fiber systems terminate in the rat (Nord, 1967; Waite, 1973; Shipley, 1974). We decided to investigate the pattern of trigeminal afferent terminations in the thalamus and lower brainstem in the adult and neonatal rat utilizing an assay for the localization of the enzyme succinic dehydrogenase. We had previously found this assay to be particularly sensitive in delimiting the afferent organization in the thalamocortical afferents (Killackey, Belford, Ryugo, and Ryugo, 1976). Like previous investigators we found no obvious signs of a detailed anatomical organization in the adult rat. However, in both the principal trigeminal

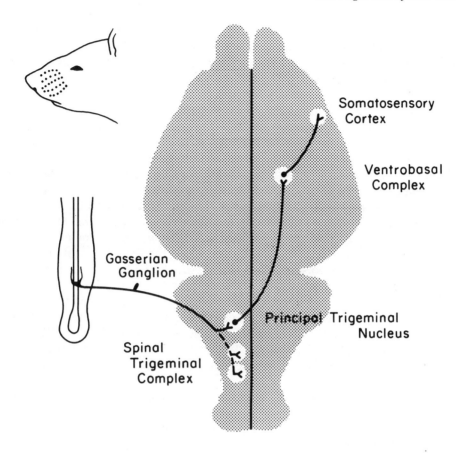

FIGURE 1 Diagrammatic illustration of the trigeminal lemniscus of the rat.

nucleus (as well as other brainstem trigeminal nuclei) and the ventrobasal complex of the neonatal rat, there are obvious and exquisitely detailed patterns of afferent segmentation which replicate the topography of the receptor surface in a manner similar to the pattern of thalamocortical afferents (Killackey and Belford, 1976; Belford and Killackey, 1979). However, unlike the pattern of thalamocortical terminations in cortex, the patterns in the subcortical trigeminal centers are ephemeral. They are most obvious at approximately the end of the first week of life, but are essentially undetectable by the end of the third postnatal week. Other structural aspects of the ventrobasal complex undergo similar changes. At the same time that the discreteness of the trigeminothalamic afferents is most obvious, the corticothalamic fibers to the ventrobasal complex are distributed in a lattice-like fashion which surrounds individual clusters of trigeminothalamic afferents (Akers and Killackey, 1978, 1979). Over the next ten days or so, this distribution changes to one in which there is a relatively uniform distribution of corticothalamic afferents in the arcuate subdivision of the ventrobasal complex. The distribution of the perikarya of thalamocortical relay cells (as defined by horseradish peroxidase labeling) undergoes similar shifts (Ivy and Killackey, 1978). Five days after birth the thalamocortical relay cells are arranged into groups which mimic the distribution of trigeminothalamic afferent terminations. Once again, this is a type or organization which is not visible in the adult animal. Further, none of these morphological peculiarities are evident in Nissl stained material from either the neonatal or adult rat. This is particularly interesting in light of the finding that the neurons of the ventrobasal complex of the adult mouse are arranged into "barreloids" which replicate the vibrissae follicle topography (Van Der Loos, 1976). However, it should be noted that this is not the only species difference in the trigeminal system between the mouse and the rat. For example, in rat somatosensory cortex the "barrels" associated with the mystacial vibrissae are filled with neurons and generally less clearly delineated than the same "barrels" in the mouse (Welker and Woolsey, 1974). Finally, it should be emphasized that this apparent loss in anatomical organization is more apparent than real. We have found an extremely close correspondence between our anatomical map of the forelimb in the ventrobasal complex of the neonatal rat and the physiological map found in the adult rat (Angel and Clarke, 1975; Belford and Killackey, 1978b). However, the biological significance of this anatomically discrete organization which is present in the neonatal, but not the adult, animal is unclear.

Thus, we have found that the afferent terminals associated with the vibrissae representation are arranged into discrete clusters in the principal trigeminal nucleus, the ventrobasal complex, and layer IV of primary somatosensory cortex. These findings alone would suggest that the periphery may play a potent role in sculpting central trigeminal structures. Indeed, this suggestion is strengthened by our finding that these central correlates of peripheral organization develop in a sequential fashion from periphery to cortex during the first five days following birth. Segmentation in the principal sensory nucleus develops between birth and postnatal day 3, in the thalamus between postnatal days 1 and 3, and finally, in cortex between postnatal days 3 and 5 (Belford, 1978; Belford and Killackey, 1978a; Killackey and Belford, 1979). This ordered sequence of appearance suggests that an initial pattern is first established in the principal trigeminal nucleus by the central processes of the gasserian ganglion cells whose peripheral processes are in direct contact with the vibrissae. This initial pattern may serve as a template for later developing patterns in the more rostrally located trigeminal structures.

The influence of the periphery on central trigeminal structure has been even more dramatically illustrated by experiments which have assayed the effect of peripheral receptor damage on central structure. Soon after describing the "barrels" and hypothesizing a unique relationship between the mystacial vibrissae and certain "barrels" of somatosensory cortex, Van Der Loos and Woolsey (1973) demonstrated that neonatal vibrissae damage in the mouse resulted in a failure of the associated cortical "barrels" to form. It was later demonstrated that this effect is restricted to a sensitive period which ends approximately four days after birth (Weller and Johnson, 1975). We have been able to detect similar changes in cortical cellular organization in the rat following neonatal vibrissae damage. However, just as normal organization is less clear in the rat than in the mouse, so is the anomalous organization of cell bodies consequent to such damage. It should be emphasized that this effect is a change in the normal cytoarchitectonics of the cortex, and the relationship between these cytoarchitectonic changes and functional ones is murky, a point I will return to shortly. It has been demonstrated that thalamocortical axons terminate at least partially on the perikarya of layer IV stellate cells (White, 1978), and given that the thalamocortical projections show the same discrete organization as these cells, an obvious question of interest is the effect of vibrissae damage on the organization of thalamocortical projections. We have found that neonatal vibrissae damage results in an anomalous organization of thalamocortical projections (see Fig. 2). If a row of vibrissae is removed at birth, the associated thalamocortical projections are a fused thin band rather than a row of punctate clusters (Killackey, Belford, Ryugo and Ryugo, 1976). In fact, this is the same effect that removal of a row of vibrissae has on the distribution of the cells of layer IV. Thus, cell distribution in layer IV of both the normal and neonatally injured animal mimics the distribution of thalamocortical afferents which suggests that these afferents may play a causal role in determining the cytoarchitectonics of layer IV. We have also determined that the anomalous organization develops with the same time course as the normal discrete organization (Killackey and Belford, 1979), suggesting that in this system at least, damage is affecting the process of initial development rather than resulting in a later reorganization. I would stress that this effect is taking place extremely rapidly. These changes are easily detectable only five days after the peripheral damage, and they are permanent ones. Further, we have determined that for the thalamocortical projections at least, the event which signals the end of the sensitive period is the formation of an initial pattern. Somehow the formation of such a pattern seems to "fix" the system and bring an end to the period during which the system responds plastically to peripheral signals. We have also detected more subtle changes in somatosensory cortex following vibrissae damage. Following damage to all the vibrissae follicles at birth there is a diminution in spine density along the portion of the apical dendrite of deep layer V pyramidal cells which traverses layer IV (Ryugo, Ryugo, and Killackey, 1975). Whether this effect on fine cortical morphology is also a change in the process of initial development or a later reorganization is at present unclear. However, data obtained by Valverde (1971) on spine densities in dark-reared mice suggest that in this instance the latter may be the case. This would argue that neonatal receptor damage may produce at least two different classes of anatomical effects: (1) the initial organization of a system may be affected, and this may happen very rapidly; and (2) there may be longer-term changes associated with the "disuse" of a damaged system. While the two types of effect can be easily separated on the intellectual level, it is often difficult to do so in any given experiment, particularly as the effects of neonatal receptor damage are often only assessed after animals reach adulthood.

The physiological correlates of these anatomical changes in cortex have been the subject of two recent studies. Both my laboratory (Killackey, Ivy and Cunningham, 1978) and Waite and Taylor (1978) have reported functional changes in the topographic map of primary somatosensory cortex following neonatal vibrissae damage. Roughly speaking, these changes can be characterized as a loss of the strict specificity seen in the normal animal. Both studies reported that units with anomalous receptive fields could be found in the regions associated with the damaged vibrissae, although both the extent of vibrissae damage and consequent anomalous organization reported by Waite and Taylor (1978) was greater than what we found in our own study. However, the nature of the relationship between these functional changes and the previously reported anatomical changes is puzzling. The anomalous organization in thalamocortical projections which follow vibrissae removal as determined anatomically seems to be largely restricted to the projections associated with the damaged vibrissae. There is no anatomical evidence for changes in the long ascending fiber systems which could mediate the reported functional changes. We have found anatomical changes in the subcortical trigeminal centers which are essentially the same as those in cortex (see Fig. 2 and Belford and Killackey, 1979). Even removal of all vibrissae does not result in a complete loss of anatomical specificity in the terminations of thalamocortical and trigeminal thalamic afferents (Killackey and Belford, 1979; Belford and Killackey, 1979). Following such damage there is still anatomical evidence for inter-row specificity, although intra-row boundaries are lost. Further, we have found no anatomical evidence for a major reduction in the amount of neuronal tissue devoted to the representations of the vibrissal pad after removal of all the vibrissae. In total, this evidence would argue against an explanation of the physiological phenomena based on "sprouting" of intact portions of the trigeminal lemniscal system into the affected regions. An alternative explanation is that the anomalous functional organization is mediated by altered local circuits and is not related to the changes in the lemniscal pathway which are detected with our anatomical methods. There is some recent evidence for this point of view. Hand et al. (1978) have reported that in the normal animal stimulation of a single vibrissae results in a "candlepin" shaped focus of activity in cortex as mapped by the metabolic indicator 2-deoxyglucose. The activity is densest in layer IV and tapers off in width and density through both the superficial and deep cortical layers—and thus is a "candlepin" shaped column of activity extending through all cortical layers. This pattern of activity could be altered by neonatally removing vibrissae which flank a given normal vibrissae. In this case, stimulation of the normal vibrissae in the adult resulted in an "hour glass" column of activity. There was still a focus of activity in layer IV, but now there was also a spreading out of activity in the other cortical layers, hence, the "hour glass" shape. This finding is suggestive of changes in local intracortical circuitry and consistent with the functional changes which have been found with more traditional physiological techniques.

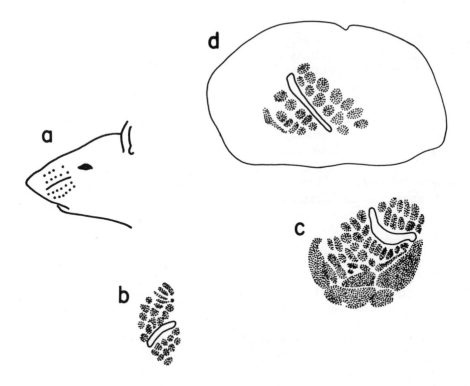

FIGURE 2 Summary drawings of the anomalous changes in the afferent terminations of the trigeminal lemniscal system consequent to vibrissae removal. Removal of the middle row of vibrissae (A) results in a fusion of the associated afferent terminations in (B) the principal sensory nucleus, (C) the ventrobasal complex, and (D) layer IV of somatosensory cortex.

In summary, I have briefly reviewed evidence which suggests that the periphery plays an important role in the formation of connectivity in the rat trigeminal system. I have suggested that these effects may take place with two different time courses. First, the periphery may influence the initial developmnent of this system, and this effect may be exerted very rapidly. Second, in addition to these initial effects it is likely that following peripheral damage there are secondary effects consequent to "disuse." Such effects may develop over a longer time course. In a similar fashion, the type of connections which are affected by peripheral damage can be divided into two subclasses. First, there is a definite and substantial effect on the anatomical organization of the lemniscal pathways. Second, there is evidence, albeit less conclusive, that local connections (intracortical and, perhaps, intranuclear) are also affected by peripheral damage. Lemniscal pathways are perhaps plastic in response to peripheral manipulation for a short period of time, while the intracortical connections remain plastic over a much longer period. In closing, I would like to put forth the speculation that perhaps the effect of neonatal receptor damage is a two-fold one. First, just as the periphery guides the development of central connectivity in the normal animal, it does likewise in the injured animal, but an altered periphery results in an altered central structure. This direct effect is perhaps best seen in the lemniscal system. A second effect of peripheral damage is altered patterns of activity consequent to "disuse." Such effects are more indirect and may only be exerted after altered lemniscal systems have begun to function. This more indirect effect may result in altered patterns of local circuit connectivity.

Acknowledgements This research was supported by a grant from the National Science Foundation.

References

Akers, R. M., and H. P. Killackey (1978). Development of segmented corticothalamic projections to the ventrobasal complex of the rat. Soc. for Neurosci. Abstracts 4:547.

Akers, R. M., and H. P. Killackey (1979). Segregation of cortical and trigeminal afferents to the ventrobasal complex of the neonatal rat. Brain Res. 161:527-532.

Angel, A., and K. A. Clarke (1975). An analysis of the representation of the forelimb in the ventrobasal thalamic complex of the albino rat. J. Physiol. (Lond.) 249:399-423.

Belford, G. R. (1978). Development of peripherally related segmentation in the ventrobasal complex of the rat. Anat. Rec. 190:336.

Belford, G. R., and H. P. Killackey (1978a). Normal and abnormal segmentation in the trigeminal complex of the young rat. Soc. for Neurosci. Abstracts 4:547.

Belford, G. R., and H. P. Killackey (1978b). Anatomical correlates of the forelimb in the ventrobasal complex and the cuneate nucleus of the neonatal rat. Brain Res. 158:450-455.

Belford, G. R., and H. P. Killackey (1979). Vibrissae representation in subcortical trigeminal centers of the neonatal rat. J. Comp. Neur. 185:305-322.

Hand, P. J., J. H. Greenberg, R. R. Miselis, W. L. Weller, and M. Reivich (1978). A normal and altered cortical column: A quantitative and qualititative (^{14}C)–2 deooxyglucose (2DG) mapping study. Soc. for Neurosci. Abstracts 4:553.

Ivy, G., and H. P. Killackey (1978). Developmental changes in the distribution of the thalamocortical relay cells of the ventrobasal complex of the rat. Soc. for Neurosci. Abstracts 4:554.

Killackey, H. P. (1973). Anatomical evidence for cortical subdivisions based on vertically discrete thalamic projections from the ventral posterior nucleus to cortical barrels in the rat. Brain Res. 51:326-331.

Killackey, H. P., and G. Belford (1976). Discrete afferent terminations in the trigeminal pathway of the neonatal rat. Anat. Rec. 184:446.

Killackey, H. P., and G. R. Belford (1979). The formation of afferent patterns in the somatosensory complex of the neonatal rat. J. Comp. Neur. 183:285-304.

Killackey, H. P., G. Belford, R. Ryugo, and D. K. Ryugo (1976). Anomalous organization of thalamocortical projections consequent to vibrissae removal in the newborn rat and mouse. Brain Res. 104:309-315.

Killackey, H. P., G. O. Ivy, and T. J. Cunningham (1978). Anomalous organization of SM1 somatotopic map consequent to vibrissae removal in the newborn rat. Brain Res. 155:136-140.

Killackey, H. P., and S. Leshin (1975). The organization of specific thalamocortical projections to the posteromedial barrel subfield of the rat somatic sensory cortex. Brain Res. 86:469-472.

Lorente de N'o, R. (1922). La corteza cerebral del raton. Trab. Lab. Invest. Biol. 20:41-78.

Lorente de N'o, R. (1938). Cerebral cortex: Architecture, intracortical connections and motor projections. In: Physiology of the Nervous System, J. F. Fulton (ed.), Oxford University Press, New York, pp. 291-339.

Nord, S. G. (1967). Somatotopic organization in the spinal trigeminal nucleus, the dorsal column nuclei and related structures in the rat. J. Comp. Neur. 130:343-355.

Ryugo, D. K., R. Ryugo, and H. P. Killackey (1975). Changes in pyramidal cell density consequent to vibrissae removal in the newborn rat. Brain Res. 96:82-87.

Shipley, M. T. (1974). Response characteristics of single units in the rat's trigeminal nuclei to vibrissa displacements. J. Neurophysiol. 37:73-90.

Smith, R. L. (1973). The ascending fiber projections from the principal sensory trigeminal nucleus in the rat. J. Comp. Neur. 148:423-441.

Torvik, A. (1956). Afferent connections to the sensory trigeminal nuclei, the nucleus of the solitary tract and adjacent structures. An experimental study in the rat. J. Comp. Neur. 106:51-132.

Valverde, F. (1971). Rate and extent of recovery from dark rearing in the visual cortex of the mouse. Brain Res. 33:1-11.

Van Der Loos, H. (1976). Barreloids in mouse somatosensory thalamus. Neurosci. Lett. 2:1-6.

Van Der Loos, H., and T. A. Woolsey (1973). Somatosensory cortex: Structural alterations following early injury to sense organs. Science 179:395-398.

Vincent, S. B. (1912). The function of the vibrissae in the behavior of the white rat. Behavior Monog. 1:7-85.

Waite, P. M. E. (1973). Somatotopic organization of vibrissal responses in the ventrobasal complex of the rat thalamus. J. Physiol. 228:527-540.

Waite, P. M. E., and P. K. Taylor (1978). Removal of whiskers in young rats causes functional changes in cerebral cortex. Nature 274:600-604.

Welker, C. (1971). Microelectrode delineation of fine grain somatotopic organization of SMI cerebral neocortex in albino rat. Brain Res. 26:259-275.

Welker, C., and T. A. Woolsey (1974). Structure of layer IV in the somatosensory neocortex of the rat: Description and comparison with the mouse. J. Comp. Neur. 158:437-454.

Welker, W. I. (1964). Analysis of sniffing of the albino rat. Behavior 22:223-244.

Weller, W. L., and Johnson, J. I. (1975). Barrels in cerebral cortex altered by receptor disruption in newborn, but not in five-day-old mice (Cricetidae and Muridae). Brain Res. 83:504-508.

White, E. L. (1978). Identified neurons in mouse SMI cortex which are postsynaptic to thalamocortical axon terminals: A combined Golgi-electron microscopic and degeneration study. J. Comp. Neur. 181:627-662.

Woolsey, T. A. (1967). Somatosensory, auditory and visual cortical areas of the mouse. Johns Hopkins Medical Journal 121:91-112.

Woolsey, T. A., and H. Van Der Loos (1970). The structural organization of layer IV in the somatosensory region (SI) of the mouse cerebral cortex. Brain Res. 17:205-242.

Woolsey, T. A., C. Welker, and R. Schwartz (1975). Comparative anatomical studies of the SMI face cortex with special reference to the occurrence of "barrels" in layer IV. J. Comp. Neur. 164:79-94.

Experimental Manipulations of the Development of Ordered Projections in the Mammalian Brain

BARBARA L. FINLAY

Department of Psychology
Cornell University
Ithaca, New York USA

Abstract The ordered representation of the retina in the tectum of hamsters can be influenced by experimental manipulation of factors intrinsic to the developmental process, such as the direction and timing of arrival of retinal fibers. The orientation of the retinotopic map in the tectum with respect to the neural axes, the amount of representation of retinal subareas on the tectal surface, and the laminar specificity of retinal terminals in tectum are differentially affected by these developmental factors.

If the amount of tectal tissue is reduced at birth in hamster by a large amount, only a portion of visual field comes to be represented. The area of visual field represented is in every case lower nasal visual field, regardless of whether caudal, rostrolateral, central, or superficial tectal tissue is removed. This asymmetry in surface representation is also reflected in the laminar distribution of retinal fibers in the tectum. A source for the inhomogeneity in visual field representation may be direction of optic tract arrival in tectum, which begins at the rostrolateral margin of the tectum, where lower nasal visual field is represented.

The polarity of the retinotopic map in tectum—its orientation with respect to the neural axes—may be dissociated from direction of fiber arrival, for if fibers are induced to enter the tectum medially, opposite to their normal entry point, a retinotopic map of normal order and polarity develops. Classes of mechanisms that account for both of these observations are discussed.

The way in which ordered topographic representations of sensory surfaces are created in the central nervous system is a central question in neuroembryology. Direct descriptions and experimental manipulations of various developing topographic representations of sensory surfaces have been made in systems ranging from insect to primate (Lopresti, Macagno, and Levinthal, 1973; Hunt and Jackson, 1973; Bunt and Horder, 1977; Straznicky and Gaze, 1971, 1972;

Chung and Cooke, 1975; Marten, this volume; LaVail and Cowan, 1971; Lund, 1978; So, Schneider, and Frost, 1977; Finlay, Wilson, and Schneider, 1979; Killackey, this volume; Rakic, 1977). Basic to all these approaches is the view that the *process* of development of orderly topographic representations of sensory surfaces in the central nervous system is a potentially rich source of information about the *mechanism* of that development.

The hamster is a convenient species for investigation of topographic map development. The hamster has the shortest gestational period of any eutherian mammal, and hamster pups are born at an usually immature state. On the day of birth in the hamster, retinal axons have distributed only sparsely in the superior colliculus, and the distribution is asymmetric; arborizations of retinal fibers are first found only in the rostral portion of the superficial gray layer of the superior colliculus, and are distributed only dorsally in that layer (So et al., 1977). Various postnatal manipulations of tectal tissue may thus be made before substantial retinal innervation has occurred, causing major reorganizations of the pattern of retinal termination (Schneider, 1973; Jhaveri and Schneider, 1974; Finlay, Wilson, and Schneider, 1979). The reorganizations observed provide evidence about the normal process of creation of an orderly retinotopic map in the tectum.

All of the experiments described in this paper involve the removal of varying amounts of tectal or retinal tissue at birth, by heat lesion, or direct excision (Schneider, 1973). After surgery, the hamster pup is allowed to mature normally, and at maturity the representation of the visual field on the remaining tectum is assessed electrophysiologically, using both single and multineuronal postsynaptic visual evoked responses (Finlay, Schneps, Wilson, and Schneider, 1978). Visual cortical areas are removed just prior to recording to eliminate a major source of non-retinal visual input; other thalamic sources remain.

After recording, the animal is perfused and the brain is removed, sectioned, and stained to permit complete reconstruction of the extent of the neonatal lesion and the position of all electrode penetrations.

Preferential Sparing of Nasal Visual Field After Partial Tectal Ablations

We have now investigated the pattern of retinal representation in over 30 hamsters in which extensive lesions of caudal, rostral, central, or superficial tectum had been made on the day of birth. The pattern of representation of the retina observed in the remaining tectum after each type of lesion was similar, an unexpected finding.

Shown in Fig. 1 is the representation of the visual field on the surface of a normal superior colliculus (A), and the representation in an animal subjected to a partial caudal lesion on the day of birth (B). In this case, the representation of the visual field is compressed onto a smaller than normal tectal surface (characteristic for lesions of this size; Finlay, 1976). The interesting feature of the retinal representation in this animal for this discussion is that the compression of the retinal representation is not homogeneous throughout. In this case,

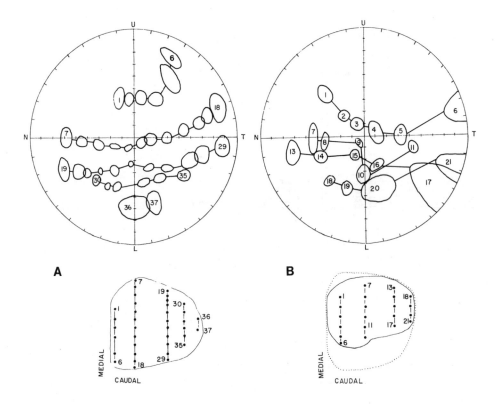

FIGURE 1 Colliculi with reconstructed electrode penetrations and associated visual receptive fields from a normal hamster (A) and from a hamster with an ablation of caudal superior colliculus at birth (B). The colliculi are shown in dorsal view; the dotted line in (B) indicates the expected normal extent of the superior colliculus.

Electrode penetrations are represented as black dots; for visual convenience, rostral to caudal series of penetrations and corresponding nasal-to-temporal visual receptive fields are connected with lines. The map of the visual field is centered about the projection of the optic disc. The nasal (N) to temporal (T) axis is determined by the attachments of the medial and lateral rectus muscles and is not the same as the horizontal plane defined by gravity. The upper (U) to lower (L) axis is defined by the insertions of the superior and inferior rectus muscles.

the representation of the nasal 70° of visual field is found on a tectal rostral to caudal surface area that is 75% of normal, while the temporal 90° of visual field is represented on an area that is 48% of normal. This inhomogeneity is also reflected in the size of visual receptive fields, which are generally larger than normal, but comparatively even larger temporally than nasally (Fig. 2). If the caudal lesion is made still larger, the entire visual field is no longer represented, and preferential sparing of the representation of nasal visual field is the result (Fig. 3). Since this neonatal lesion removed the normal terminal area of temporal visual field, the causal tectum, this result was not unexpected.

What are the results if the normal area for representation of nasal visual field, rostral tectum, is removed at birth? An example of a very large neonatal

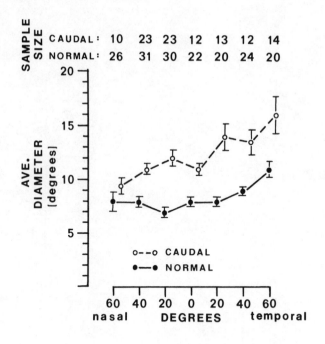

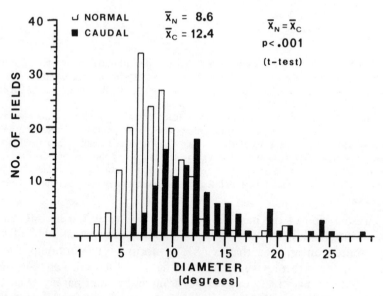

FIGURE 2 Receptive field sizes in various locations in the visual field taken from the data from seven normal hamsters and four hamsters with neonatal caudal lesions in which the representation of the visual field was compressed into a smaller than normal tectal area. Receptive field size was defined as the average of the major and minor axis of the typical elliptical collicular receptive field. Shown in the upper diagram is receptive field size, variability, and number of observations for divisions of visual field, from nasal to temporal. In the bottom diagram is the total, summed over all receptive field locations.

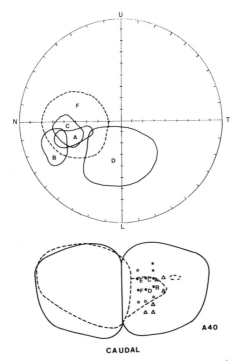

FIGURE 3 Dorsal view of the remaining right superior colliculus (dotted line) in a hamster with a large caudal tectal ablation at birth. In solid lines is the expected extent of the superior colliculus. Open circles indicate electrode penetrations where auditory responses were found; triangles indicate penetrations in identified pretectal nuclei. Visual evoked responses in the residual superior colliculus are indicated with black dots and are numbered; corresponding visual fields can be located in the receptive field diagram.

The map of the visual field is centered about the projection of the optic disc. The nasal (N) to temporal (T) axis is determined by the attachments of the medial and lateral rectus muscles. The upper (U) to lower (L) axis is defined by the insertions of the superior and inferior rectus muscles. Visual receptive fields recorded at the surface of the superior colliculus are outlined with solid lines; visual receptive fields recorded deep to the surface of the colliculus are indicated with dotted lines.

rostral tectum lesion is shown in Fig. 4. As in the case of the neonatal caudal lesion, *nasal* visual field is preferentially represented, even though the part of tectum in which nasal field is normally represented has been removed.

We have replicated this preferential sparing of nasal visual field, with a variety of lesion locations and sizes, in over 30 hamsters. Two types of lesions are of particular interest. If a slit is made across the tectum on the day of birth, it creates a transitory block to fiber passage before healing, and also a loss of tectal cells, presumably from central tectum where the slit is made. Although the tissue loss is presumably from central tectum, it is temporal visual field, and not central visual field, that shows evidence of loss of representation (Finlay and So, 1979).

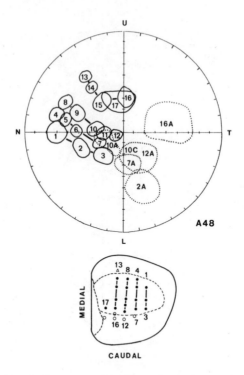

FIGURE 4 Dorsal view of the remaining right superior colliculus (dotted line) in a hamster with a large rostral tectum lesion at birth. Solid lines represent the expected extent of the superior colliculus. Triangles indicate penetrations where auditory receptive fields were found; squares indicate penetrations in identified pretectal nuclei; open circles indicate penetrations where no visual or auditory response could be evoked; and lettered, filled circles mark penetrations where visual responses were found.

The map of the visual field is centered about the projection of the optic disc. The nasal (N) to temporal (T) axis is determined by the attachments of the medial and lateral rectus muscles. The upper (U) to lower (L) axis is defined by the insertions of the superior and inferior rectus muscles.

Finally, if a large lesion is made that removes all of the superficial gray layer of the tectum, the normal retinal terminal area, anomalous retinal terminals in the residual deep layers of the colliculus can be demonstrated (Schneider, 1973). We have examined this anomalous projection area in five hamsters with large bilateral lesions of the entire superficial gray layer; a representative case appears in Fig. 5. After a lesion of this size, the midbrain collapses around the residual deep tectal area, where visual responses can still be demonstrated. Because of the alteration of gross morphology, we do not know what area of visual field would have been expected to be represented in this residual tissue, although we can be sure it is not a normal retinal terminal zone. Again, nasal and lower nasal field is preferentially represented.

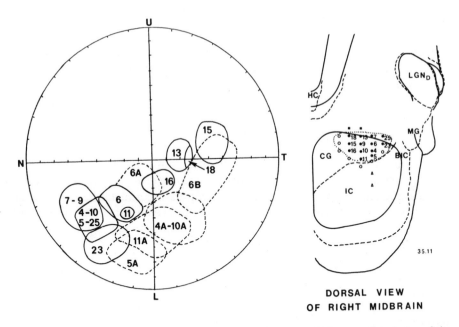

FIGURE 5 Dorsal view of the right midbrain in a hamster in which a complete lesion of the superficial gray layer of the superior colliculus had been made on the day of birth. Solid lines indicate the normal appearance of the midbrain; dotted lines indicate the appearance of the remaining midbrain in this animal. Triangles indicate penetrations where auditory evoked responses were found; squares indicate penetrations located in pretectal nuclei; open circles indicate penetrations where no visual or auditory response could be elicited. Filled numbered circles indicate penetrations in the superior colliculus where visual responses could be obtained; corresponding visual receptive fields can be seen on the visual receptive field diagram.

The map of the visual field is centered about the projection of the optic disc. The nasal (N) to temporal (T) axis is determined by the attachments of the medial and lateral rectus muscles. The upper (U) to lower (L) axis is defined by the insertions of the superior and inferior rectus muscles. Visual receptive fields recorded at the surface of the superior colliculus are outlined with solid lines; visual receptive fields recorded deep to the surface of the colliculus are indicated with dotted lines.

Changes in Laminar Specificity of Retinal Terminals Under Conditions of Increased or Decreased Retinal Convergence on Tectum

In the normal hamster, the retina projects throughout the entire extent of the superficial gray layer of the superior colliculus (Schneider, 1973). During development, however, retinal arbors are first found in the superficial part of the superficial gray and only later arborize more ventally (So, et al., 1977). The effects of various manipulations that either increase or decrease the amount of retinal convergence on tectal tissue (the number of retinal ganglion cells per postsynaptic tectal cell) appear to reflect this developmental process.

If the retinal input is made more convergent than normal, by the partial tectal lesions described in the preceding sections, the preferential representation of nasal visual field on the tectal surface is also reflected in its laminar representation. The areas of temporal visual field not represented on the tectal surface can often be found represented deep in the tectum, in the more ventral parts of the superficial gray and the stratum opticum. This tendency becomes progressively more pronounced as one moves farther from the point of fiber entry, i.e., moves caudally in the colliculus. This is shown in Fig. 2 where the dotted fields represent subsurface fields. This retinotopic order in the depth dimension was observed in all of the animals with various large tectal lesions that spared some part of superficial gray, and we could not account for it by factors such as anomalous angles of electrode penetration, or morphology changes in the neonatally lesioned midbrains (Finlay, 1976).

The projection of the retina can be made more sparse than normal, by a variety of methods. The most straightforward of these methods was one used by Frost (1976); hamsters were given partial retinal lesions at birth, and the laminar extent of the projection of the retina in the tectum was described at maturity using anatomical methods. Under less than normal convergence, retinal projections did not become simply uniformly more sparse throughout the laminar extent of their projection, but were found preferentially distributed in the most dorsal part of the superficial gray.

The projection from the retina to one tectum may also be reduced by the following procedure: if one eye and the tectum ipsilateral to it are removed at birth, the optic tract from the remaining eye will travel to the lesioned tectum, where some fibers will terminate. The rest will continue to grow medially, eventually recrossing the midline, and innervating the ipsilateral tectum with a topographic representation mirror-symmetric to the normal contralateral representation (Schneider, 1973; Finlay et al., 1979). When the fiber bundle recrosses the midline it is reduced in bulk due to the progressive loss of fibers through prior termination. The distribution of this projection is abnormally superficial, as shown by both anatomical (Schneider, 1973; So and Schneider, 1978) and physiological techniques (Finlay et al., 1979). This abnormal superficial distribution is most pronounced in the lateral tectum, at the pole opposite from the direction of fiber entry. In terms of retinotopic organization, the most striking result of this abnormal superficial distribution is that extreme lower visual field, represented in lateral tectum, is represented on a surface area in colliculus that is two to five times its normal extent.

After monocular enucleation at birth, the projection from the remaining eye to the colliculus ipsilateral to it is another example of an unusually sparse retinotectal projection. Interpretation of this case is difficult, however, since the retinal axons projecting to the ipsilateral tectum now probably represent two groups: an augmented normal ipsilateral projection, and a second projection consisting of sprouted or diverted fibers that normally would have gone contralaterally (Finlay et al., 1979; Cunningham, 1976). The ipsilateral projection in the hamster, like other rodents, has a patchy representation confined to the ventralmost part of rostral superficial gray (Schneider, 1973; Drager and Hubel, 1975). After monocular enucleation at birth, this projection expands, both in its laminar and its rostrocaudal extent. In rostralmost colliculus, this projection occupies the entire depth of the superficial gray; more caudally, the projection, like other sparse projections, is found only superficially.

The fact that incoming retinal fibers arborize first in the most superficial part of the tectum can account for all the anomalous laminar patterns under increased or decreased conditions of convergence. It is interesting to speculate on the possible effect of such a preferential termination pattern in normal development. If earlier and later maturing retinal ganglion cells represent different functional subgroups, passive ordering by time of arrival in tectum would give functional subdivisions in tectum by lamina. The ipsilateral retinal projection to the tectum may represent such a functional subgroup; it is represented deep in the tectum, and it appears 8-12 days later than the contralateral projection (So et al., 1977). Likewise, the various cortical projections to tectum might be organized by time of arrival.

Implications for the Process and Mechanism of the Development of Ordered Rectinotectal Connections

The abnormal retinotopic maps we observe after these experimental manipulations of development could be accounted for by a variety of mechanisms. One hypothesis that systematizes all the results described is the following: all of the asymmetric retinotopic maps observed represent features in the normal process of development of the tectal retinotopic map. This developmental process has the following characteristics:

1. The tectum supplies a polarity cue to incoming retinal fibers independent of the direction they enter the tectum (amphibian, Chung and Cooke, 1975; mammal, Finlay et al.).

2. Incoming retinal fibers arborize exclusively in the most superficial part of the superficial gray. If this area is occupied, fiber will then secondarily terminate deep to the tectal surface (So et al., 1978; Frost, 1976).

3. The retinotopic map is "unrolled" beginning at the first tectal boundary contacted. So while polarity is independent of the direction of fiber entry, the initial quadrant or sector of visual field to be represented in tectum during development depends directly on the direction of fiber arrival.

This hypothesis is a statement of process, not of mechanism. Characteristics (1) and (2) have been demonstrated both by direct description and experimental manipulation. Characteristic (3) is presently under investigation. Two types of results might be anticipated: (1) the area of retina represented in lateral tectum on first innervation is the same as that represented at maturity, or (2) the area represented is not the same; in particular, since maturation in the retina proceeds from the center to the periphery, a prediction that central retina would be first represented in rostral tectum is plausible.

How does this description of process relate to mechanism, and to other investigations of the development of ordered topographic maps of sensory surfaces? Two aspects are of interest: the nature of the polarity cue provided in the tectum to order incoming fibers, and the ability of explanations based on axon-sorting processes, as opposed to fiber-tectum recognition, to account for these observations.

Various investigators have indicated that polarity cues are independent of the direction of fiber arrival (Chung and Cooke, 1975; Attardi and Sperry, 1963; Gaze, 1970; Finlay et al., 1979). Investigators have differed, however, about the nature of the polarity cue. In chemospecificity models, where optic tract axons read labels on tectal cells, attaching only to cells with a matching label (Meyer and Sperry, 1976), polarity of the retinotopic map emerges naturally from the matching process. A variety of other mechanisms of generating polarity have been proposed that depend either on gradients of various types that do not supply unique location cues (Prestige and Willshaw, 1975; Chung and Cooke, 1975; Wolpert, 1971; Gaze, 1970) or require specification of a starting point for the retinotopic map (Hope, Hammond, and Gaze, 1976). The results of these experiments conform more easily to generation of polarity cues by mechanisms other than chemospecificity, since we could find no evidence that incoming optic tract fibers are seeking particular tectal locations. "Respecification" of the tectum consequent to the early lesion as an explanation for these results is made somewhat less plausible in that asymmetries in the hamster retinotopic map, unlike regenerating topographic maps (Meyer, 1977) are unrelated to lesion location.

Knowledge of the developmental sequence of the laying down of a retinotopic map in tectum, taken together with the experimental results already described, will define the nature of the polarity generator further. If central retina is first represented in rostral tectum, and is later selectively displaced by a segment of peripheral retina, a strong argument can be made for some asymmetry in the interaction of retinal axons and postsynaptic cells according to retinal and tectal location. Retinal fiber interaction models, and time-position models are made unlikely. If, however, the first area to be represented in tectum is topographically appropriate, "first-come, first-served" time-position models, or selective axonal sorting models remain viable hypotheses.

In some simpler systems, a painstaking description of development has served to directly define mechanism (Lopresti et al., 1973). In more complex vertebrate systems, description may not uniquely define a mechanism, but can clearly constrain the types of mechanisms possible.

Acknowledgements This work was supported by National Science Foundation Grant BNS77-07066. I wish to thank Sara Cairns and Anne Berg for their help with various parts of this study and Tedd Judd and Harold Zakon for their criticisms of this manuscript.

References

Attardi, D. G., and R. W. Sperry (1963). Preferential selection of central pathways of regenerating optic fibers. Exp. Neurol. 7:46-64.

Bunt, S. M., and T. J. Horder (1977). A proposal regarding the significance of simple mechanical events such as the development of the choroid fissure, in the organization of central visual connections. J. Physiol. (Lond.) 272:10-11.

Chung, S. H., and Cooke (1975). Polarity of structure and of ordered nerve connections in the developing amphibian brain. Nature (Lond.) 258:126-132.

Cunningham, T. J. (1976). Early eye removal produces excessive bilateral branching in the rat: application of cobalt filling method. Science 194:857-859.

Drager, U. C., and D. H. Hubel (1975). Responses to visual stimulation and relationship between visual, auditory and somatosensory units in mouse superior colliculus. J. Neurophysiol. 39:690-713.

Finlay, B. L. (1976). Neuronal specificity and plasticity in hamster superior colliculus: electrophysiological studies. Doctoral dissertation, Department of Psychology, Massachusetts Institute of Technology.

Finlay, B. L., S. E. Schneps, K. G. Wilson, and G. E. Schneider (1978). Topography of visual and somatosensory projections to the superior colliculus of the golden hamster. Brain Res. 142:223-235.

Finlay, B. L., and K. F. So (1979). Altered retinotectal topography in hamsters with neonatal tectal slits. Neuroscience (in press).

Finlay, B. L., K. G. Wilson, and G. E. Schneider (1979). Anomalous ipsilateral retinotectal projections in hamsters with early lesions: topography and functional capacity. J. Comp. Neurol. 183:721-740.

Frost, D. O. (1975). Factors influencing the development and plasticity of retinal projections in the Syrian hamster. Doctoral dissertation, Department of Psychology, Massachusetts Institute of Technology.

Gaze, R. M. (1970). The Formation of Nerve Connections. Academic Press, New York.

Hope, R. A., B. J. Hammond, and R. M. Gaze (1976). The arrow model: retinotectal specificity and map formation in the goldfish visual system. Proc. Roy. Soc. Lond. B 194:447-466.

Hunt, R. K., and M. Jacobson (1973). Specification of positional information in retinal ganglion cells of *Xenopus* Assays for analysis of the unspecified state. Proc. Natl. Acad. Sci. (USA) 70:507-511.

Jhaveri, S. R., and G. E. Schneider (1974). Retinal projections in Syrian hamsters: normal topography and alterations after partial tectum lesions at birth. Anat. Rec. 178:383.

LaVail, J. H., and W. M. Cowan (1971). The development of the chick optic tectum. I. Normal morphology and cytoarchitectonic development. Brain Res. 28:391-419.

Lopresti, V., R. E. Macagno, and C. Levinthal (1973). Structure and development of neuronal connections in isogenic organisms: Cellular interactions in the development of the optic lamina of *Daphnia* Proc. Natl. Acad. Sci. (USA) 70:433-437.

Lund, R. D. (1978). Development and Plasticity of the Brain. Oxford University Press, London.

Meyer, R. L. (1977). Eye-in-water mapping of goldfish with and without tectal lesions. Exp. Neurol. 56:23-41.

Meyer, R. L., and R. W. Sperry (1976). Retinotectal specificity: Chemoaffinity theory. In: Studies on the development of behavior and the nervous system, Vol. 3, Neural and behavioral specificity. G. Gottlieb (ed.). Academic Press, New York, pp. 111-149.

Prestige, M. C., and D. H. Willshaw (1975). On a role for competition in the formation of patterned neural connexions. Proc. Roy. Soc. B 190:77-98.

Rakic, P. (1977). Prenatal development of the visual system in rhesus monkey. Phil. Trans. B 278:245-260.

Schneider, G. E. (1973). Early lesions of the superior colliculus: factors affecting the formation of abnormal retinal projections. Brain Behav. Evol. 8:73-109.

So, K. F., G. E. Schneider, and D. O. Frost (1977). Normal development of the retinofugal projections in Syrian hamsters. Anat. Rec. 187:719.

Straznicky, K., and R. M. Gaze (1971). The growth of the retina in *Xenopus laevis:* An autoradiographic study. J. Embryol. Exp. Morphol. 26:67-79.

Straznicky, K., and R. M. Gaze (1972). The growth of the tectum in *Xenopus laevis:* An autoradiographic study. J. Embryol. Exp. Morphol. 28:87-115.

Wolpert, L. (1971). Positional information and pattern formation. Curr. Top. Develop. Biol. 6:183-224.

Development and Plasticity of Neuronal Connections in the Lamb Visual System

P. G. H. CLARKE
K. A. C. MARTIN
V. S. RAMACHANDRAN
V. M. RAO
D. WHITTERIDGE
University Laboratory of Physiology
Parks Road
Oxford, England

Abstract The visual system of the sheep has a number of features in common with that of the cat, but the wide interocular distance of the sheep makes it more suitable for the study of binocular receptive field disparities of cortical cells. On the day of birth the physiology of the lamb cortex is similar to that of the adult, with the exception of the limited amount of facilitation seen in cells with binocular fields. Monocular deprivation in the first three or four months of life results in most cortical cells being driven by the experienced eye alone and a relative shrinkage of cells in the lateral geniculate nucleus and medial interlaminar nucleus supplied by the deprived eye. Exposure of the deprived eye to stimulation of only one hour by slowly rotating square wave gratings increases the number of cortical cells which can be driven by the deprived eye.

Introduction

Given the extensive amount of data that has accumulated from studies on the visual systems of the cat and monkey, it may seem strange that we have chosen to study the visual system of the sheep. In this paper, however, we will attempt to show that the sheep is useful for more than *souvlakia* and Cretan rugs.

Compared to the cat and monkey the neonatal lamb is far more advanced developmentally, being able to follow its mother and avoid obstacles by sight within a few hours of birth. Unlike those of the kitten the optic media are quite clear at birth and the fundus can be easily seen (Ramachandran et al.,

403

1977). One of the principal features of their visual system is the wide interocular distance of about 12 cm in the adult compared with 6 cm in man, 4 cm in the monkey, and 3 cm in the cat. Thus, for equally distant visual targets the angle of parallax for the sheep is four times that of the cat. The consequence is that the range of disparities between the receptive fields of binocular cortical cells is far greater than that in the cat or monkey. These binocular disparities are probably part of the mechanism responsible for stereoscopic vision, and large disparities may be particularly important in the sheep which probably cannot converge but which may need to localize grass at a short distance (Clarke et al., 1976).

The principal disadvantages of using the sheep for developmental studies are the restricted lambing seasons and the small number of animals per birth compared to the cat.

Anatomy

All ungulates have three approximately parallel sulci (the entolateral, lateral, and ectolateral sulci), which run anteromedially over the posterior third of the cerebral hemispheres. Of these the middle or lateral sulcus is the deepest and contains most of the striate area, visual area 1 (V1; see Fig. 1). The boundary between V1 and visual area 2 (V2) is at the bottom of the ectolateral sulcus. As can be seen in Fig. 1 this boundary is found 10-15° out in the ipsilateral visual field. In V2 the extensive representation of the peripheral visual field (up to 130°) extends over the lateral cortical surface to the suprasulvian gyrus (Clarke and Whitteridge, 1976).

The retina in the ungulate has a very marked horizontal streak with an increased ganglion cell density which is reflected in the cortex to the extent that the upper and lower peripheral fields have very little cortical representation. The optic axes in the sheep diverge 48° from the midline, but although the midline falls on the two area centrales, very little binocular representation is found more than 25° from the centre of the visual axis. This gives a total binocular field of 50° and a total visual field of 276° (Hughes and Whitteridge, 1973).

The lateral geniculate nucleus (LGN) has three layers (A, A_1, and C) which are similar to those of the cat. Layers A and C are supplied by fibres from the contralateral eye, layer A_1 from the ipsilateral eye. There is a group of large cells in layers A and A_1 which do not appear in C, but the cell sizes of C overlap with those of cells in A and A_1. As would be expected from the small ipsilateral field, layer A_1 has the smallest area of the three layers. The LGN is roughly rectangular, about three times as long as it is wide, with its long axis running downward, forward, and medially. Vertical meridians in the visual field are roughly parallel to the long axis of the nucleus. Rather surprisingly, the horizontal meridian is represented along its short lateral axis, and the effect of the increased ganglion cell concentration of the horizontal streak is to increase its long vertical axis, not its lateral axis.

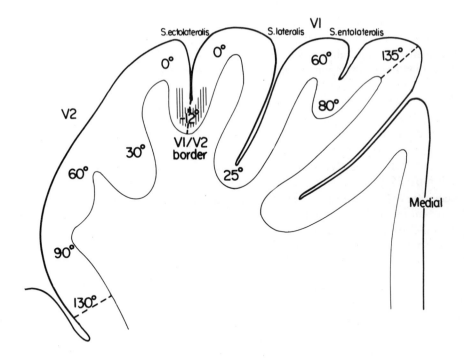

FIGURE 1 Diagrammatic view of a coronal section of the left visual cortex of the sheep at the level of the inter-aural plane indicating the main sulci and the topographic representation of the visual field along the horizon. The reveral of the map at the V1/V2 border occurs at the bottom of the ectolateral sulcus, 10-15° in the ipsilateral visual field. The striped region indicates the region of termination and the cells of origin of the callosal fibres as shown by horseradish-peroxidase and Fink-Heimer degeneration techniques.

There is also a medial interlaminar nucleus (MIN) which seems to contain areas corresponding to layers A, A$_1$, and C (Clarke et al., in preparation). Recent work with horseradish peroxidase has established that the main laminae of the nucleus project both to V1 and to V2. The MIN projects to V2 and possibly to V1 as well (V. M. Rao, in preparation).

Callosal connections exist between the cortex of the ectolateral gyrus of one hemisphere and the same area of the other hemisphere. Surprisingly, the results of both horseradish peroxidase and Fink-Heimer methods agree in that the main callosal connection is to the *depth* of the ectolateral sulcus—the V1/V2 border which is 10-15° ipsilateral in the field (Fig. 1)—and not to the representations of the area centralis at 0° in the visual field (V. M. Rao, in preparation).

Physiology

Cells in V1 have sharply tuned orientation specificity with regular sequences of different preferred orientations as one progresses across the cortex. Ocular dominance columns also occur. Simple, complex, and hypercomplex cells are all to be found in V1. There are more cells driven by the contralateral than by the ipsilateral eye alone (see Fig. 2). In V2 the receptive field disparities of binocular neurones cover a wider range than in V1 and are more sharply tuned to disparity. In layers II and III one very frequently finds "true binocular" cells, i.e., cells which cannot be driven by either eye alone. Other cells can be driven weakly by one eye, but a sharply facilitated response is given by stimulating both eyes at the appropriate disparity. There are indications of disparity "columns" about 400 μm wide in which cells of the same disparity of 3-5° crossed, zero disparity, or 3-5° uncrossed are grouped. We have so far not seen columns of cells responding to stimuli in the same direction but at different disparities (Clarke et al., 1976, Clarke et al., 1979).

In lambs born in the dark and examined a few hours later, having not more than 1-2 minutes of visual experience, we found cells sharply tuned to orientation, in regular sequences (Fig. 2), and having simple, complex, and hypercomplex type receptive field organization. Cells with receptive field disparities comparable to those found in the adult were also found. The principal difference

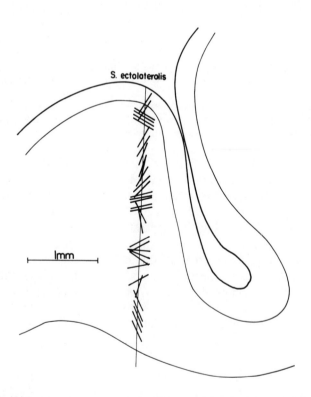

FIGURE 2 Changes in the orientation preference of cells in a typical single penetration through V2, recorded on the day of birth of the lamb.

was that these binocular cells showed far less binocular facilitation in the newborn than in the adult. Only three "true binocular" cells have been seen in the newborn. The adult pattern in binocular cells of brisk facilitation when the eyes were stimulated at the optimal disparity is established by 3 months (Ramachandran et al., 1977; Clarke et al., 1979).

Studies on Deprivation

Although genetic factors have produced everything except facilitated binocular cells without visual experience, there is a critical period which extends from the day of birth for at least three to four months (Fig. 3). Closing both eyes for about six to seven weeks (in two animals) has been found to have little effect. Cells with normal orientation specificity and binocularity are found, but the total number of binocular cells is somewhat reduced. During the first few months of life, closing one eye results in almost complete disappearance of cells in the opposite hemisphere driven by the deprived eye. Binocular cells similarly disappear with early closures. The great reduction in the number of cells driven by the deprived eye is limited to the binocular field. In the monocular field the deprived eye drives cells apparently normally.

This severe reduction in the number of cortical cells driven by the deprived eye can be reversed to some extent by an hour's exposure to slowly revolving black and white gratings (Banks et al., 1978) with the good eye closed (Martin et al., 1979). Such exposure gives an increase of 15-20% in the cells driven by the deprived eye which persists for at least seven to eight hours. On some occasions the recovery after stimulation has been even greater (Fig. 4). Preliminary results suggest that application of amphetamines, which are catecholamine agonists, further increases the improvement produced by the revolving grating.

As with the cat and monkey, monocular deprivation from birth results in shrinkage of cells in the deprived eye's laminae in the LGN. This effect is clearly seen in the binocular segment and seems to be more marked in layer A_1 (driven by the ipsilateral eye). In addition to the changes in the LGN, shrinkage effects have also been observed in the binocular segment of the MIN (V. M. Rao, in preparation). Among other questions being investigated are the effects of deprivation on the monocular segment (larger in comparison with the cat), which will throw more light on the morphological changes caused by deprivation *per se* and by competitive effects.

Discussion

Partly because the ungulates produce young which are more mature at birth than any other mammal, the lamb is the only animal so far whose visual system has been studied on the day of birth. It provides unequivocal evidence that the cortical visual cells at birth have very similar properties as those of adults, with the exception of the limited amount of facilitation seen in cells with binocular fields. Clearly, genetic mechanisms without the help of visual experience can generate the neural connections needed for these mechanisms. It is somewhat

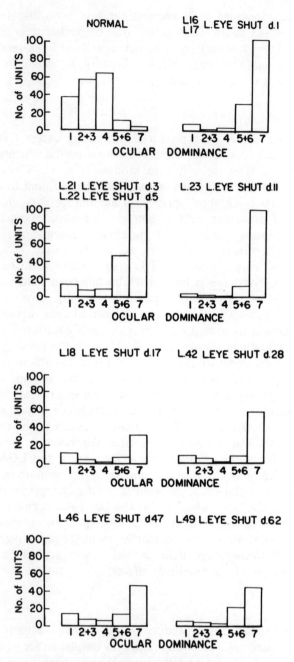

FIGURE 3 The effect of closing one eye, at the indicated days after birth, on the ocular dominance distribution of cells recorded in V1. Similar histograms were obtained from cells in V2. The dominance histograms were generally obtained six to eight weeks after closing the eye. Note the strong dominance of the contralateral eye in the normal animal.

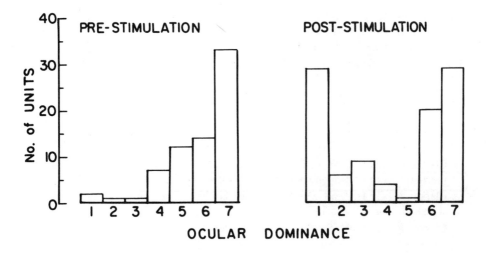

FIGURE 4 The most striking example of the recovery of cells in V1 driven by the deprived eye after one hour of stimulation by revolving gratings in the acute preparation. This lamb had the left eye shut six days after birth and was mapped two days later. The marked effect of deprivation of only two days on the ocular dominance distribution can be seen by comparing the pre-stimulation sample with that of the normal animals shown in Fig. 3. In all lambs the recovery was mainly of cells driven solely by the contralateral deprived eye.

surprising to find that these connections can, however, be readily disrupted at three to four months after birth. Brief periods of interim visual stimulation can re-establish the working connections of the deprived eye in the visual cortex. The effect of the short grating stimulation is similar to that of closing the experienced eye and opening the deprived eye ("reverse suturing") in that the cells which recover are mainly those driven only by the deprived eye. Obviously, it would be interesting to see whether, by altering the stimulus conditions, more binocularly driven cells could be produced. We are also investigating whether stimulation of a portion of the field would induce any recovery in the unstimulated portion.

The recovery after grating stimulation appears to be much quicker than after reverse suturing. This may be related to the fact that a progressive sequence of orientation columns is stimulated as the grating revolves, which may have a stronger cumulative effect than the intermittent stimulation of non-sequential columns which would occur during random visual stimulation. There are clear advantages of using the gratings in an acute preparation rather than reverse suturing if one is studying the mechanism of the plasticity. The details of this recovery process are still far from clear, but a mechanism by which the cells driven by an active eye can "inhibit" corresponding cells driven by a closed eye, and a hypothetical process by which such an "inhibition," if maintained for 24-48 hours, can produce a permanent decrease in excitability of the inhibited cells, would cover most of the facts.

In addition to providing some clear answers to the question of the role of experience in the development of stereoscopic mechanisms, work on the lamb visual cortex may also provide some clues as to the mechanisms whereby experience modifies the original genetically determined neuronal connections.

References

Banks, R. V., F. W. Campbell, R. Hess, and P. G. Watson (1978). A new treatment for amblyopia. Brit. orthopt. J. 35:1-12.

Clarke, P. G. H., I. M. L. Donaldson, and D. Whitteridge (1976). Binocular visual mechanisms in cortical areas I and II of the sheep. J. Physiol. 256:509-526.

Clarke, P. G. H., and D. Whitteridge (1976). The cortical visual areas of the sheep. J. Physiol. 256:497-508.

Clarke, P. G. H., V. S. Ramachandran, and D. Whitteridge (1979). The development of the binocular depth cells in the secondary visual cortex of the lamb. Proc. R. Soc. B (in press).

Hughes, A., and D. Whitteridge (1973). The receptive fields and topographical organization of goat retinal ganglion cells. Vision Res. 13:1101-1114.

Ramachandran, V. S., P. G. H. Clarke, and D. Whitteridge (1977). Cells selective to binocular disparity in the cortex of newborn lambs. Nature 268:333-335.

Martin, K. A. C., V. S. Ramachandran, V. M. Rao, and D. Whitteridge (1979). Changes in ocular dominance induced in monocularly deprived lambs by stimulation with rotating gratings. Nature 277:391-393.

Refractive Changes in the Chicken Eye Following Lid Fusion

URI YINON
Electrophysiological Laboratory
Goldschleger Eye Institute, Tel Aviv University
Sheba Medical Center
Tel Hashomer, Israel

LIONEL ROSE
AMIRAM SHAPIRO
MENACHEM M. GOLDSCHMIDT
Department of Ophthalmology
Hadassah University Hospital
Jerusalem, Israel

TATANYA Y. STEINSCHNEIDER
Vision Research Laboratory
Hadassah University Hospital
Jerusalem, Israel

Abstract Lid fusion was surgically performed in newly hatched and grown-up chicks for various periods of up to three-and-one-half months. This led to an unusual enlargement of the eyeball in the first group of chicks in all equators, resulting in average myopia of -8.02 D. The value that was obtained in normal eyes was +0.42 D. Simultaneously, an antagonistic procedure, although much smaller, took place, leading to flattening of the cornea and decrease of its refractive power.

Hypermetropia was found in the second group of late-operated chicks if illumination conditions were poor. This was expressed almost only in a corneal change.

In the myopic chicks the two eyes grew independently, while in the hypermetropic chicks the same change, although smaller, was found also for the open eye.

Introduction

Lid fusion in monkeys, whether monocularly or binocularly performed, leads to an axial enlargement of the eye, resulting in high myopia (Wiesel and Hubel, 1977a, 1977b; Raviola and Wiesel, 1978). Similar results were obtained in tree

411

shrews similarly operated on, although the myopia that was found was not attributed only to an axial change (Sherman et al., 1977). Slight myopia (-1.00 to -2.00 D) has been preliminarily reported also for the deprived eye of monocularly sutured cats (Wilson and Sherman, 1977). On the other hand, inconsistency of the relationship between lid closure and refractive errors in monkeys was found by von Noorden and Crawford (1978).

The present report on chicks is part of a long-term study on mechanisms controlling refractive changes during development. Although we have obtained a high myopia and hypermetropia in chicks, in comparison to myopic monkeys there are some important differences which will be reported here.

Methods

Thirty-eight chicks of the domestic fowl were operated binocularly or monocularly. In the first group of these chicks the eyes were closed from the first day until the age of 3 to 3.5 months; in the second group they were opened at various times after hatching; and in the third group they were closed at late ages. For comparison, data from 8 control chicks of the same batch and from 14 laying hens on a farm (most of the latter were between 1 and 2 years old) were also taken.

The operation was performed by partitioning the upper and lower eyelids following sectioning of 1 mm of their margins. Then the internal aspects of the outer sides were attached and fused by the aid of four to six single stitches of silk thread No. 6-0. After healing, the fused eyelids produced a thin translucent whitish membrane on the eye through which the corneal reflex could be seen, thus preventing patterned input only. The membrane was very loose and did not produce any pressure on the cornea. All eyes were perfectly intact and the optical media was clear under the fused eyelids.

Immediately following the operation the chicks were labeled on the legs with numbered rings. They were raised for three weeks in the presence of a low-intensity lamp (50 foot candles) for warmth in a 30-cm high, 70-cm long, and 50-cm wide box. They were then transferred to a chicken coop. The first and second groups ("myopic") received only sunlight (250-300 foot candles at noon), while the third group ("hypermetropic") received low illumination (about 50 foot candles).

Following incision of the completely fused lids the chicks were examined. At first the curvature (radius) of the cornea was determined by using a keratometer. The corneae in most of the chicks were too steep for direct keratometry using the American Optical Co. keratometer. This was overcome by applying lenses of +1.00 D to +2.75 D in the optical pathway. We therefore corrected our readings by measuring the radius of metal balls 9.52-14.29 mm in diameter through these lenses. A curve was produced showing the ratio between the real radius of the ball to its new radius when viewed through each lens. The readings of the chicks were then corrected in accordance with the curve.

The refractive error was determined with a streak retinoscope and hand-held trial case lenses. Since the avian ciliary body and iris contain striated muscles

(Pumphrey, 1961), cycloplegia could not be obtained. This was true in our chicks even by applying D-Tubocurarine chloride 2.25 mg/ml in Benzalkonium Chloride 0.025% which was successfully used for pigeons (Campell and Lawton-Smith, 1962). Therefore, refraction was performed at maximum relaxation of the accommodation. The values that were obtained were similar to those obtained in several chicks under deep Nembutal anesthesia or immediately after death. Furthermore, these values were consistently similar in the same chick within ±1.00 D even at intervals of several weeks. The chicks were also independently refracted by two investigators in alternate weeks with good agreement; the direction of error was consistently the same (in view of the fact that the refractive errors that were found express mainly changes in the size of the eyeball and the curvature of the cornea, as will be shown later, accommodative changes at the moment of refraction contribute only minor effects).

The direction of beak was taken as the horizontal meridian for each eye for purposes of retincoscopy and keratometry. We calculated the average of the horizontal and vertical values which were practically similar.

The length of the anterior-posterior axis of the eyeball was measured using an ultrasound apparatus. The ultrasound measurement gives a value which is 2-3 mm smaller than the actual value obtained by measuring with a caliper the length of the enucleated eye. This difference is partly attributed to the width of the avian sclera with its ossicles and the considerable thickness of the avian retina (Pumphrey, 1961). However, the relative difference between the eyes was preserved.

The difference in the means for the retinoscopy, keratometry, and ultrasound readings in all chicks have been statistically tested by comparing left and right eyes and the control and operated groups. The appropriate degrees of freedom, *t* and *P* values, were calculated (Snedecor, 1964; Arkin and Cotton, 1963).

Results

Our normal chicks with both eyes open were either slightly hypermetropic or emmetropic with their refractive error ranging between 0.00 D and +1.25 D (average: +0.42 D). This is not very different from the value obtained for laying hens (n = 28 eyes), where retinoscopy findings ranged between -0.25 D and +1.50 D (average: +0.90 D). In the normal chicks the refractive power of the cornea was almost 65 D and the length of the anterior-posterior axis about 12 mm (Table 1). These values are very homogeneous within the control group, giving, as expected, nonsignificant differences between right and left eyes (p > 0.5).

For the open eye of monocularly closed chicks of the first and second groups, the same refractive values and eye size were obtained as for the controls. On the other hand, the bulging of the cornea of the closed eye could be easily seen by the naked eye immediately following opening of the fused lids. Figures 1(A,B) show the two eyes following enucleation. The closed eyes became extremely large, as also reflected by the length of the anterior-posterior axis (14

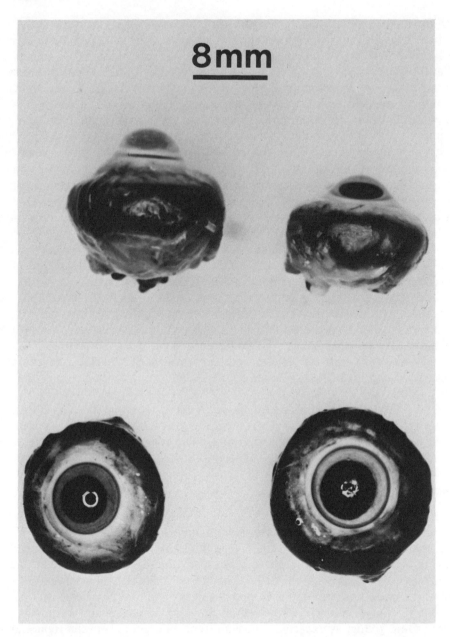

FIGURE 1 The closed (left) and open (right) eye of a 3-month-old chick in which the eyelids were fused immediately following hatching. Note the relative increase in the axial length of the left eye (upper) and in the diameter of the globe and cornea (lower).

to 16.5 mm, average 15 mm) and, consequently, became myopic (-3.00 D to -12.25 D, average -8.02 D) (Table 1). With respect to axial length and refractive error, the difference between the formerly closed and the normal eyes is statistically highly significant (p < 0.005).

In addition to the enlargement of the globe, which produces the myopia in the first group of chicks, optically an antagonistic change was found for the cornea, which becomes flattened under the closed lid. This leads to statistically significant (P < 0.025) lower values when the refractive power of the cornea of the closed eye is compared with the open eye. The fact that the antagonistic processes occur simultaneously disturbed the constant relationship expected between eye size and refractive error as in the mammalian eye. Furthermore, the two changes are not linearly related as could be judged from the individual results. The refractive power of the cornea could not even be approximated on the basis of the eye globe dimensions, although the direction of the change was always as expected (the large eyeball having the flatter cornea). For instance, in one chick the difference between the two eyes in the anterior-posterior axis was 1.0 mm, while the difference in refractive power of the cornea was 10 D; in another chick from the same batch with the same difference in refractive power between the corneae there was a difference of 4 mm in axial length. Therefore, in chicks early operated for eye closure, the refractive error caused by flattening of the cornea was unable to "compensate" for the myopia obtained.

The closed eye of the chicks monocularly operated on the first day had properties similar to the closed eye of the binocularly closed chicks (P > 0.5 for the refractive error). Since the open eye in these chicks was not affected by the fact that the other eye was permanently closed, we conclude that the two eyes are independent with respect to the myopia produced.

On the other hand, the two eyes seem to be dependent, or at least simultaneously affected, when the closure is monocularly performed for short periods at late ages and if the chicks are raised under poor illumination conditions (third group, Table 2). In this case only hypermetropia is produced consistently higher for the closed eye. However, the eye size remained almost equal for the two eyes, indicating that the main change observed was in the corneal curvature. It is interesting to note here that the corneae in all these chicks were very

TABLE 1 Effects of permanent eye closure[a] immediately following hatching on refraction and eye size in chicks.

Eye closed	Number of chicks	Right eye Retinoscopy (diopters)	Keratometry (diopters)	Anterior-posterior axis (mm)	Left eye Retinoscopy (diopters)	Keratometry (diopters)	Anterior-posterior axis (mm)
Left	6	+0.96	65.42	12.25	-8.02	58.57	14.75
Both	4	-6.25	59.75	13.94	-6.50	59.65	13.69
None	4	+0.50	64.87	12.08	+0.34	64.62	12.21
Control	14 (adults)	+0.87			+0.92		

[a]Based on one measurement of refraction and ultrasound for each eye.

TABLE 2 Effects of temporary eye closure[a] for 11-64 days within the first 79 days following hatching.

Eye closed	Number of chicks	Right eye			Left eye		
		Retinoscopy (diopters)	Keratometry (diopters)	Anterior-posterior axis (mm)	Retinoscopy (diopters)	Keratometry (diopters)	Anterior-posterior axis (mm)
Left	3	+2.35	49.46	13.89	+ 8.60	42.69	14.32
Both	2	+9.9	39.12	14.00	+10.41	40.48	14.75
None	4	+0.23	60.00	13.30	+ 0.48	60.00	13.20

[a] Total number of measurements: 50 retinoscopy, 18 keratometry, 28 ultrasound.

flat, even in the control unoperated group (compare Tables 1 and 2). This might indicate that another factor also takes place, probably the illumination level.

Discussion

It is interesting to note that while neonatal lid fusion in monkeys caused an elongation of the eye globe confined to the posterior segment (Wiesel and Raviola, 1977a, 1977b; Raviola and Wiesel, 1978), in the chick under similar conditions the enlargement was of the whole eye. Since the chick's eye is ellipsoid in is sagittal plane, it is not yet clear whether the 30% increase we found for the anterior-posterior axis and the 12% and 18% increases in the diameter of the globe and cornea, respectively, indicate linearity of growth. Another difference was that while corneal refraction was not affected significantly in monkeys (Wiesel and Raviola, 1977a), in chicks there was a significant difference, causing relative flattening of the cornea under the fused eyelids.

The myopia that was obtained in early operated chicks was age-dependent, i.e., it was obtained only if the lid fusion was performed early in life and for a period of at least six weeks after hatching; periods longer than three months after hatching were not effective (Table 3). Therefore, there is a critical period for the myopia to take place, the exact extent of which is now being studied. On the other hand, the "corneal" flattening found in late operated chicks (Table 2) seems to be age-independent, since, more recently, we were also able to obtain it in adult chickens.

When developing monkeys and cats were raised in animal rooms and in cages under poor visual conditions, myopia was found to develop (Young, 1963, 1964; Young et al., 1973; Rose et al., 1974; Belkin et al., 1977). Whether this finding is related from a mechanistic point of view to the results of the present and other studies with pattern deprivation where myopia was found (Wiesel and Raviola, 1977a, 1977b; Raviola and Wiesel, 1978; Sherman et al., 1977; Wilson and Sherman, 1977) is still an open question.

The results of the present study on myopia and hypermetropia in developing chicks have also been presented at the second European Neurosciences Meeting in Florence, Italy in September 1978. Later, Wallman et al. (Science

TABLE 3 Effects of eye closure[a] for various periods immediately following hatching on refraction and eye size in chicks.

Age at eye opening (days after hatching)	Retinoscopy (diopters)	Anterior-posterior axis (mm)
4-12	+0.19	12.31
20-43	-1.68	12.04
80	-8.22	14.00
94-101	-7.41	14.15
Control (both eyes open)	+0.42	12.22

[a]Based on one measurement of refraction and ultrasound for each eye; 34 eyes were examined in chicks either monocularly or binocularly sutured.

201:1249-1251 (1978)) confirmed our results on myopia in chicks. However, these authors have not performed the keratometric measurements which have been proved by us to be very crucial for determination of the optical condition in chicks.

Acknowledgements This work was supported by the Office for the Absorption of Scientists, Ministry of Absorption and Immigration, Government of Israel.

References

Arkin, H., and R. R. Colton (1963). Tables for Statisticians. Barnes and Noble, Inc., New York.

Belkin, M., U. Yinon, L. Rose, and I. Reisert (1977). Effect of visual environment on refractive error of cats. Doc. Ophthal. 42:433-437.

Campell, H. S., and J. Lawton-Smith (1962). The pharmacology of the pigeon pupil. Arch. Ophthal. 67:501-504.

von Noorden, G. K., and M. L. J. Crawford (1978). Lid closure and refractive error in macaque monkeys. Nature 272:53:54.

Pumphrey, R. J. (1961). Sensory organs: vision. In: Biology and Comparative Physiology of Birds, A. J. Marshall (ed.), Academic Press, New York, pp. 55-68.

Raviola, E., and T. N. Wiesel (1978). Effect of dark-rearing on experimental myopia in monkeys. Invest. Ophthal. 17:485-488.

Rose, L., U. Yinon, and M. Belkin (1974). Myopia induced in cats deprived of distance vision during development. Vision Res. 14:1029-1032.

The Cholinergic System in the Chicken Retina: Cellular Localization and Development

ROBERT W. BAUGHMAN

Department of Neurobiology
Harvard Medical School
Boston, Massachusetts USA

Abstract The cellular localization of the cholinergic system of the chicken retina was determined by means of freeze-drying and dry autoradiography following incubation with [³H]choline under conditions favoring high-affinity uptake. The cholinergic cells were localized to the inner nuclear and ganglion cell layers, and they extended processes in two bands in the inner plexiform layer. During embryogenesis in the chick, the cholinergic system was found to develop in two stages; the first occurs relatively early in retinal differentiation and is associated with increased ACh synthesis and storage and with a large rise in CAT activity; and the second occurs just before hatching, coincident with synaptogenesis and the appearance of visual function, and is associated with further increases in ACh synthesis and storage and with the development of high-affinity choline uptake.

Acetylcholine (ACh) appears to be a neurotransmitter in the retina of several vertebrate species (Lindeman, 1947; Hebb, 1957; Graham, 1974; Masland and Ames, 1976; and Baughman and Bader, 1977). In the present studies the presence of a cholinergic system in the chicken retina was confirmed, and by means of autoradiography, the cholinergic cells were localized (Baughman and Bader, 1977). Two important components of the system, choline acetyltransferase and high-affinity choline uptake, were found to develop during different stages of embryogenesis (Bader, Baughman, and Moore, 1978).

The localization of the cholinergic neurons was determined on the basis of high-affinity uptake of [³H]choline followed by autoradiography. The uptake of choline exhibited properties similar to those observed in other cholinergic systems (Haga and Noda, 1973; Yamamura and Snyder, 1973); both high- and low-affinity processes were observed with K_m and V_{max} values of 1.1 ± 1.0 μM (mean $\pm$ S.D.) and 8.0 ± 3.8 pmole min^{-1}mgprot^{-1} for the high-affinity and

214 ± 58 μM and 578 ± 94 pmole $\text{min}^{-1}\text{mgprot}^{-1}$ for the low-affinity uptake, respectively. The high-affinity uptake was blocked in the absence of sodium or in the presence of micromolar concentrations of hemicholinium-3. At choline concentrations in the micromolar range, a large percentage of choline transported by the high-affinity system was converted to ACh, and this appeared to be the major source of ACh since ACh synthesis almost stopped when the high-affinity uptake was blocked. The [^{3}H] ACh that was synthesized following incubation with [^{3}H]choline could be released by depolarization with increased extracellular K^+ in a Ca^{2+} dependent manner.

Since neither choline nor ACh is fixable, special procedures were required to retain the [^{3}H]choline and [^{3}H]ACh *in situ* prior to autoradiography. This was achieved by freeze-drying the tissue, fixing in the vapor phase with OsO_4, embedding directly in Epon, sectioning under dry conditions and using a dry-film autoradiographic technique (Baughman and Bader, 1977). A bright-field micrograph of a Toluidine Blue-stained section from this procedure is shown in Fig. 1A. The various layers of the retina from the photoreceptors to the optic nerve fibers are indicated. In Fig. 1B, obtained following an incubation with [^{3}H]choline, the distribution of silver grains is shown in dark field. Patches of label are seen in the inner nuclear and ganglion cell layers, and two bands are present in the inner plexiform layer. In order to test whether the localized distribution seen in Fig. 1B was associated with the high-affinity uptake system, the autoradiography was repeated following an incubation that included hemicholinium-3, which should specifically block the high-affinity uptake. As is shown in Fig. 1C, the localized uptake was completely eliminated with this procedure. A further confirmation was obtained by counting the silver grains

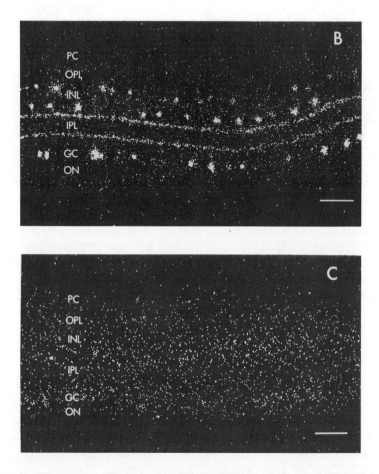

FIGURE 1 Morphology and distribution of [³H]choline uptake in chicken retina after freeze-drying. All micrographs are at the same magnification. The calibration bar equals 50 μm. PC: photoreceptor cell layer; OPL: outer plexiform; INL: inner nuclear layer; IPL: inner plexiform layer; GC: ganglion cell layer; and ON: optic nerve fibers. (A) Bright-field illumination of a Toluidine Blue-stained section. (B) Dark-field illumination of an autoradiograph obtained following incubation with 5 μm [³H]choline for 15 minutes. (C) Dark-field illumination of an autoradiograph obtained following similar [³H]choline incubation including hemicholinium-3. Sections from hemicholinium-3 incubated retinas were autoradiographed and processed together on the same slides with sections from normal [³H]choline incubated retinas which showed localization as seen in Fig. 1B.

localized over various regions of the retina. Calculations based on this procedure indicate that the grains localized in patches and bands in Fig. 1B account for essentially all of the high-affinity [³H]choline uptake and [³H]ACh content. In Fig. 2, which is a bright-field montage at higher magnification, the patches of label are seen to be present over cell bodies in two rows in the inner nuclear layer, one row approximately in the middle of the layer and the other near the inner margin, and over small cell bodies in the ganglion cell layer.

The localization that is observed for uptake of [³H]choline can be compared with that obtained with other means of histochemically visualizing components of the cholinergic system. With acetylcholinesterase staining there is relatively good agreement; there are two heavy bands of staining in the inner plexiform layer that exactly overlap with the two bands seen with [³H]choline uptake (Shen, et al., 1956). These two bands incidentally coincide with Cajal's bands 2 and 4 of the chicken retina (Cajal, 1972). With α-bungarotoxin binding, which should label nicotinic cholinergic receptor sites, the agreement is not as good in that the α-bungarotoxin binding in the inner plexiform layer is more diffuse and, unlike either [³H]choline uptake or acetylcholinesterase staining, labeling is seen in the outer plexiform layer as well (Vogel and Nirenberg, 1976). With [³H]quinuclidinyl benzilate, which should label muscarinic receptor sites, the labeling is more similar to the [³H]choline or acetylcholinesterase pattern; two bands are seen in the inner plexiform layer and little labeling occurs in the outer plexiform layer (Sugiyama, et al., 1977).

Another general question concerning the cholinergic system in retina is how it develops during embryogenesis. The retina is well suited for such developmental studies in that it has no extrinsic inputs other than the centrifugal fibers (Cowan, 1970). Lindeman (1947) found that in the chick retina, although acetylcholinesterase levels increased steadily throughout embryogenesis, ACh levels rose sharply just before hatching. A likely site for controlling ACh content is the synthetic enzyme choline acetyltransferase (CAT), and therefore the activity of this enzyme was measured at various times throughout the developmental period. As is shown in Fig. 3, at embryonic day 5, the earliest stage studied, the activity was low, but during the period from day 6 to day 11 it increased more than 100-fold. The level then continued to rise slightly until hatching at day 20, after which it increased by an additional twofold. Three control experiments were carried out to test whether other acetylating enzymes, e.g., carnitine acetyltransferase (White and Wu, 1973), contributed to the choline-acetylating activity plotted in Fig. 3. Firstly, the developmental time course of carnitine-acetylating activity was determined and compared with that of choline-acetylating activity. As is indicated by the dotted line in Fig. 3, the carnitine-acetylating activity increased by only about twofold during the entire period from embryonic day 5 to hatching. Secondly, the kinetic parameters for choline acetylation were determined at different developmental stages. At embryonic days 6, 13, and 19, and in the adult the K_m values obtained were 0.81 ± 0.27, 0.93 ± 0.25, 1.02 ± 0.12, and 1.08 ± 0.27. The fact that the K_m, which reflects the affinity of the enzyme for choline, is essentially constant

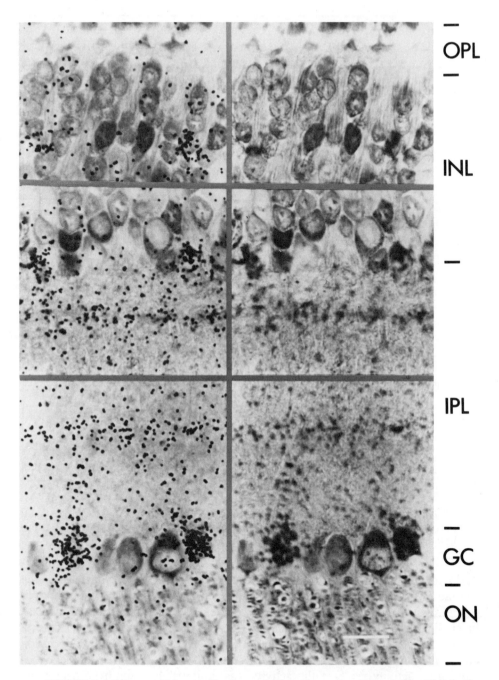

OPL

INL

IPL

GC

ON

FIGURE 2 Montage at high magnification made from the section shown in Fig. 1B following counterstaining with Toluidine Blue. On the right focus is on the section and on the left focus is on both the grains and the section. The calibration bar equals 10 μm. Abbreviations as in Fig. 1.

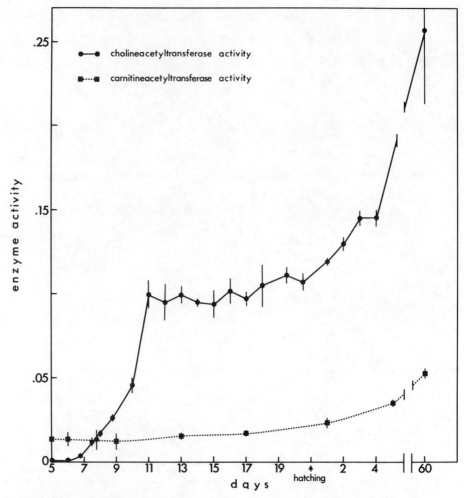

FIGURE 3 Development of CAT (filled circles) and carnitine acetyltransferase (filled squares) activity. Each point is the mean (± S.D.) of three to six determinations. The arrowhead indicates the time of hatching.

during development suggests that the same enzyme may be responsible for the choline acetylation throughout. The increase in the V_{max}, which is indicative of the number of enzymatic sites, however, closely parallels the observed increase in choline-acetylating activity. Thirdly, the effect of naphthylvinylpyridine, a specific inhibitor of CAT (White and Wu, 1973), was determined during development. Even at embryonic day 6, before the large increase in choline-acetylating activity, 96% of the activity was inhibited by naphthylvinylpyridine (in the adult an inhibition of 95% was observed). These results suggest that only one enzyme, with kinetics and pharmacological properties appropriate for CAT, accounts for the increase in choline-acetylating activity shown in Fig. 3.

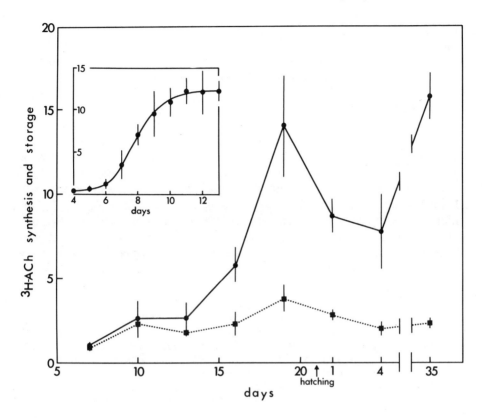

FIGURE 4 Development of synthesis and storage of [³H]ACh measured in the absence (filled circles) or presence (filled squares) of 5 μM HC-3. Pieces of tissue were incubated with 0.5 μM [³H]choline (10 μM in the inset) for 10 minutes (30 minutes in the inset) and analyzed for [³H]ACh. Each point is the mean (± S.D.) of five or more determinations in the absence of three determinations in the presence of HC-3. The arrowhead indicates the time of hatching.

The developmental time course for CAT activity shows no increase near the time of hatching that would account for the rise in ACh levels reported by Lindeman (1947). The CAT assays, however, were carried out on homogenates. Another way to study mechanisms underlying increased ACh content is to look at synthesis and storage in intact tissue. Such experiments measure the ability of the tissue to take up [³H]choline from the medium, to convert it to ACh, and to store the newly synthesized ACh. In addition to an early increase that coincides with that of CAT activity (Fig. 4), the time course of development of such ACh synthesis and storage, in contrast to CAT activity, showed another marked rise just before hatching. This suggested that some additional component of the cholinergic system present in intact tissue was maturing during this second rise. A possible candidate was identified when the synthesis and storage experiments were repeated in the presence of 5 μM hemicholinium-3, which, as indicated before, should block high-affinity choline uptake. Although this reagent had little effect up to embryonic day 13, at embryonic day 19 it reduced the rise in ACh synthesis and storage by more than 70%, presumably as a result of interfering with high-affinity choline uptake.

To pursue this possibility, the development of the high-affinity uptake was measured by determining the kinetics of choline uptake at embryonic day 13, after the large increase in CAT activity was complete, at embryonic day 19, when the increase in ACh synthesis occurred, and in the adult. The uptake experiments were done with the P_2 synaptosomal fraction (Gray and Whittaker, 1962) from retinal homogenates, and the kinetic parameters were determined both with a nonlinear least squares curve-fitting program and on the basis of hemicholinium-3 dependent uptake. As is shown in Table 1, the K_m of the high-affinity uptake, as determined with both methods, was not significantly different at any of these developmental stages. The V_{max}, which reflects the number of uptake sites, however, increased approximately sixfold between embryonic days 13 and 19.

TABLE 1 Kinetic parameters for the rate of choline uptake with respect to choline concentration. Each value is the mean ($\pm$ S.D.) of the V_{max}(pmole/μg$_{prot}$·hr) and K_m(μM). The subscripts H and L refer to the high- and low-affinity components. (A) Parameters calculated with a nonlinear least squares program assuming two independent Michaelis-Menten processes. (B) HC-3 sensitive uptake defined as uptake blocked in the presence of 10 μM HC-3 at 0.1, 0.3, 1.0, and 3.0 μM [^{3}H]choline (uptake at higher concentrations was unaffected by HC-3). Parameters calculated with a linear least squares treatment assuming a single Michaelis-Menten process.

		EMBRYONIC DAY 13	EMBRYONIC DAY 19	ADULT
V_{MAX_H}	A	6.2 ± 3.5	40 ± 5	49 ± 6
	B	8.2 ± 2.7	47 ± 12	36 ± 13
K_{M_H}	A	0.6 ± 0.3	1.0 ± 0.1	0.8 ± 0.1
	B	1.2 ± 0.6	1.1 ± 0.5	0.6 ± 0.4
V_{MAX_L}	A	1062 ± 99	2906 ± 882	2327 ± 668
K_{M_H}	A	94 ± 15	690 ± 256	503 ± 188

A) COMPUTER ANALYSIS OF CHOLINE UPTAKE

B) HC-3 SENSITIVE CHOLINE UPTAKE

An apparent increase in ACh synthesis and high-affinity choline uptake might occur if sequestration and protection of ACh newly synthesized from choline was more efficient at embryonic day 19 than day 13. Such an effect would probably be detectable kinetically, but a direct test is also possible. To do this, incubations with [³H]choline in the presence of eserine, which should block both intracellular and extracellular acetylcholinesterase, were carried out with tissue from these two embryonic days, and the levels of [³H]ACh were measured in both the medium and tissue. The result was that the tissue levels of [³H]ACh in both cases were not significantly higher than those of the controls, and although somewhat more [³H]ACh was released into the medium of day 13, this amount when added to the tissue levels was far from sufficient to bring the day 13 total [³H]ACh synthesis and [³H]choline uptake up to the day 19 levels.

The time course of biochemical development just described can be compared with other aspects of development. Morphologically, the initial increase in ACh synthesis and storage and large rise in CAT activity coincides with the appearance of the inner plexiform layer (Coulombre, 1955) and the second increase, which is accompanied by the development of high-affinity choline uptake, coincides with synaptogenesis in the same layer (Hughes and LaVelle, 1974). These results are consistent with previous findings indicating that in chicken retina the cholinergic system is localized primarily in the inner plexiform layer (Baughman and Bader, 1977). The large increase in CAT activity, which occurs soon after most retinal cells have withdrawn from the cell cycle (Kahn, 1974), appears to precede the morphological development of synaptic specialization. Other studies have reported a similar increase in CAT activity preceding synaptic specialization in the developing spinal cord (Burt, 1968) and ciliary ganglion (Chiappinelli, et al., 1976). It is difficult to judge the effect of neuronal activity on the development of the cholinergic system. The first sign of the electroretinogram appears on embryonic day 18 (Witkovsky, 1963), and thus the large rise in CAT activity, which occurs several days earlier, appears not to be associated with the appearance of a response to light stimulation. The development of the high-affinity uptake, however, does coincide approximately with the appearance of retinal function.

In summary, the chicken retina contains a cholinergic system consisting of cells in the inner nuclear and ganglion cell layers that extend processes in two bands in the inner plexiform layer. This cholinergic system appears to develop in two stages; the first occurs relatively early in retinal differentiation and is associated with increased ACh synthesis and storage and with a large rise in CAT activity; the second occurs just before hatching, coincident with synaptogenesis and the appearance of the light response.

Acknowledgements The work described here was done in collaboration with Dr. Charles Bader and Janet Moore. The encouragement and support of Torsten Wiesel are gratefully acknowledged. Funding was provided by NIH grants EYO2317, EYOO606, and EYOO82.

References

Bader, C. R., R. W. Baughman, and J. L. Moore (1978). Different time course of development for high-affinity choline uptake and choline acetyltransferase in the chick retina. Proc. Natl. Acad. Sci. USA 75:2525-2529.

Baughman, R. W., and C. R. Bader (1977). Biochemical characterization and cellular localization of the cholinergic system in the chicken retina. Brain Res. 138: 469-485.

Burt, A. M. (1968). Acetylcholine esterase and choline acetyltransferase activity in the developing chick spinal cord. J. Exp. Zool. 169:107-112.

Cajal, S. R.y (1972). The Structure of the Retina. S. A. Thorpe and M. Glickstein (translators). Thomas, Springfield, Ill., pp. 76-121.

Chiappinelli, V., E. Giocobini, G. Pilar, and H. Uchimura (1976). Induction of cholinergic enzymes in chick ciliary ganglion and iris muscle cells during synapse formation. J. Physiol. (Lond.) 257:749-766.

Coulombre, A. J. (1955). Correlations of structural and biochemical changes in the developing retina of the chick. Am. J. Anat. 96:153-189.

Cowan, W. M. (1970). Centrifugal fibers to the avian retina. Brit. Med. Bull. 26:112-118.

Graham, L. T. (1974). Comparative aspects of neurotransmitters in the retina. In: The Eye. H. Davson and L. T. Grahams (eds.). Academic Press, New York, vol. 6, pp. 283-342.

Gray, E. G., and V. P. Whittaker (1962). The isolation of nerve endings from brain: an electron microscope study of cell fragments derived by homogenization and centrifugation. A. Anat. (Lond.) 96:79-87.

Haga, T., and H. Noda (1973). Choline uptake systems of rat brain synaptosomes. Biochim. Biophys. Acta (Amst.) 291:564-575.

Hebb, C. (1957). Biochemical evidence for the neural function of acetylcholine. Physiol. Rev. 37:169-220.

Hughes, W. F., and A. LaVelle (1974). On the synaptogenic sequence in the chick retina. Anat. Rec. 179:297-302.

Kahn, A. J. (1974). An autoradiographic analysis of the time of appearance of neurons in the developing chick neural retina. Dev. Biol. 38:30-40.

Lindeman, V. F. (1947). The cholinesterase and acetylcholine content of the chick retina, with special reference to functional activity as indicated by the pupillary constrictor reflex. Am. J. Physiol. 148:40-44.

Masland, R. H., and A. Ames III (1976). Responses to acetylcholine of ganglion cells in an isolated mammalian retina. J. Neurophysiol. 39:1220-1235.

Shen, S. C., R. Greenfield, and E. J. Boell (1956). Localization of acetylcholinesterase in chick retina during histogenesis. J. Comp. Neurol. 106:433-461.

Sugiyama, H., M. P. Daniels, and M. Nirenberg (1977). Muscarinic acetylcholine receptors of the developing retina. Proc. Natl. Acad. Sci. USA 74:5224-5528.

Vogel, Z., and M. Nirenberg (1976). Localization of acetylcholine receptors during synaptogenesis in retina. Proc. Natl. Acad. Sci. USA 73:1806-1810.

White, H. L., and J. C. Wu (1973). Choline and carnitine acetyltransferases of heart. Biochemistry 12:841-846.

Witkovsky, P. (1963). An ontogenic study of retinal function in the chick. Vision Res. 3:341-355.

Yamamura, H. I., and S. H. Snyder (1973). High-affinity transport of choline into synaptosomes of rat brain. J. Neurochem. 21:1355-1374.

Characterization of Neural Enzyme Development in Dissociated Chick Embryo Brain Cell Cultures

ANTONIA VERNADAKIS
ELLEN B. ARNOLD

Departments of Psychiatry and Pharmacology
University of Colorado School of Medicine
Denver, Colorado USA

Abstract The present study characterizes the *in vitro* development of certain neural marker enzymes in dissociated cell cultures of embryonic chick brain. In cultures of cerebral hemispheres (CH) from 6- and 10-day-old embryos, acetylcholinesterase (AChE) activity declined with age in culture, a phenomenon which may be related to increasing growth of non-neuronal cells in culture. Choline acetyltransferase (CAT) activity in cultures from 10-day-old embryos, except for a decline between 7 to 11 days, increased with age in culture. The development of CAT activity suggests non-neuronal influences on the expression of CAT activity. Tyrosine hydroxylase (TH) activity increases at later periods in culture than does CAT and may reflect enhanced differentiation of adrenergic neurons by cholinergic influences, or by glial or other non-neuronal cells. The differential maturational patterns of CAT and TH *in vitro* are also observed *in vivo*, suggesting that the biochemical development of dissociated brain cell cultures exhibits similarities to the biochemical development of the brain.

Introduction

Neural tissue culture has served as a useful tool for the study of regulatory processes in neural growth and differentiation. There are several types of neural tissue culture (see review, Vernadakis and Culver, 1979). These include *organ culture,* in which the neural explant maintains its *in vivo* cytoarchitecture for at least 24 hours. This culture system has been used by several investigators, including ourselves, to study acute effects of drugs and hormones on neural growth (see review, Vernadakis and Culver, 1979). Another culture system is the *organotypic culture* (Maximow double coverslip assembly technique), in which the explant loses its *in vivo* cellular organization but can be maintained in culture for several weeks. This culture system can be used to conduct electrophysiological studies of neural growth and electron microscopic studies of synaptic development and myelination (see reviews, Crain, 1976; and Murray, 1965). More recently, the *dissociated brain cell culture* system has been used for morphological and biochemical studies of neural growth (see review, Vernadakis and Culver, 1979). We have used this culture system to study the maturation of several enzymes associated with neural growth in general, such as acetylcholinesterase (AChE) (Vernadakis and Gibson, 1974), and the enzymes choline acetyltransferase (CAT) and tyrosine hydroxylase (TH), which are associated with cholinergic and catecholaminergic neurons, respectively.

Materials and Methods

Preparation of cultures A modification of the method described by Sensenbrenner and her associates (Sensenbrenner et al., 1971) was used to dissociate neural cells from chick embryos. Tissue (either cerebral hemispheres or whole brain, without optic lobes) was removed, dissected free of meningeal membranes, and dispersed through a single thickness of nylon mesh (73 μ pore size) using a glass rod. The cell suspension was prepared in Dulbecco's Modified Eagle Medium (GIBCO), fortified with 20% fetal calf serum (Reheis Division, Armour Pharmaceuticals). Approximately 3×10^6 cells, in a volume of 4 ml, were plated per 25cm^2 plastic tissue culture flask, and the cultures were incubated at 37°C in a humidified atmosphere of 95% air-5% CO_2. The growth medium was replaced on the fourth day of cultivation and on every third day thereafter.

Biochemical determinations To harvest cells, growth medium was decanted and the cells were removed from the surface of the flask using a rubber policeman. The cells were harvested in serum-free medium, pelleted by low-speed centrifugation at 4°C, washed with cold 0.67 M phosphate buffer, pH 7.0, and pelleted again. Cells were frozen and enzyme activities were determined within one to two weeks from the time of harvest, except for cells for AChE assay which were not frozen.

Acetylcholinesterase activity was determined colorimetrically, using a Beckman DU spectrophotometer to measure the rate of hydrolysis of acetylthiocholine (ATCh), according to the method of Ellman et al. (1961). Acetylcholinesterase activity was expressed as μmoles ATCh hydrolysed per minute per gram protein.

Choline acetyltransferase activity was determined using a modification (Chiappinelli et al., 1976) of the radiochemical method of Fonnum (1975). This method is based on the principle of liquid cation exchange, which specifically separates acetylcholine (ACh) from anions and uncharged compounds such as acetylcarnitine. Labeled ACh was extracted from the reaction mixture with the liquid ion exchanger sodium tetraphenylboron (Kalignost) in 3-heptanone into a toluene-based scintillation fluid. Radioactivity was counted and results were expressed as μmoles of ^{14}C–ACh synthesized per hour per culture.

Tyrosine hydroxylase activity was determined as described by Waymire et al. (1971), by measuring the enzymatic decarboxylation of ^{14}C–carbosyl–labeled DOPA formed from ^{14}C–carboxyl–labeled tyrosine. Enzyme activity was expressed as pmoles ^{14}C–CO_2 formed per mg protein per hour.

In all experiments protein content was determined by the method of Lowry et al. (1951).

Results

Development of cholinesterase activity In cultures of cerebral hemispheres from 6-day-old (E6) chick embryos, AChE activity declined by 65% from initially high levels over a two-week time period (days 3 through 18 in culture) (Fig. 1). In similar cultures from 10-day-old (E10) embryos, AChE activity on day 3 in culture was approximately twice as high as in the E6 cultures, and the subsequent decline in enzymatic activity was more rapid and more pronounced, falling by 90% on day 11 in culture (Fig. 2).

Development of choline acetyltransferase activity In contrast to cholinesterase activity, CAT activity in E10 cerebral hemisphere cultures, expressed as the rate of ^{14}C–acetylcholine synthesis per hour per gm protein or per culture, increased from eight- to thirteen-fold, respectively, between days 3 and 7 in culture (Fig. 3). A slight but statistically significant decline in CAT activity per gm protein occurred between days 7 and 11 in culture; this was followed by an increase in activity throughout the duration of the culture period.

Development of tyrosine hydroxylase activity In E8 cultures of either cerebral hemispheres or whole brain without optic lobes, TH activity remained at a low level throughout the first ten days in culture (Fig. 4). There was then a marked increase in enzymatic activity, which was maximal on day 14 in cerebral hemisphere cultures and on day 15 in whole brain cultures.

Discussion

Dissociated cell cultures obtained from six to ten days of embryonic life, a period during which the chick brain rapidly develops its characteristic gross morphology and cytoarchitecture (Burdick and Strittmatter, 1965), exhibit age-related differences in their patterns of growth and enzyme development. In cultures of both E6 and E10 cerebral hemispheres, the specific activity of AChE declined with age in culture. This corresponds to observations of others who have found a decline of AChE activity during cultivation of dissociated brain cell cultures from chick embryos (Werner et al., 1971; Ebel et al., 1974), newborn mice (Wilson et al., 1972), and fetal rats (Shapiro and Schrier, 1973). Ebel et al. (1974) attributed the fall of AChE activity in their cultures to a progressive proliferation of astroblasts and mesenchymal cells in which AChE activity was negligible. The possibility that a decrease in AChE activity reflects dilution by proliferating cells not containing AChE is also suggested by the experiments of Werner et al. (1971), where the addition of 5-fluorouracil prevented overgrowth of non-neuronal cells and significantly elevated AChE

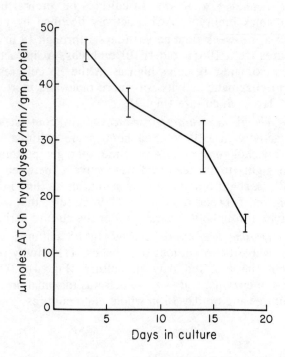

FIGURE 1 Acetylcholinesterase activity in dissociated cerebral hemisphere cultures from 6-day-old chick embryos. Activity is expressed as μmoles of acetylthiocholine hydrolysed per minute per gm protein. Each point represents the mean ±standard error of the mean of values obtained from four to eight cultures.

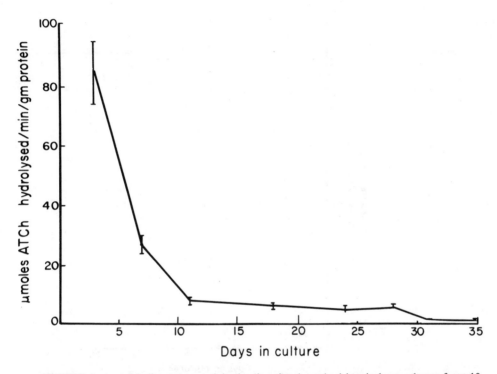

FIGURE 2 Acetylcholinesterase activity in dissociated cerebral hemisphere cultures from 10-day-old chick embryos. Activity is expressed as μmoles of acetylthiocholine hydrolysed per minute per gm protein. Each point represents the mean ±standard error of the mean of values obtained from four to eight cultures.

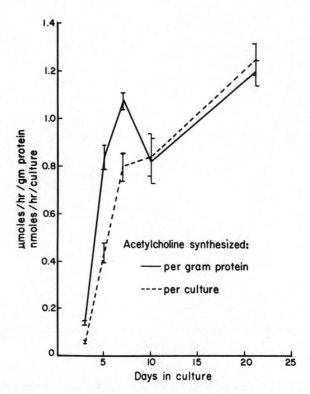

FIGURE 3 Choline acetyltransferase activity in dissociated cerebral hemisphere cultures from 10-day-old chick embryos. Activity is expressed as μmoles acetylcholine synthesized per hour per culture. Each point represents the mean ± standard error of the mean of values obtained from six to eight cultures.

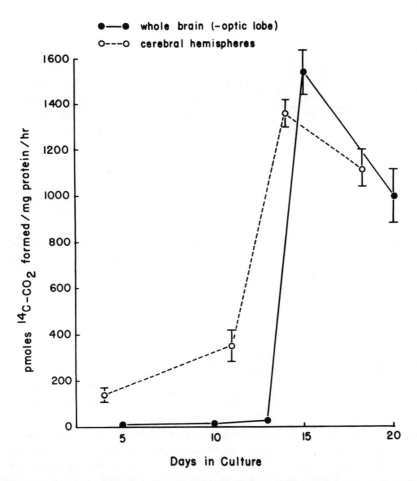

FIGURE 4 Tyrosine hydroxylase activity in dissociated cerebral hemisphere or whole brain (optic lobes removed) cultures from 8-day-old chick embryos. Activity is expressed $p$moles $^{14}C-CO_2$ formed per milligram protein per hour. Each point represents the mean ±standard error of the mean of values obtained from six to eight cultures.

activity in comparison to untreated cultures. In neuroblastoma cell cultures, AChE decreases with increasing cell density (Vernadakis et al., 1976). Blume et al. (1970) reported that while the specific activity of AChE did not change appreciably during the period of rapid cell division, the specific activity during the stationary phase of growth increased by 25-fold. It has therefore been suggested that AChE responds to a regulatory mechanism which is coupled to the rate of cell division.

In contrast to cholinesterase activity, CAT activity increased with age in culture. Although this increase is probably due in part to the growth and differentiation of cholinergic neuroblasts in culture, other factors may also be involved. In E10 cultures, for example, the second peak in CAT activity occurs at a time (21-23 days) when neuronal growth is no longer prominent and the majority of the neuronal population has begun to degenerate. Werner et al. (1971) observed a marked increase in CAT activity in dissociated chick embryo brain cultures which was correlated with the synthesis of DNA and protein and with the proliferation of flat cells, suggesting that CAT was either contained in flat cells or that these cells were in some way required for the full expression of CAT activity in neuronal cells. Likewise, Shapiro and Schrier (1973) found that, in dissociated cultures of fetal rat brain, CAT activity was related to cell division. These investigators raised the question as to whether CAT-producing cells were capable of multiplication in culture. Since neurons are not known to divide in culture, while non-neuronal cells readily proliferate, their observations suggest a non-neuronal contribution to CAT activity. In our culture system, for example, the later increase in CAT activity may be related to the proliferation and maturation of glial cells in the cultures. Contact with glial cells is known to enhance and maintain the differentiation of neurons in culture. Monard et al. (1973) have demonstrated the release by cultured glial cells of a macromolecular factor which can induce morphological differentiation of neuroblastoma cells, and Murphy et al. (1977) have reported that culture glial cells secrete a factor which is biologically and immunologically similar to nerve growth factor (NGF). Primary cultures of chick fibroblasts also secrete NGF (Young et al., 1975). There may thus be various non-neuronal influences on the expression of CAT activity in dissociated cultures, as well as influences attributable to other, undefined, neuronal elements in the culture system.

The demonstration of TH activity in both cerebral hemispheres and whole brain cultures (E8) suggests that dissociated neural cells possess at least some characteristics associated with adrenergic functioning. The pattern of development of TH differs from that of CAT or AChE, both of which could be detected during very early periods in culture: TH activity rose sharply near the end of the second week in culture and reached a maximum on day 14-15. The specific activity of TH increased 4.5-fold over a period of three days in cerebral hemisphere cultures and 30-fold over a period of two days in whole brain cultures. This sudden increment in enzymatic activity was not correlated with any observable morphological changes in the cultures and, in fact, occurred at a time when neuronal growth appeared to have ceased. This suggests that the increase in TH activity was related to some intrinsic influence which facilitated expression of TH activity in cells having adrenergic characteristics, rather than to an active proliferation of any specific neuronal cell type. One possibility is that cholinergic activity, which is high during the first two weeks, stimulates adrenergic activity in culture. In the peripheral nervous system, presynaptic cholinergic neurons regulate the development of postsynaptic adrenergic neurons, as has been demonstrated by Black and his associates (Black et al., 1971). Whether similar interactions exist between cholinergic and adrenergic neurons in dissociated brain cell cultures remains to be determined. As discussed previously, there is evidence to suggest that both glial cells and fibroblasts can secrete NGF, and perhaps other macromolecular growth factors, in culture (Monard et al., 1973; Murphy et al., 1977; Young et al., 1975). The increase in TH activity may reflect an enhanced differentiation of adrenergic neurons by one or more such factors elaborated by glial or other non-neuronal cells. The maturational profiles of CAT and TH observed in dissociated embryonic brain cultures are somewhat analogous to the time course of development of these enzymes *in vivo*. For example, CAT activity has been detected in early neuroblasts (Giacobini and Filogamo, 1973), while we have found (Waymire et al., 1974) that TH activity cannot be detected in the chick brain until 14 days of embryonic age. This neural culture system therefore seems to reflect at least some significant characteristics of *in vivo* neural growth.

Acknowledgements This work was supported by USPHS Training Grant T32 HD 07072-02, a Developmental Psychobiology Research Group Endowment Fund, and a Research Scientist Career Development Award KO2 MH 42479 from the National Institute of Mental Health (A. Vernadakis).

References

Black, I. B., I. A. Hendry, and L. L. Iversen (1971). Transynaptic regulation of growth and development of adrenergic neurons in a mouse sympathetic ganglion. Brain Res. 34:229-240.

Blume, A., F. Gilbert, S. Wilson, J. Farberg, R. Rosenberg, and M. Nirenberg (1970). Regulation of acetylcholinesterase in neuroblastoma cells. Proc. Soc. Nat. Acad. Sci. 67:786-792.

Burdick, C. J., and C. F. Strittmatter (1965). Appearance of biochemical components related to acetylcholine metabolism during the embryonic development of chick brain. Arch. Biochem. Biophys. 109:293-301.

Chiappinelli, V., E. Giacobini, G. Pilar, and H. Uchimura (1976). Induction of cholinergic enzymes in chick ciliary ganglion and iris muscle cells during synapse formation. J. Physiol. 257: 749-766.

Crain, S. M. (1976). Neurophysiologic Studies in Tissue Culture. Raven Press, New York.

Ebel, A., R. Massarelli, M. Sensenbrenner, and P. Mandel (1974). Choline acetyltransferase and acetylcholinesterase activities in chicken brain hemispheres *in vivo* and in cell cultures. Brain Res. 76:461-472.

Ellman, G. L., K. D. Courtney, V. Andres, and R. M. Featherstone (1961). A new and rapid colorimetric determination of acetylcholinesterase activity. Biochem. Pharmacol. 7:88-95.

Fonnum, F. (1975). A rapid radiochemical method for the determination of choline acetyltransferase. J. Neurochem. 24: 407-409.

Giacobini, G., and G. Filogamo (1973). Changes in the enzymes for the metabolism of acetylcholine during development of the central nervous system. In Central Nervous System: Studies on Metabolic Regulation and Function, E. Genazzani and H. Herken (eds.), Springer-Verlag, Berlin, pp. 153-157.

Lowry, O. H., N. F. Rosebrough, A. L. Farr, and R. J. Randall (1951). Protein measurement with the Folin phenol reagent. J. Biol. Chem. 193:265-275.

Monard, D., F. Solomon, M. Rentsch, and R. Gysin (1973). Glia- induced morphological differentiation in neuroblastoma cells. Proc. Nat. Acad. Sci. USA 70:1894-1897.

Murphy, R. A., J. Oger, J. D. Saide, M. H. Blanchard, G. W. Aranson, C. Hogan, N. J. Pantazis, and M. Young (1977). Secretion of nerve growth factor by central nervous glioma cells in culture. J. Cell Biol. 72:769-773.

Murray, M. R. (1965). Nervous tissues *in vitro* In: Cells and Tissues in Culture: Methods, Biology, and Physiology, E. N. Willmer (ed.), Academic Press, New York, pp. 371-435.

Sensenbrenner, M., J. Booher, and P. Mandel (1971). Cultivation and growth of dissociated neurons from chick embryo cerebral cortex in the presence of different substrates. Z. Zellforsch. 117:559-569.

Shapiro, D. L., and B. K. Schrier (1973). Cell cultures of fetal rat brain: growth and marker enzyme development. Exp. Cell Res. 77:239-247.

Vernadakis, A., and B. Culver (1979). Neural tissue culture: a biochemical tool. In: The Biochemistry of Brain, S. Kumar (ed.), Pergamon Press, Ltd. (in press).

Vernadakis, A., and D. A. Gibson (1974). Role of neurotransmitter substances in neural growth. In: Perinatal Pharmacology: Problems and Priorities, J. Dancis and J. C. Hwang (eds.), Raven Press, New York, pp. 65-77.

Vernadakis, A., R. Nidess, M. L. Timiras, and R. Schlesinger (1976). Responsiveness of acetylcholinesterase and butyrylcholinesterase activities in neural cells to age and cell density in culture. Exp. Cell Res. 97:453-457.

Waymire, J. C., R. Bjur, and N. Weiner (1971). Assay of tyrosine hydroxylase by coupled decarboxylation of DOPA formed from $1-{}^{14}C-L-$tyrosine. Anal. Biochem. 43:588-600.

Waymire, J. C., A. Vernadakis, and N. Weiner (1974). Studies on the development of tyrosine hydroxylase, monoamine oxidase, and aromatic-L-amino acid decarboxylase in several regions of the chick brain. In: Drugs and the Developing Brain, A. Vernadakis and N. Weiner (eds.), Plenum Press, New York, pp. 149-170.

Werner, I., G. R. Peterson, and L. Shuster (1971). Choline acetyltransferase and acetylcholinesterase in cultured brain cells from chick embryos. J. Neurochem. 18:141-151.

Wilson, S. H., B. K. Schrier, J. L. Farber, E. J. Thompson, R. N. Rosenberg, A. J. Blume, and M. W. Nirenberg (1972). Markers for gene expression in cultured cells from the nervous system. J. Biol. Chem. 247:3159-3169.

Young, M., J. Oger, M. H. Blanchard, H. A. Asdourian, and B. G. W. Aranson (1975). Secretion of a nerve growth factor by primary chick fibroblast cultures. Science 187:361-362.

Index

accommodation, 292
acetylcholine, 421
acuity, *see visual acuity*
area 17, *see striate cortex*
area 18, *see parastriate cortex*
area centralis, 24, 205
astigmatism, 295

binocular competition, 81
binocular deprivation, *see deprivation*

callosal connections, 405
cat, 163, 175, 185, 195, 205, 227, 235
cells, retinal ganglion (X/Y), 73, 83, 185, 205
chemoaffinity theory, 332
chicken, 411, 421, 433
choline, 421
choline acetyltransferase, 424, 434
columnar organization, 268
consolidation, 48, 99
contrast sensitivity, 290, 301
cornea, 412
corpus callosum, 228
critical period, 56, 116, 123, 135, 149, 164, 304, 407
cytoarchitecture, 361, 433

dark rearing, 43, 136
    recovery from, 153
dendritic field, 21